Study Guide

to accompany

Prescott, Harley, and Klein's Microbiology

Seventh Edition

Joanne M. Willey
Hofstra University
Linda M. Sherwood
Montana State University
Christopher J. Woolverton
Kent State University

Prepared by
Linda M. Sherwood
Montana State University
and
Mark Schneegurt
Wichita State University

 **Higher Education**

Boston   Burr Ridge, IL   Dubuque, IA   New York   San Francisco   St. Louis
Bangkok   Bogotá   Caracas   Kuala Lumpur   Lisbon   London   Madrid   Mexico City
Milan   Montreal   New Delhi   Santiago   Seoul   Singapore   Sydney   Taipei   Toronto

The **McGraw-Hill** Companies

Study Guide to accompany
PRESCOTT, HARLEY, AND KLEIN'S MICROBIOLOGY, SEVENTH EDITION
JOANNE M. WILLEY, LINDA M. SHERWOOD, AND CHRISTOPHER J. WOOLVERTON

Published by McGraw-Hill Higher Education, an imprint of The McGraw-Hill Companies, Inc., 1221 Avenue of the Americas, New York, NY 10020. Copyright © 2008 by The McGraw-Hill Companies, Inc. All rights reserved.

1 2 3 4 5 6 7 8 9 0 QPD/QPD 0 9 8 7 6

ISBN: 978-0-07-299323-3
MHID: 0-07-299323-5

www.mhhe.com

Contents

To The Student...A Word from Lansing Prescott

One of the most important factors contributing to success in college, and in microbiology courses, is the use of good study techniques. The textbook, *Microbiology, 7e,* is organized to help you study more effectively. But even a text with many learning aids is not effective unless it is used properly. Thus this section briefly outlines some practical study skills that will help ensure your success in microbiology and make your use of this textbook more productive. Many of you already have the study skills mentioned here and will not need to spend time reviewing familiar material. These suggestions are made in the hope that they may be useful to those who are unaware of approaches like the SQ4R technique for studying textbooks.

Time Management and Study Environment

Many students find it difficult to study effectively because of a lack of time management skills and a proper place to study. Often a student will do poorly in a course because he or she has not spent enough time studying outside class. For best results, you should plan to spend at least an average of four to eight hours a week outside class working on each course. There is sufficient time in the week for this, but it does require that you use your time wisely. If you spend a few minutes early in the morning planning how to use your day, and allow adequate time for studying, you will accomplish much more. Students who make efficient use of every moment find they have plenty of time for recreation.

A second important factor is a proper place to study so that you can concentrate and efficiently use your study time. Try to find a quiet location with a desk and adequate lighting. If possible, always study in the same place and use it only for studying. In this way you will be mentally prepared to study when you are at your desk This location may be in the dorm, the library, a special study room, or somewhere else. Wherever it is, your study area should be free from distractions—including friends who drop by to socialize. You will accomplish much more if you really study during your designated study times.

Make the Most of Lectures

Attendance at lectures is essential for success. Students who chronically miss classes usually do not do well. To gain the most from lectures, it is best to read any relevant text material beforehand. Be prepared to concentrate during lectures; do not simply sit back passively and listen to the instructor. During the lecture record your notes in a legible way so that you can understand them later. It is most efficient to employ an outline or simple paragraph format. The use of abbreviations or some type of shorthand notation often is effective. During lecture, concentrate on what is being said and be sure to capture all of the main ideas, concepts, and definitions of important terms. Do not take sketchy notes assuming that you will remember things because they are easy or obvious; you won't. Diagrams, lists, and terms written on the board are almost always important, as is anything the instructor clearly emphasizes by tone of voice. Feel free to ask questions during class when you don't understand something or wish the instructor to pursue a point further. Remember that if you don't understand, it is very likely that others in the class don't either but simply aren't willing to show their confusion. As soon as possible after a lecture, carefully review your notes to be certain that they are complete and understandable. Refer to the textbook when you are uncertain about something in your notes; it will be invaluable in clearing up questions and amplifying major points. When studying your notes for tests, it is a good idea to emphasize the most important points with a highlighter just as you would when reading the textbook.

Studying the Textbook

Your textbook is one of the most important learning tools in any course and you should use it very carefully and conscientiously. Many years ago, Francis P. Robinson developed a very effective study technique called SQ3R (survey, question, read, recite, and review). More recently, L.L. Thistlethwaite and N.K. Snouffer have slightly modified it to yield the SQ4R approach (survey, question, read, revise, record, and review). This latter approach is summarized here:

1. *Survey.* Briefly scan the chapter to become familiar with its general content. Quickly read the title, introduction, summary, and main headings. Record the major ideas and points that you think the chapter will make. If there is a list of chapter concepts and a chapter outline, pay close attention to these. This survey should give you a feel for the topic and how the chapter is approaching it.
2. *Question.* As you reach each main heading or subheading, try to compose an important question or two that you believe the section will answer. This preview question will help focus your reading of the section. It is also a good idea to keep asking yourself questions as you read. This habit facilitates active reading and learning.
3. *Read.* Carefully read the section. Read to understand concepts and major points, and try to find the answer to your preview question(s). You may want to highlight very important terms or explanations of concepts, but do not indiscriminately highlight everything. Be sure to pay close attention to key terms printed in color or boldface since the author(s) considered these to be important.
4. *Revise.* After reading the section, revise your question(s) to more accurately reflect the section's contents. These questions should be concept-type questions that force you to bring together a number of details. You can write them in the margins of your text.
5. *Record.* Underline the information in the text that answers your questions, if you have not already done so. You may wish to write down the answers in note form as well. This process will give you good material to use in preparing for exams.
6. *Review.* Review the information by trying to answer your questions without looking at the text. If the text has a list of key words and a set of study questions, be sure to use these in your review. You will retain much more if you review the material several times.

Preparing for Examinations

It is extremely important to prepare for examinations properly so that you will not be rushed and tired on examination day. You should complete all textbook reading and lecture note revision well ahead of time so that you can spend the last few days in mastering the material, not in trying to understand the basic concepts. Cramming at the last moment for an exam is no substitute for daily preparation and review. By managing your time carefully and keeping up with your studies, you will have plenty of time to review thoroughly and clear up any questions. This will allow you to get sufficient rest before the test and to feel confident in your preparation. Because both physical condition and general attitude are important factors in test performance, you will automatically do better. Proper reviewing techniques also aid retention of material.

Our website (www.mhhe.com/prescott7) contains many useful study aids, including, flash cards with terminology, self-quizzes, microbiology career links, clinical case studies, and animations.

1 The History and Scope of Microbiology

CHAPTER OVERVIEW

This chapter introduces the field of microbiology and discusses the importance of microorganisms not only as causative agents of disease, but also as important contributors to food production, antibiotic manufacture, vaccine development, and environmental management. It presents a brief history of the science of microbiology, an overview of the microbial world, a discussion of the scope and relevance of microbiology in today's society, and predictions about the future of microbiology.

CHAPTER OBJECTIVES

After reading this chapter you should be able to:

- define the science of microbiology and describe some of the general methods used in the study of microorganisms
- discuss the historical concept of spontaneous generation and the experiments that were performed to disprove this erroneous idea
- discuss how Koch's postulates are used to establish the causal link between a suspected microorganism and a disease
- describe some of the various activities of microorganisms that are beneficial to humans
- describe procaryotic and eucaryotic morphology and the distribution of microorganisms among the three domains in which living organisms are categorized
- discuss the importance of the field of microbiology to other areas of biology and to society

CHAPTER OUTLINE

I. Microbiology—Introduction
   A. Microbes are found everywhere and are indispensable for cycling of essential elements on earth.
   B. Most microbes are beneficial to society by producing foods, oxygen, and commercial products, and by enhancing human health; some microbes cause disease.
II. Members of the Microbial World
   A. Microbiology is the study of organisms too small to be clearly seen by the unaided eye (i.e., microorganisms); these include viruses, bacteria, archaea, protozoa, algae, and fungi
   B. Some microbes (e.g., algae and fungi) are large enough to be visible, but are still included in the field of microbiology; it has been suggested that microbiology be defined not only by the size of the organisms studied but by techniques employed to study them (isolation, sterilization, culture in artificial media)
   C. Procaryotes have a relatively simple morphology and lack a true membrane-delimited nucleus
   D. Eucaryotes are morphologically complex and have a true, membrane-enclosed nucleus
   E. A classification scheme consisting of three domains (*Bacteria*, *Archaea*, and *Eucarya*) has become widely accepted; this scheme is followed in this textbook

1

F. Bacteria are single-celled procaryotes with peptidoglycan cell walls that are found throughout the environment

G. Archaea are single-celled procaryotes with different cell walls and unique membrane lipids that are often found in extreme environments

H. *Eucarya* includes the protists (algae, protozoans, slime molds, and water molds), fungi, plants, and animals

I. Viruses are not composed of cells and not part of the domain classification scheme

III. The Discovery of Microorganisms *germ theory of disease*

A. Invisible living creatures were thought to exist and were thought to be responsible for disease long before they were observed

B. Antony van Leeuwenhoek (1632–1723) constructed microscopes and was the first person to observe and describe microorganisms accurately *& extensively*

IV. The Conflict over Spontaneous Generation

A. The proponents of the concept of spontaneous generation claimed that living organisms could develop from nonliving or decomposing matter

B. Francesco Redi (1626–1697) challenged this concept by showing that maggots on decaying meat came from fly eggs deposited on the meat, and not from the meat itself

C. John Needham (1713–1781) showed that mutton broth boiled in flasks and then sealed could still develop microorganisms, which supported the theory of spontaneous generation

D. Lazzaro Spallanzani (1729–1799) showed that flasks sealed and then boiled had no growth of microorganisms, and he proposed that air carried germs to the culture medium; he also commented that external air might be needed to support the growth of animals already in the medium; the latter concept was appealing to supporters of spontaneous generation

E. Theodore Schwann (1810–1882) and others carried out experiments testing the need for air if growth was to occur

F. Louis Pasteur (1822–1895) heated the necks of flasks, drawing them out into long curves, sterilized the media, and left the flasks open to the air; no growth was observed because particles carrying organisms did not reach the medium, instead they were trapped in the neck of the flask; if the necks were broken, particles would enter the flasks and the organisms would grow; in this way Pasteur disproved the theory of spontaneous generation

G. John Tyndall (1820–1893) demonstrated that dust carried microbes and that if dust was absent, the broth remained sterile—even if it was directly exposed to air; Tyndall also provided evidence for the existence of heat-resistant forms of bacteria

V. The Golden Age of Microbiology

A. Recognition of the relationship between microorganisms and disease

1. Agostino Bassi (1773–1856) showed that a silkworm disease was caused by a fungus

2. M. J. Berkeley (ca. 1845) demonstrated that the Great Potato Blight of Ireland was caused by a fungus

3. Heinrich de Bary in 1853 showed that smut and rust fungi caused cereal crop diseases

4. Louis Pasteur showed that the pébrine disease of silkworms was caused by a protozoan parasite

5. Joseph Lister (1872–1912) developed a system of surgery designed to prevent microorganisms from entering wounds; his patients had fewer postoperative infections, thereby providing indirect evidence that microorganisms were the causal agents of human disease; his published findings (1867) transformed the practice of surgery

B. Koch's postulates

1. Robert Koch (1843–1910), using criteria developed by his teacher, Jacob Henle (1809–1895), established the relationship between *Bacillus anthracis* and anthrax; his criteria became known as Koch's Postulates and are still used to establish the link between a particular microorganism and a particular disease:

a. The microorganism must be present in every case of the disease, but absent from healthy individuals

b. The suspected microorganism must be isolated and grown in pure culture

c. The same disease must result when the isolated microorganism is inoculated into a healthy host

2

        d.   The same microorganism must be isolated again from the diseased host

     2.   Koch's work with anthrax was independently confirmed by Pasteur   (TB)

C.   The development of techniques for studying microbial pathogens
1. Koch and his associates developed techniques, reagents, and other materials for culturing bacterial pathogens on solid growth media; agar plates enable microbiologists to isolate microbes in pure culture
2. Charles Chamberland (1851–1908) constructed a bacterial filter that removed bacteria and larger microbes from specimens; this led to the discovery of viruses as disease-causing agents

D.   Immunological studies
1. Edward Jenner used a vaccination procedure to protect individuals from smallpox
2. Louis Pasteur developed other vaccines including those for chicken cholera, anthrax, and rabies
3. Emil von Behring (1854–1917) and Shibasaburo Kitasato (1852–1931) induced the formation of diphtheria-toxin antitoxins in rabbits; the antitoxins were effectively used to treat humans and provided evidence for humoral (antibody) immunity
4. Elie Metchnikoff (1845–1916) demonstrated the existence of phagocytic cells in the blood, thus demonstrating cell-mediated immunity

VI. The Development of Industrial Microbiology and Microbial Ecology

A.   Louis Pasteur demonstrated that alcoholic fermentations were performed by yeasts under anaerobic w/out oxygen conditions and that different fermentation products were produced by different microorganisms; he also developed the process of pasteurization used to preserve wine and milk during storage

B.   Sergei Winogradsky (1856–1953) worked with soil bacteria and discovered that they could oxidize iron, sulfur, and ammonia to obtain energy; he also studied anaerobic nitrogen fixation and cellulose decomposition

C.   Martinus Beijerinck (1851–1931) isolated aerobic nitrogen-fixing bacteria, a root-nodule bacterium capable of fixing nitrogen, and sulfate-reducing bacteria

D.   Beijerinck and Winogradsky pioneered the use of enrichment cultures and selective media to isolate microorganisms that can eat and/or breathe inorganic minerals

VII. The Scope and Relevance of Microbiology

A.   Microorganisms were the first living organisms on the planet, live everywhere life is possible, are more numerous than any other kind of organism, and constitute the largest component of the Earth's biomass

B.   The entire ecosystem depends on the activities of microorganisms, and microorganisms influence human society in countless ways

C.   Microbiology has an impact on many fields including medicine, agriculture, food science, ecology, genetics, biochemistry, and molecular biology

D.   Microbiologists may be interested in specific types of organisms:
1. Virologists—viruses
2. Bacteriologists—bacteria
3. Phycologists or Algologists—algae
4. Mycologists—fungi
5. Protozoologists—protozoa

pathology - study + diagnosis of disease thru examination of organs, tissues, bodily fluids etc

E.   Microbiologists may be interested in various characteristics or activities of microorganisms:
1. Microbial morphology structure + form
2. Microbial cytology structure, function, multiplication, pathology + life history of cells
3. Microbial physiology mechanical, physical + biochemical functions of living organisms
4. Microbial ecology environment
5. Microbial genetics and molecular biology
6. Microbial taxonomy classification

F.   Microbiologists may have a more applied focus:
1. Medical microbiology
2. Public health microbiology
3. Immunology
4. Agricultural microbiology

5. Microbial ecology
6. Food microbiology
7. Industrial microbiology

VIII. The Future of Microbiology
    A. Microbiology has had and will continue to have a profound influence on society; there is a clear mission and practical significance
    B. Future microbiologists will continue to:
       1. Try to better understand and control existing, emerging, and reemerging infectious diseases
       2. Learn more about host defenses and host-pathogen interactions
       3. Develope new uses for microbes in industry, agriculture, and environmental clean-up
       4. Understand microbial diversity and isolate new microorganisms with unique features
       5. Understand microbial communities, biofilms, and microbial interactions
       6. Use bioinformatics to analyze growing genomic and proteomic databases
       7. Assess and communicate potential impacts of new discoveries and technologies on society

AREAS OF MICROBIOLOGY

From the list below, select the area that best matches the description in each of the numbered statements.

_f_ 1. Deals with diseases of humans and animals

_j_ 2. Endeavors to control the spread of communicable diseases

_D_ 3. Deals with the mechanisms by which the human body protects itself from disease-causing organisms

_a_ 4. Deals with microorganisms that cause damage to crops or live in herds of domestic animals; also deals with ways of increasing soil fertility

_g_ 5. Studies the relationship between microorganisms and their habitats

_b_ 6. Investigates the causes of spoilage of products for human consumption and the use of microorganisms in the production of cheese, yogurt, pickles, beer, etc.

_e_ 7. Employs microorganisms to make products such as antibiotics, vaccines, steroids, alcohols, vitamins, amino acids, and enzymes

_h_ 8. Investigates the synthesis of antibiotics and toxins, microbial energy production, the ways in which microorganisms survive harsh environmental conditions, etc.

_i_ 9. Focuses on the nature of genetic information and how genes regulate the development and function of cells and organisms

_c_ 10. Involves the insertion of new genes into organisms in order to investigate the genes' functions or to produce useful organisms with new properties

a. agricultural microbiology
b. food and dairy microbiology
c. genetic engineering
d. immunology
e. industrial microbiology
f. medical microbiology
g. microbial ecology
h. microbial physiology
i. molecular biology
j. public health microbiology

ok

FILL IN THE BLANK

1. The concept that living organisms could develop from nonliving or decomposing matter is referred to as ___spontaneous generation___. An Italian physician named ___Francesco Redi___ challenged this by showing that maggots developed from fly eggs, not from decaying meat as had previously been thought. However, even after this, many thought that simple microorganisms could develop from nonliving material, even if more complex organisms could not. This was finally disproved by the work of a French scientist named ___Louis Pasteur___ and an English physicist named ___John Tyndall___, who both demonstrated that organisms only developed in sterile broth exposed to air if dust particles carrying living organisms dropped into the broth.

2. Indirect evidence that microorganisms were agents of human disease came from the work of an English surgeon named ___Joseph Lister___, who developed a procedure for antiseptic surgery in which he heat-sterilized his surgical instruments before use and also sprayed the compound phenol, which kills bacteria, over the surgical area. These procedures lowered the incidence of postoperative infections.

3. ___Pasteur___ and ___Roux___ discovered that older cultures of the bacterium that caused chicken cholera lost their ability to cause disease, or were said to be ___attenuated___. If chickens were injected with these older strains, they remained healthy but developed the ability to resist disease. They called this culture a ___vaccine___ in honor of the English physician ___Edward Jenner___, who developed the procedure in order to protect against smallpox.

4. Inactivated toxin from the microorganism that causes diphtheria was injected into rabbits by ___Emil Von Behring___ and ___Shibasaburo Kitasato___, inducing the rabbits to produce a soluble substance in the blood that would neutralize the toxin and protect against disease. This is referred to as ___humoral___ immunity. Taking a different approach, Elie Metchnikoff discovered that some blood cells were capable of ingesting bacteria and thereby destroying them. He called this process ___phagocytosis___, and the cells that can do this, ___phagocytes___. This is the basis for cell-mediated immunity.

5. The theory that yeast cells were responsible for converting sugar to alcohol was first proposed by ___Theodore Schwann___ and others. However, it was ___Pasteur___ who did the definitive work that clearly demonstrated this to be true. One of his most important discoveries was that some organisms carried out these processes ___aerobically___ (in the presence of oxygen) while others carried them out ___anaerobically___ (in the absence of oxygen).

6. Two of the most important contributors in the field of microbial ecology were ___Sergei Winogradsky___, who discovered that soil bacteria could oxidize iron, sulfur, and ammonia to obtain energy, and ___Martinus Beijerinck___ who isolated a root-nodule bacterium capable of nitrogen fixation. These individuals also pioneered the use of ___enrichment-culture___ techniques and ___selective___ media.

7. Scientists begin their attempts to understand natural phenomena by gathering observations on a topic and then formulating an ___hypothesis___, which is subsequently tested experimentally. If this initial educated guess about the phenomenon of interest survives the testing, it may become a ___theory___, which is a set of propositions and concepts that provides a reliable, systematic, and rigorous account of an aspect of nature.

MULTIPLE CHOICE

For each of the questions below select the *one best* answer.

1. Which of the following is (are) used to define the field of microbiology?
    a. the size of the organism studied
    b. the techniques employed in the study of organisms regardless of their size
    c. Both (a) and (b) are correct. Both size + techniques
    d. Neither (a) nor (b) is correct.

2. Which of the following developed a set of criteria that could be used to establish a causative link between a particular microorganism and a particular disease?
    a. van Leeuwenhoek
    b. Fracastoro
    c. Pasteur
    d. Koch

3. Which of the following was the first to observe and accurately describe microorganisms?
   a. van Leeuwenhoek
   b. Fracastoro
   c. Pasteur
   d. Koch
4. Which of the following provided strong evidence against the concept of spontaneous generation?
   a. van Leeuwenhoek
   b. Fracastoro
   c. Pasteur
   d. Koch
5. Which of the following provided evidence that a microorganism could be responsible for a particular disease?
   a. Bassi *silkworm disease*
   b. Berkeley *water mold – Potato blight*
   c. Koch *anthrax, TB*
   d. All of the above are correct.

TRUE/FALSE

___T___ 1. Although Robert Koch used and published the criteria for establishing a causative link between a particular microorganism and a particular disease, the criteria were actually first developed by his former teacher, Jacob Henle.

___T___ 2. The major difference between procaryotic cells and eucaryotic cells is that eucaryotic cells have a membrane-bound nucleus but procaryotic cells do not.

___T___ 3. Agar is used instead of gelatin to solidify microbial media because agar is not digested by many bacteria while gelatin is.

___F___ 4. The first disease to be identified as being caused by a virus was anthrax.

___T___ 5. The discovery of viruses and their role in disease was made possible when Charles Chamberland constructed a porcelain filter that would retain bacteria. Using these filters, others found that some disease-causing organisms were not retained by the filter; these were referred to as viruses.

___F___ 6. Louis Pasteur was the first to document the use of a vaccination procedure to prevent disease.

___T___ 7. Microorganisms can be used in bioremediation to reduce pollution effects.

___T___ 8. Genetically engineered microorganisms are used to produce a variety of products including hormones, antibiotics, and vaccines.

___T___ 9. Recent advances in classification led to the proposal that organisms should be divided into three domains: *Archaea*, *Bacteria*, and *Eucarya*.

CRITICAL THINKING

1. Discuss the spontaneous generation theory. Present the evidence that was used to support it and present the evidence that was used to discredit it. How was this theory finally discredited? *Pasteur + Redi*

definition – living organisms develop from non-living matter

2. Describe in detail at least two ways in which microorganisms have a direct and substantial impact on your life (other than as causative agents of disease). Include a discussion of the role(s) of the microorganisms in these processes.

3. In the 1980s when AIDS was first recognized, scientists were faced with the task of identifying the causative agent. Suppose you were one of those scientists and that you observed a virus in the tissues of AIDS patients. Which steps of Koch's postulate could you use to support your belief that the virus was responsible for the disease? Which steps couldn't you use? Explain your answers.

ANSWER KEY

Areas of Microbiology

1. f, 2. j, 3. d, 4. a, 5. g, 6. b, 7. e, 8. h, 9. i, 10. c

Fill in the Blank

1. spontaneous generation; Francesco Redi; Louis Pasteur; John Tyndall 2. Joseph Lister 3. Pasteur; Roux; attenuated; vaccine; Jenner 4. Emil von Behring; Shibasaburo Kitasato; humoral; phagocytosis; phagocytes 5. Theodore Schwann; Louis Pasteur; aerobically; anaerobically 6. Sergius Winogradsky; Martinus Beijerinck; enrichment culture; selective 7. hypothesis; theory

Multiple Choice

1. c, 2. d, 3. a, 4. c, 5. d

True/False

1. T, 2. T, 3. T, 4. F, 5. T, 6. F, 7. T, 8. T, 9. T

2 The Study of Microbial Structure: Microscopy and Specimen Preparation

CHAPTER OVERVIEW

This chapter provides a relatively detailed description of the bright-field microscope and its use. Other common types of light microscopes are also described. Following this, various procedures for the preparation and staining of specimens are introduced. The chapter continues with a description of the two major types of electron microscopes and the procedures associated with their use. It concludes with descriptions of recent advances in microscopy: confocal microscopy and scanning probe microscopy.

CHAPTER OBJECTIVES

After reading this chapter you should be able to:

- describe how lenses bend light rays to produce enlarged images of small objects
- describe the various parts of the light microscope and how each part contributes to the functioning of the microscope
- describe the preparation and simple staining of specimens for observation with the light microscope
- describe the Gram-staining procedure and how it is used to categorize bacteria
- describe the basis for the various staining procedures used to visualize specific structures associated with microorganisms
- compare the operation of the transmission and scanning electron microscopes with each other and with light microscopes
- describe confocal microscopy and scanning probe microscopy
- compare and contrast light microscopes, electron microscopes, confocal microscopes, and scanning probe microscopes in terms of their resolution, the types of specimens that can be examined, and the images produced

CHAPTER OUTLINE

I. Lenses and the Bending of Light
   A. Light is refracted (bent) when passing from one medium to another; the refractive index is a measure of how greatly a substance slows the velocity of light; the direction and magnitude of refraction is determined by the refractive indexes of the two media forming the interface
   B. Convex lenses bend parallel light rays from a distant light source and focus the light rays at a specific place known as the focal point; the distance between the center of the lens and the focal point is the focal length
II. The Light Microscope
   A. The bright-field microscope produces a dark image against a brighter background; the total magnification of the image is the product of the magnification of the objective lens and the magnification of the ocular (eyepiece) lens
   B. Resolution
      1. Microscope resolution refers to the ability of a lens to separate or distinguish small objects that are close together
      2. The major factor determining resolution is the wavelength of light used; the shorter the wavelength, the greater the resolution

3.   The numerical aperture of the objective lens also impacts resolution; the larger the numerical aperture, the greater the resolution and the shorter the working distance of the lens

4.   The unaided eye has a resolution of 0.2 mm; the typical light microscope has a resolution 1000 times greater

   C.   The dark-field microscope produces a bright image of the object against a dark background and is used to observe living, unstained preparations

   D.   The phase-contrast microscope converts slight differences in refractive index and cell density into easily detected variations in light intensity; it is an excellent way to observe living cells

   E.   The differential interference contrast microscope detects differences in refractive indices and thickness using two beams of polarized light, which form brightly colored, three-dimensional images of living, unstained specimens

   F.   The fluorescence microscope excites a specimen with a single color of light and shows a bright image of the object resulting from the fluorescent light emitted by the specimen; specimens are usually stained with a fluorochrome

III. Preparation and Staining of Specimens

   A.   Fixation refers to the process by which internal and external structures are preserved and fixed in position; it usually kills the organism and firmly attaches it to the microscope slide

      1.   Heat fixing preserves overall morphology but not internal structures

      2.   Chemical fixing is used to protect fine cellular substructure and the morphology of larger, more delicate microorganisms

   B.   Dyes and simple staining   *conjugated system*

      1.   Dyes have two common features: chromophore groups and the ability to bind cells by ionic, covalent, or hydrophobic bonding

         a. + Basic dyes bind negatively charged molecules and cell structures

         b. – Acid dyes bind to positively charged molecules and cell structures

      2.   Simple staining uses a single staining agent to stain a specimen

   C.   Differential staining is used to divide bacteria into separate groups based on their different reactions to an identical staining procedure

      1.   Gram staining is the most widely used differential staining procedure because it divides bacterial species into two groups—gram positive and gram negative—based on cell wall characteristics

gram + purple

gram – pink to Red

         a.   The smear is first stained with crystal violet, which stains all cells purple

         b.   Iodine is used as a mordant to increase the interaction between the cells and the dye

         c.   Ethanol or acetone is used to decolorize; this is the differential step because gram-positive bacteria retain the crystal violet whereas gram-negative bacteria lose the crystal violet and become colorless

         d.   Safranin is then added as a counterstain to turn the gram-negative bacteria pink while leaving the gram-positive bacteria purple

      2.   Acid-fast staining is a differential staining procedure for cell walls that can be used to identify two medically important species of bacteria—*Mycobacterium tuberculosis,* the causative agent of tuberculosis, and *Mycobacterium leprae,* the causative agent of leprosy

   D.   Staining specific structures

      1.   Negative staining is widely used to visualize diffuse capsules surrounding the bacteria; those capsules are unstained by the procedure and appear colorless against a stained background

      2.   Endospore staining is a double-stain technique by which bacterial endospores stain one color and vegetative cells stain a different color

      3.   Flagella staining is a procedure in which mordants are applied to increase the thickness of flagella to make them easier to see

IV. Electron Microscopy

   A.   The transmission electron microscope (TEM)

      1.   The TEM has a resolution about 1,000 times better than that of the light microscope (0.5 nm versus 0.2 μm) due to the short wavelength of the electron beam used to create the image

2.   In TEM, electrons scatter when they pass through thin sections of a specimen; the transmitted electrons (those that do not scatter) are used to produce an image on a fluorescent screen

3.   Specimen preparation for TEM involves procedures for cutting thin sections, chemical fixation, drying, and staining with electron-dense materials (analogous to the procedures used for the preparation of specimens for light microscopy); other preparation methods include negative staining, shadowing, and freeze-etching

B.   The scanning electron microscope (SEM) uses electrons reflected from the surface of a specimen to produce a three-dimensional image of its surface features; many SEMs have a resolution of 7 nm or less; specimen preparation usually involves chemical fixation, drying, and coating

V.   Newer Techniques in Microscopy

A.   Confocal microscopy

1.   A focused laser beam is used to illuminate a point on a specimen (usually fluorescently stained); light from the illuminated spot is focused by an objective lens; an aperture above the lens blocks out stray light from parts of the specimen that lie above and below the plane of focus; a detector measures the amount of illumination from each point, creating a digitized signal

2.   After examining many points (optical z-sections), a computer combines all the digitized signals to form a three-dimensional image with excellent contrast and resolution, especially valuable for examining living biofilms

B.   Scanning Probe Microscopy

1.   Scanning probe microscopy measures surface features by moving a short probe over the object's surface

2.   The scanning tunneling electron microscope creates an image using a tunneling electrical current, held at a steady level by the vertical movement of the probe across the surface of the specimen; the vertical motion of the probe is used to create a three-dimensional image of the specimen's surface atoms; resolution is such that individual atoms can be observed

3.   The atomic force microscope uses a very small amount of force on the probe tip to maintain a constant distance between the tip and the specimen; the vertical movement of the probe across the surface of the specimen is used to create a three-dimensional image; because it does not make use of a tunneling current, it is useful for surfaces that do not conduct electricity well

TERMS AND DEFINITIONS

Place the letter of each term in the space next to the definition or description that best matches it.

_q_ 1. The bending of light rays at the interface of one medium with another

_r_ 2. A measure of how greatly a substance changes the velocity of light, and a factor determining the direction and magnitude of the bending of light rays

_k_ 3. The point at which a lens focuses rays of light

_j_ 4. The distance between the center of a lens and the focal point

_b_ 5. Conventional microscope that produces a dark image against a brighter background

_o_ 6. Describes a microscope whose image remains in focus when the objectives are changed

_s_ 7. The ability of a lens to separate or distinguish small objects that are close together

_d_ 8. A microscope that uses two beams of polarized light to form three-dimensional images of living, unstained specimens

_z_ 9. The distance between the front surface of the lens and the surface of the cover glass (or the specimen) when the specimen is in sharp focus

_c_ 10. A microscope that produces a bright image of the specimen against a dark background

_p_ 11. A microscope that converts slight differences in refractive index and cell density into easily detected variations in light intensity

_i_ 12. A microscope that exposes specimens to ultraviolet, violet, or blue light and forms an image from the resulting light emitted, which has a different wavelength

_g_ 13. The process by which the internal and external structures of cells and organisms are preserved and maintained in position

_w_ 14. A staining process in which a single staining agent is used

_e_ 15. A staining process that divides organisms into two or more separate groups depending on their reaction to the same staining procedure

_m_ 16. A substance that accelerates the reaction of cell structures with a dye so that the cell is more intensely stained

_n_ 17. A staining procedure in which the background is dark and the organism remains unstained

_x_ 18. A staining process in which heat is used to increase the affinity of bacterial endospores to dye; endospores are usually resistant to simple staining procedures

_h_ 19. A staining process that enables the observation of thin, threadlike flagella by increasing their thickness and then staining them

a. atomic force microscope
b. bright-field microscope
c. dark-field microscope
d. differential interference contrast (DIC) microscope
e. differential staining
f. electron microscope
g. fixation
h. flagella staining
i. fluorescence microscopy
j. focal length
k. focal point
l. freeze-etching
m. mordant
n. negative staining
o. parfocal
p. phase-contrast microscope
q. refraction
r. refractive index
s. resolution
t. scanning electron microscope (SEM)
u. scanning tunneling electron microscope
v. shadowing
w. simple staining
x. spore staining
y. transmission electron microscope (TEM)
z. working distance

11

20. A microscope that forms an image by focusing a beam of electrons on a specimen

21. An electron microscope that creates an image from transmitted electrons (those not scattered when they pass through a thin section of a specimen) *TEM*

22. A staining process in which heavy metals are applied to specimens at an approximately 45° angle; this provides a three-dimensional image similar to shadowing with light

23. An electron microscope that creates an image from electrons emitted from the surface of a specimen that has been excited by a beam of focused electrons

24. A procedure in which frozen specimens are broken along lines of greatest weakness, usually down the middle of internal membranes; exposed surfaces are then shadowed for production of a better image

25. A scanning probe microscope that uses voltage flow between the tip of the probe and the electron clouds of the surface atoms of the specimen

26. A scanning probe microscope that is useful for surfaces that do not conduct electricity well

MICROSCOPE IDENTIFICATION

Using the terms listed below, label the parts of the microscope indicated on the accompanying picture. In the space beside each term provide a *brief* description of its function.

1. Ocular (eyepiece) lens: 10x magnification

2. Arm: connects base · barrel

3. Objective lens:

4. Coarse focusing adjustment knob:

5. Fine focusing adjustment knob:

6. Base: Supports the microscope

7. Nosepiece:

8. Stage: — supports slide

9. Substage condenser:

10. Diaphragm lever: regulates light

11. Mechanical stage:

12. Lamp: w/ Base

13. Body:

14. Field diaphragm lever:

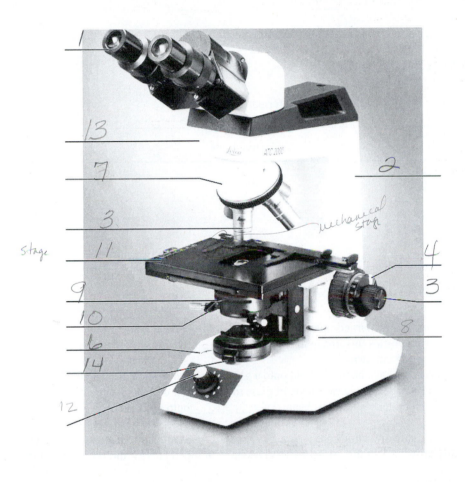

1

13

7

3

Stage   11

9

10

6

14

12

2

mechanical stage

4

3

8

GRAM-STAINING PROCEDURE

Complete the table for the Gram-staining procedure by supplying the missing information.

Procedure Step	Reagent	Color after Completion of Step	
		Gram Positive	Gram Negative
1. Primary stain	crystal violet	Purple	Purple
2. _mordant_	Iodine	Purple	Purple
3. Decolorizer	ethanol or acetone	Purple	Colorless
4. Counterstain	safranin	Purple	red

LENSES AND MAGNIFICATION

Complete the table below by filling in the missing information.

	Ocular Lens	Objective Lens	Magnification
1.	10×	40×	400×
2.	10×	100×	1000×
3.	15×	40×	600×

FILL IN THE BLANK

1. The objective lens forms an enlarged image within the microscope called the _"real"_ image. The eyepiece lens further magnifies this image to form the _"virtual"_ image, which appears to lie just beyond the stage about 25 cm away.
2. Thin films of bacteria that have been air-dried onto a glass microscope slide are called _smears_.
3. A _light_ microscope uses lenses made of glass to focus light onto a specimen, while an electron microscope uses magnetic lenses to focus beams of _electrons_ onto a specimen.
4. Special dyes called _fluorochrome_ are used in fluorescence microscopy. These dyes are excited by light with a specific wavelength and emit light with a _longer_ wavelength, thus having less energy than the light originally absorbed. In this way the dye gives up its trapped energy and returns to a more stable state.
5. The presence of diffuse capsules surrounding many bacteria is commonly revealed by _negative stain or capsule_ staining, in which the background is stained dark. The cell can also be counterstained for greater visibility, leaving the capsule colorless.
6. Although there are many types of dyes, all share two common features: _chromophore_ groups, which give dyes their colors, and the ability to bind to cells by ionic, covalent, and hydrophobic bonds. Those that bind by ionic bonds are called ionizable dyes. They can be divided into two broad classes based on the nature of their charged group. Dyes such as methylene blue and crystal violet have positively charged groups and are called _basic_ dyes. Dyes such as eosin have negatively charged groups and are called _acidic_ dyes.

MULTIPLE CHOICE

For each of the questions below select the *one best* answer.

1. Acid-fast organisms such as *Mycobacterium tuberculosis* resist decolorization by acid-alcohol solutions because of the high concentration of _____ in their cell walls.
   a. proteins
   b. carbohydrates
   c. lipids
   d. peptidoglycan

2. Why are smears heat-fixed prior to staining?
   a. to kill the organism
   b. to preserve the internal structures
   c. to attach the organism firmly to the slide
   d. All of the above are correct.

3. Which type of microscope is best for visualizing small morphological features within the cell interior?
   a. light microscope
   b. transmission electron microscope
   c. dark-field microscope
   d. scanning electron microscope

4. Transmission electron microscopy requires the use of thin slices of a microbial specimen. What should the thickness of the specimen be?
   a. 20 to 100 mm
   b. 100 to 200 nm
   c. 20 to 100 nm
   d. 0.2 to 10 nm

5. For transmission electron microscopy, a specimen can be spread out in a thin film with uranyl acetate, which does not penetrate the specimen. What is this procedure called?
   a. negative staining
   b. shadowing
   c. freeze-etching
   d. simple staining

6. Which type of microscope best reveals surface features of an organism?
   a. fluorescence microscopy
   b. phase-contrast microscopy
   c. scanning electron microscopy
   d. transmission electron microscopy

7. As the magnification of a series of objective lenses increases, what happens to the working distance?
   a. It increases.
   b. It decreases.
   c. It stays the same.
   d. It cannot be predicted.

8. The Gram-staining procedure differentiates bacteria based on the chemical composition of which cell structure?
   a. cytoplasmic membrane
   b. cell wall
   c. cytoplasm
   d. chromosome

9. What is the distance between the focal point of a lens and the center of the lens called?
   a. working distance
   b. numerical aperture
   c. focal length
   d. parallax distance

10. A microscope is able to keep objects in focus when the objective lens is changed. What term is used to describe this property?
    a. equifocal
    b. parfocal
    c. optically constant
    d. focally constant

11. Which microscope is especially useful for examining thick specimens such as biofilms?
    a. transmission electron microscope
    b. dark-phase-contrast microscope
    c. confocal scanning laser microscope
    d. bright-field light microscope

TRUE/FALSE

__T__ 1. Light is refracted at the interface between two materials with different refractive indexes because the velocity of light is altered.

__T__ 2. Resolution becomes greater as the wavelength of the illuminating light decreases.

__T__ 3. Phase-contrast microscopy enhances density differences among internal cellular structures and therefore allows these structures to be visualized without stains or dyes.

__T__ 4. Basic dyes are cationic (positively charged) and are commonly used to stain bacteria since the surfaces of these organisms are usually negatively charged.

__T__ 5. The Gram-staining procedure is one of the most widely used differential stains because it divides bacterial species into two groups: gram-positive and gram-negative.

__F__ 6. Since transmission electron microscopy uses electrons rather than light, it is not necessary to stain biological specimens before observation.

__T__ 7. Freeze-etching minimizes the production of artifacts since the cells are not subjected to chemical fixation, dehydration, or plastic embedding.

__T__ 8. The resolution of a microscope is not related to its magnification. Therefore, although it is possible to build a light microscope capable of 10,000× magnification, it would only be magnifying a blur.

__T__ 9. Scanning tunneling electron microscopes can be used to visualize individual atoms.

__T__ 10. Denser regions of a specimen scatter more electrons and therefore appear darker in the image projected onto the screen of a transmission electron microscope.

__T__ 11. The larger the numerical aperture of the objective lens, the greater the resolution of the microscope.

CRITICAL THINKING

1. Explain why it is possible to increase the magnification of the light microscope above 1,500× and yet not be able to see any additional details.

2. Compare and contrast electron microscopy with light microscopy. Include in your answer the operation of the instruments, the degree of magnification and resolution possible, and the procedures used for preparation, fixation, and staining of specimens. What major advances in our knowledge of cell structure were made possible by the invention of the electron microscope that were not possible with the light microscope?

ANSWER KEY

Terms and Definitions

1. q, 2. r, 3. k, 4. j, 5. b, 6. o, 7. s, 8. d, 9. z, 10. c, 11. p, 12. i, 13. g, 14. w, 15. e, 16. m, 17. n, 18. x, 19. h, 20. f, 21. y, 22. v, 23. t, 24. l, 25. u, 26. a

Gram-Staining Procedure

1. crystal violet; purple; purple 2. mordant; purple 3. ethanol or acetone; purple 4. safranin; purple; red

Lenses and Magnification

1. 400×, 2. 100×, 3. 40×

Fill in the Blank

1. real; virtual 2. smears 3. light; electrons 4. fluorochromes; longer 5. negative 6. chromophore; basic; acid

Multiple Choice

1. c, 2. d, 3. b, 4. c, 5. a, 6. c, 7. b, 8. b, 9. c, 10. b 11. c

True/False

1. T, 2. T, 3. T, 4. T, 5. T, 6. F, 7. T, 8. T, 9. T, 10. T, 11. T

3 Procaryotic Cell Structure and Function

CHAPTER OVERVIEW

This chapter provides a description of the procaryotic cell, including bacterial and archaeal cells. Throughout this chapter, the term *procaryote* is used to mean both bacteria and archaea. The discussion begins with the general features of size, shape, and arrangement of procaryotic cells. Then the general features of biological membranes and the specific features of procaryotic membranes are given. Important internal structures of procaryotes, such as the cytoplasmic matrix, ribosomes, inclusion bodies, and the nucleoid are described, as well as structures external to the cell, such as the cell wall, glycocalyx, fimbriae, and flagella. The differences between the cell walls of gram-positive bacteria and gram-negative bacteria are discussed and the mechanism of this differential staining reaction is explained. The chapter concludes with a discussion of bacterial chemotaxis and bacterial endospores.

CHAPTER OBJECTIVES

After reading this chapter you should be able to:

* identify major procaryotic cell structures in a drawing or photomicrograph
* describe the various sizes, shapes, and cellular arrangements exhibited by procaryotes
* describe the bacterial plasma membrane and the limited internal membrane structures found in procaryotes
* describe the appearance, composition, and function of the various internal structures found in procaryotic organisms (such as inclusion bodies, ribosomes, and the nucleoid)
* compare the structure of gram-positive and gram-negative bacterial cell walls and explain how the differences between the two contribute to their Gram reaction
* compare and contrast the major protein secretion pathways: sec-dependent protein secretion pathway and the five types of gram-negative protein secretion pathways
* describe external structures such as capsules, fimbriae, and flagella
* diagram and describe the various arrangements of bacterial flagella
* describe how bacteria use their locomotive ability to swim toward chemical attractants and away from chemical repellents
* describe the production of the bacterial endospore and how it enables endospore-forming bacteria to survive harsh environmental conditions and renew growth when the environment becomes conducive to growth

CHAPTER OUTLINE

I. An Overview of Procaryotic Cell Structure
    A. Size, shape, and arrangement
        1. Most prokaryotes are spheres (cocci) or rods (bacilli)
            a. During the reproductive process, some cocci remain attached to each other to form pairs (diplococci), chains, clusters, square planar configurations (tetrads), or cubic configurations (sarcinae)
            b. Some rods are so short and wide that they appear to be ovals (coccobacilli); most rods occur singly, but some form pairs or chains
        2. Other shapes include: curved rods (vibrios); rigid helices (spirilla), and flexible helices (spirochetes); a few bacteria are flat; filaments, which can produce a network, are called a mycelium; bacteria that exhibit more than one form are called pleomorphic

3. Procaryotic cells vary in size (generally 1 to 5 μm) although they are typically smaller than most eucaryotic cells; there have been reports of nanobacteria (0.2 μm to less than 0.05 μm in diameter) and of very large procaryotes (up to 750 μm in diameter)
   B. Procaryotic cell structure
   1. A variety of structures (cell wall, periplasmic space, plasma membrane, nucleoid, ribosomes, inclusion bodies, flagella, capsules, and slime layers) are observed in procaryotic cells
   2. Not all structures are found in every genus; procaryotic cells are morphologically simpler than eucaryotic cells

II. Procaryotic Cell Membranes
   A. Membranes contain both lipids and proteins; most membrane lipids are amphipathic molecules, having both hydrophilic (interact with water) and hydrophobic (insoluble in water) components; the amphipathic nature of these lipids allows them to form lipid bilayers
   B. The plasma membrane serves several functions:
      1. It retains the cytoplasm and separates the cell from its environment
      2. It serves as a selectively permeable barrier
      3. It contains transport systems used for nutrient uptake, waste excretion, and protein secretion
      4. It is the location of a variety of crucial metabolic processes including respiration, photosynthesis, lipid synthesis, and cell wall synthesis
      5. It contains special receptor molecules that enable detection of and response to chemicals in the surroundings
   C. The fluid mosaic model of membrane structure
      1. This model, proposed by Singer and Nicholson, states that membranes are lipid bilayers with floating proteins
      2. Cell membranes are very thin (5–10 nm thick); the lipids are amphipathic, having hydrophilic head groups and long hydrophobic tails; the head groups face out of the membrane while the tails are buried in the membrane
      3. Two types of proteins are associated with the fluid lipid portion of the membrane: peripheral (loosely associated and easily removed) and integral (embedded within the membrane and not easily removed)
   D. Bacterial membranes
      1. The plasma membrane of bacteria consists of a phospholipid bilayer with hydrophilic surfaces and a hydrophobic interior; bacterial membranes lack sterols, but many contain sterol-like molecules called hopanoids that help stabilize the membrane
      2. Bacteria do not have membranous organelles, but can have internal membrane systems with specialized functions such as photosynthesis or respiration
   E. Archaeal membranes
      1. Lipids have branched hydrocarbons attached to glycerol by ether links rather than straight-chain fatty acids attached to glycerol by ester links, as seen in bacteria and eucaryotes
      2. Long tetraether structures that form monolayer membranes also are found
      3. Different lipids can be combined to yield membranes of different fluidity and stability

III. The Cytoplasmic Matrix
   A. The cytoplasmic matrix is the substance between the membrane and the nucleoid; it is featureless in electron micrographs but is often packed with ribosomes and inclusion bodies; although lacking a true cytoskeleton, the cytoplasmic matrix of bacteria does have a cytoskeleton-like system of proteins
   B. The procaryotic cytoskeleton has homologs of the elements seen in eucaryotes, filaments of actin, tubulin, and intermediate filament proteins
   C. Inclusion bodies
      1. Many inclusion bodies are granules of organic or inorganic material that are stockpiled by the cell for future use; some are not bounded by a membrane, but others are enclosed by a single-layered membrane
      2. Organic inclusion bodies include glycogen (carbon storage), poly-β-hydroxybutyrate (carbon storage), cyanophycin granules (nitrogen storage in cyanobacteria), carboxysomes (ribulose-1, 5-bisphosphate carboxylase reserves)

19

3. Gas vacuoles are composed of hollow protein sacs that are filled with gases and used for buoyancy control in aquatic environments
4. Inorganic inclusion bodies include polyphosphate granules (phosphate storage), sulfur granules, and magnetosomes (for magnetic orientation)

D. Ribosomes
1. Ribosomes are complex structures consisting of protein and RNA
2. They are responsible for the synthesis of cellular proteins
3. Procaryotic ribosomes are similar in structure to, but smaller (70S with 50S and 30S subunits) than, eucaryotic ribosomes (80S)

IV. The Nucleoid
A. The nucleoid is an irregularly shaped region in which the chromosome of the procaryote is found
1. In most procaryotes, the nucleoid contains a single circular chromosome, though some have more than one chromosome or have one or more linear chromosomes
2. The nucleoid is not membrane bound, but it is sometimes found to be associated with the plasma membrane
B. The bacterial chromosome is an efficiently packed DNA molecule that is looped and coiled extensively

V. Plasmids
A. Plasmids are small, circular DNA molecules that are not part of the bacterium's chromosome
1. Plasmids have their own replication origins, replicate autonomously, and are stably inherited
2. Episomes are plasmids that can exist either with or without being integrated into the host chromosome
3. Plasmids can be eliminated from a cell by a process called curing, which can occur either spontaneously or as a result of treatments that inhibit plasmid replication but do not affect host cell reproduction
B. Conjugative plasmids (fertility factors, F factors, or F plasmids) are episomes that usually have genes for sex pili and can transfer copies of themselves to other bacteria during conjugation
C. Resistance factors (R plasmids) have genes for resistance to various antibiotics; some are conjugative; however, they usually do not integrate into the host chromosome; R factors can be transferred to other cells even across species lines, spreading antibiotic resistance, a major public health concern
D. Col plasmids carry genes for the synthesis of bacteriocins (e.g., colicins), proteins that kill other bacterial species
E. Other types of plasmids include virulence plasmids, which may carry a gene for toxin production, and metabolic plasmids, which carry genes for enzymes that utilize certain substances as nutrients (aromatic compounds, pesticides, etc.)

VI. The Bacterial Cell Wall
A. The cell wall is a rigid structure that lies just outside the plasma membrane; it creates characteristic shapes for the bacteria and protects from osmotic lysis (bursting)
B. Overview of bacterial cell wall structure
1. The cell walls of most bacteria contain peptidoglycan
2. The cell walls of gram-positive bacteria and gram-negative bacteria differ greatly, but both have a periplasmic space between the cell wall material and the plasma membrane
C. Peptidoglycan structure
1. Peptidoglycan (murein) is a polysaccharide polymer composed of two sugar derivatives and several amino acids; the polysaccharide polymer is a linear chain of alternating N-acetylglucosamine and N-acetylmuramic acid molecules
2. Polysaccharide chains of peptidoglycan are cross-linked via a peptide interbridge attached to the sugar backbone via a short peptide chain; these peptides contain amino acids not found in proteins; cross-linking adds strength to the peptidoglycan mesh
D. Gram-positive cell walls
1. They consist of a thick wall composed of many layers of peptidoglycan and large amounts of teichoic acids

AA - all l amino acids

2. Techoic acids are polymers with a glycerol and phosphate backbone that span the cell wall and likely enhance its structural stability
3. The periplasmic space of gram-positive cells is usually thin and contains only a few secreted proteins (exoenzymes)
4. Most gram-positive bacteria also have a layer of proteins (S-layer proteins) on the outer surface of the peptidoglycan that have a role in wall synthesis and virulence

E. Gram-negative cell walls
1. The gram-negative cell wall is more complex than the gram-positive cell wall; and has a thin layer of peptidoglycan surrounded by an outer membrane
2. The periplasmic space is often wide and contains many different proteins; some are involved with energy conservation or nutrient acquisition
3. The outer membrane is composed of lipids, lipoproteins, and lipopolysaccharides (LPS); Braun's lipoprotein attaches the outer membrane to the peptidoglycan
4. LPSs are large complex molecules composed of lipid A, core polysaccharides, and O antigen carbohydrate side chains; LPSs stabilize the outer membrane, protect against some toxins, and can cause strong host immunological responses, acting as an endotoxin ?
5. The outer membrane is more permeable than the plasma membrane because of porin proteins that form channels through which molecules smaller than 600–700 Daltons can pass

F. The mechanism of Gram staining
1. constriction of the thick peptidoglycan layer of gram-positive cells, thereby preventing the loss of crystal violet during the brief decolorization step
2. the thinner, less cross-linked peptidoglycan layer of gram-negative bacteria cannot retain the stain as well, and these bacteria are thus more readily decolorized when treated with alcohol

G. The cell wall and osmotic protection
1. The cell wall prevents swelling and lysis of bacteria in hypotonic solutions; in hypertonic habitats, the plasma membrane shrinks away from the cell wall in a process known as plasmolysis
2. Bacteria without cell walls (by removal with lysozyme or through peptidoglycan synthesis inhibition by penicillin) called spheroplasts are osmotically sensitive
3. Mycoplasmas lack a cell wall and tend to be pleomorphic   stain gram-

VII. Archaeal Cell Walls
A. Archaea can stain either gram positive or gram negative, but their cell wall structure differs significantly from that of bacteria
1. Many archaea that stain gram positive have a cell wall made of a single homogeneous layer
2. The archaea that stain gram negative lack the outer membrane and complex peptidoglycan network associated with gram-negative bacteria
B. Archaeal cell wall chemistry is different from that of bacteria
1. Cell walls lack muramic acid and D-amino acids and therefore is resistant to lysozyme and β-lactam antibiotics
2. Some have pseudomurein, a peptidoglycanlike polymer that has L-amino acids in its cross-links and different monosaccharide subunits and linkage
3. Others have different polysaccharides
C. The archaea that stain gram negative have a layer of protein or glycoprotein outside their plasma membrane

VIII. Protein Secretion in Procaryotes
A. Overview of bacterial protein secretion
1. Proteins are located outside the plasma membrane, outside the cell wall and in the periplasmic space
2. Gram-positive cells secrete proteins across the plasma membrane using the Sec-dependent pathway; gram-negative cells use this pathway and others to secrete proteins across the plasma membrane and the outer membrane; all of these pathways require energy from ATP or GTP
B. The Sec-dependent pathways
1. This pathway (also called the general secretion pathway) translocates proteins from the cytoplasm across or into the plasma membrane

21

2. Proteins secreted by this pathway are synthesized as preproteins having a signal peptide at their amino terminal; the signal peptide delays protein folding
3. Chaperone proteins keep the preproteins unfolded and help them reach the transport machinery
4. The protein is transferred across or into the plasma membrane; this is accompanied by the removal of the signal peptide and subsequent folding of the protein

C. Protein secretion in gram-negative bacteria
   1. The type II protein secretion pathways
      a. This pathway is observed in a number of pathogens; it transports proteins that passed from the periplasmic space across the outer membrane after the proteins pass through the plasma membrane via the Sec-dependent or Tat pathways
      b. These systems are very complex and consist primarily of integral membrane proteins
   2. Type V protein secretion pathways also start with Sec-dependent transport across the plasma membrane, but the proteins then form a channel in the outer membrane through which they autotransport
   3. The type I protein secretion pathways (ABC protein secretion pathway)
      a. This pathway moves proteins from the cytoplasm across both the plasma membrane and the outer membrane
      b. Proteins secreted by this pathway contain C-terminal secretion signals
      c. Proteins that comprise type I systems form channels through the membranes; translocation is driven by both ATP hydrolysis and proton motive force
   4. Type III protein secretion pathway
      a. This pathway is used by several gram-negative pathogens to secrete virulence factors from the cytoplasm, across both the plasma membrane and outer membrane, and into host cells
      b. Some bacteria form needle-shaped type III secretion machinery and the secreted proteins are thought to move through a translocation channel
   5. Type IV protein secretion pathways use a syringelike structure as in the type III system, but can transport DNA during conjugation in addition to proteins

IX. Components External to the Cell Wall
   A. Capsules, slime layers and S-layers
      1. Capsules and slime layers (also known as glycocalyx) are layers of polysaccharides lying outside the cell wall; they protect the bacteria from phagocytosis, desiccation, viral infection, and hydrophobic toxic materials such as detergents; they also aid bacterial attachment to surfaces and gliding motility; capsules are well organized, whereas slime layers are diffuse and unorganized
      2. S-layers are regularly structured layers of protein or glycoprotein observed in both bacteria and archaea; in archaea, the S-layer may be the only structure outside the plasma membrane; S-layers protect against ion and pH fluctuations, osmotic stress, hydrolytic enzymes, or the predacious bacterium *Bdellovibrio*; they can also help maintain cell shape and envelope rigidity, promote cell adhesion, and protect against host defenses
   B. Pili and fimbriae are short, thin, hairlike appendages that mediate bacterial attachment to surfaces (fimbriae) or to other bacteria during sexual mating (sex pili); fimbriae tend to be narrower in diameter and shorter than sex pili; certain fimbriae are required for twitching motility and gliding motility observed for some bacteria
   C. Flagella and motility
      1. Flagella are threadlike locomotor appendages extending outward from the plasma membrane and cell wall; they may be arranged in various patterns:
         a. Monotrichous—a single flagellum
         b. Amphitrichous—a single flagellum at each pole
         c. Lophotrichous—a cluster (tuft) of flagella at one or both ends
         d. Peritrichous—a relatively even distribution of flagella over the entire surface of the bacterium

2. Flagellar ultrastructure—the flagellum consists of a hollow filament composed of a single protein known as flagellin; the hook is a short, curved segment that links the filament to the basal body, a series of rings that drives flagellar rotation

3. Flagellar synthesis—synthesis of flagella involves many genes for the hook and basal body, as well as the gene for flagellin; new molecules of flagellin are transported through the hollow filament so that the growth of the flagellum is from the tip, not from the base

4. The mechanism of flagellar movement—procaryotic flagella rotate; the direction of flagellar rotation determines the nature of bacterial movement: counterclockwise rotation causes forward motion (called a run) and clockwise rotation disrupts forward motion (resulting in a tumble)

5. Procaryotes can move by other mechanisms: in spirochetes, axial filaments cause movement by flexing and spinning; other procaryotes exhibit gliding motility, a mechanism by which they coast along solid surfaces

X. Chemotaxis

A. Chemotaxis is directed movement of bacteria either toward a chemical attractant or away from a chemical repellent

B. The concentrations of these attractants and repellents are detected by chemoreceptors in the periplasmic space or the plasma membrane

C. Directional travel toward a chemoattractant (biased random walk toward attractant) is caused by lowering the frequency of tumbles (twiddles), thereby lengthening the runs when traveling up the gradient, but allowing tumbling to occur at normal frequency when traveling down the gradient

D. Directional travel away from a chemorepellent (biased random walk away from repellent) involves similar but opposite responses

E. The mechanism of control of tumbles and runs is complex, involving numerous proteins and several mechanisms (conformation changes, methylation, and phosphorylation) to modulate their activity; despite this complexity, chemotaxis is fast, with responses occurring in as little as 200 milliseconds

XI. The Bacterial Endospore

A. The bacterial endospore is a special, resistant, dormant structure formed by some bacteria; it enables them to resist harsh environmental conditions

B. Endospore structure is complex, consisting of an outer covering called the exosporium, a spore coat beneath the exosporium, the cortex beneath the spore coat, and the spore cell wall, which is inside the cortex and surrounds the core

C. Endospore formation (sporulation) normally commences when growth ceases because of lack of nutrients; it is a complex, multistage process

D. Transformation of dormant endospores into active vegetative cells is also a complex, multistage process that includes activation (preparation) of the endospore, germination (breaking of the endospore's dormant state), and outgrowth (emergence of the new vegetative cell)

TERMS AND DEFINITIONS

Place the letter of each term in the space next to the definition or description that best matches it.

_g_ 1. Bacteria that are roughly spherical in shape

_c_ 2. Bacteria that are rod shaped

_m_ 3. Short, thin, hairlike structures that mediate bacterial attachment to other bacteria during mating

_qq_ 4. Curved, rod-shaped bacteria

____ 5. The vegetative cell (mother cell) in which an endospore forms

_x_ 6. Branched network of hyphae

_ll_ 7. Long, rod-shaped bacteria that are twisted into rigid helices

_nm_ 8. Long, rod-shaped bacteria that are twisted into flexible helices

____ 9. The domain of procaryotic organisms distinguished by their rRNA, cell wall composition, lipids, and other features

_ff_ 10. Bacteria that lack a single characteristic shape and that therefore vary in shape

_w_ 11. The structural component of bacterial cell walls that is also called peptidoglycan

_q_ 12. Molecules or regions of molecules that readily interact with water

_r_ 13. Molecules or regions of molecules that are insoluble in water, or do not readily react with water

_aa_ 14. Proteins that are loosely associated with a membrane and that can therefore be easily removed

_t_ 15. Proteins that are not easily extracted from membranes and that are insoluble in aqueous solutions when freed of lipids

_gg_ 16. Plasma membrane and everything contained within

____ 17. A unit of measure of the sedimentation velocity in a centrifuge

_y_ 18. The nonmembrane-bound region of a procaryotic cell in which the DNA is located

_cc_ 19. A circular, double-stranded DNA molecule in procaryotes that can exist and replicate independently of the chromosome

_bb_ 20. A gap between the plasma membrane and the cell wall of bacteria

____ 21. Organisms with a peptidoglycan-like polymer with L-amino acids instead of D-amino acids

_l_ 22. Enzymes that are secreted out of the cell to aid in the acquisition and digestion of nutrients from the environment

_z_ 23. Movement of water across selectively permeable membranes from dilute solutions (higher water concentration) to more concentrated solutions (lower water concentration)

_u_ 24. Bursting of cells that occurs when cells are placed in hypotonic solutions so that water flows in the cell

_ee_ 25. Shrinkage of the plasma membrane away from the cell wall that occurs when cells are placed in hypertonic solutions so that water flows out of the cell

a. Archaea
b. axial filament
c. bacilli
d. capsule
e. carboxysomes
f. chemotaxis
g. cocci
h. conjugative plasmid
i. cyanophycin granules
j. endospore
k. episome
l. exoenzymes
m. fimbriae
n. germination
o. glycocalyx
p. glycogen
q. hydrophilic
r. hydrophobic
s. inclusion bodies
t. integral proteins
u. lysis
v. metabolic plasmid
w. murein
x. mycelium
y. nucleoid
z. osmosis
aa. peripheral proteins
bb. periplasmic space
cc. plasmid
dd. plasmids
ee. plasmolysis
ff. pleomorphic
gg. protoplast
hh. R factor
ii. sex pili
jj. S-layer
kk. slime layer
ll. spirilla
mm. spirochetes
nn. sporangium
oo. sporogenesis (sporulation)
pp. Svedberg unit
qq. Vibrios
rr. virulence plasmid

_____ 26. Well-organized polysaccharides outside the cell wall that are not easily washed off

_____ 27. Diffuse, unorganized polysaccharides outside the cell wall that are easily removed

_____ 28. Protein or glycoprotein layer that exhibits a pattern not unlike floor tiles

_____ 29. Hairlike appendages that are thinner than flagella and are usually involved in attachment rather than motility

_____ 30. A flexible filament that is used for motility by spirochetes

_____ 31. Movement toward chemical attractants or away from repellents

_____ 32. A structure formed by some species of bacteria that is resistant to some environmental stresses

_____ 33. The process of forming endospores within a vegetative cell

_____ 34. The breaking of an endospore's dormant state as it begins to form a vegetative cell

_____ 35. Granules of organic or inorganic material that are stockpiled by the cell for future use

_____ 36. A network of polysaccharides extending from the surface of bacteria and other cells

_____ 37. Inclusion bodies that serve as a reserve of the enzyme ribulose-1,5-bisphosphate carboxylase; they may also be the site of CO_2 fixation

_____ 38. Inclusion bodies that store extra nitrogen for bacteria

_____ 39. A polymer of glucose units; it serves as a carbon storage reservoir for many bacteria

_____ 40. Small, circular DNA molecules that can exist independently of the host chromosome, that have their own replication origins, that are autonomously replicated, and that are stably inherited

_____ 41. A plasmid that can exist independent of the host chromosome or be integrated into it

_____ 42. A plasmid that has genes for pili and can transfer copies of itself to other bacteria

_____ 43. A plasmid that carries genes that encode resistance to antibiotics

_____ 44. A plasmid that carries genes for a toxin, and thus renders the host bacterium pathogenic

_____ 45. A plasmid that carries genes for enzymes that degrade environmental substances such as aromatic compounds and pesticides

IDENTIFICATION OF PROCARYOTIC STRUCTURES

Label the appropriate structures indicated on the accompanying figure. The terms to be used are listed below. In the space beside each term provide a *brief* description of its structure and/or function.

1. Capsule:

2. Cell wall:

3. Flagellum:

4. Inclusion body:

5. Nucleoid:

6. Periplasmic space: between cell wall & membrane – little in gram⁺ large in gram⁻ where it is on both side of peptidoglycan

7. Plasma membrane:

8. Ribosome:

9. S-layer:

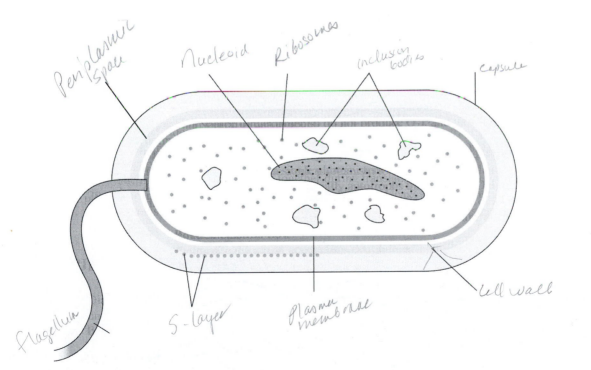

FLAGELLAR ARRANGEMENTS

Label each flagellar arrangement with the appropriate name.

1. _____

2. _____

3. _____

4. _____

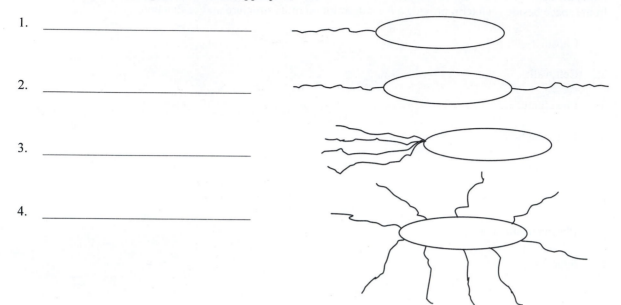

BACTERIAL SHAPES

Label each bacterial shape.

1. _____

2. _____

3. _____

4. _____

5. _____   rigid

6. _____   flexible

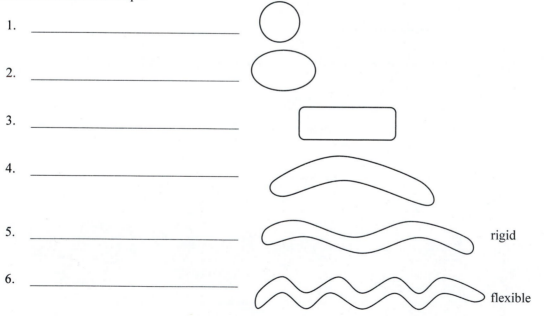

BACTERIAL ARRANGEMENTS

Label each arrangement of spherical bacteria with the appropriate term.

1. _____

2. _____

3. _____

4. _____

   planar configuration of four cells

5. _____

   cubic configuration of eight cells

FILL IN THE BLANK

1. The _____ _____ encompasses the cytoplasm of both procaryotic and eucaryotic cells. It is the chief point of contact with the cell's _____ and thus is responsible for much of its relationship with the outside world.

2. The amphipathic property of lipids enables them to form a _____ in membranes. Thus the outer surface of the membrane is _____, and the _____ ends of the lipids are buried in the interior of the membrane, away from the surrounding water.

3. The most widely accepted structural model for membranes is called the _____ _____ model. In this model, proteins are associated with or embedded in a fluid lipid bilayer. Those proteins embedded in the membrane are called _____ proteins; those proteins loosely attached to the surface of the membrane are called _____ proteins.

4. An important constituent of the cell walls of gram-negative bacteria is lipopolysaccharide (LPS). This complex molecule is composed of _____ ____, _____ _____, and _____ side chain (antigen).

5. In _____ environments, a cell will lose water so that the plasma membrane shrinks away from the cell wall. This is referred to as _____.

6. The space between the cell wall and the plasma membrane is called the_____ _____. In gram-negative bacteria, this space contains many proteins that participate in nutrient acquisition. In gram-positive bacteria, this same function is performed by _____ that are secreted out of the cell.

7. Solutes are less concentrated in _____ environments than in cells. When cells are placed in this type of environment, water will flow _____, causing the cell to swell and eventually burst if the plasma membrane is unprotected because the cell lacks a _____ _____.

8. Many bacteria are motile by means of _____, which are composed of three parts. The largest and most obvious part is the _____, which extends from the cell surface. The ____ _____ is embedded in the cell. The _____ is a short, curved segment that acts as a flexible coupling, joining the basal body to the filament.

9. The direction of flagellar rotation determines the nature of bacterial movement. For bacteria with peritrichous flagella, rotation in the _____ direction propels the bacterium forward, while rotation in the _____ direction disrupts the bundle of flagella and causes the bacterium to _____.

10. A bacterium travels in a straight line (a _____) for a few seconds; then it stops and _____; this is followed by a _____ in a different direction. When exposed to a chemical attractant gradient, a bacterium tumbles _____ frequently when traveling up the gradient, but tumbles at _____ frequency if moving down the gradient.

11. A thin delicate covering called the _____surrounds an endospore and the _____ _____ lies directly below it. The latter is composed of several protein layers and may be fairly thick. As much as half of the volume of the endospore may be occupied by the _____, which is made of peptidoglycan, and within this is the _____ _____ _____ (also called the core wall), which surrounds the protoplast or _____ of the endospore.

12. Many aquatic bacteria contain _____ _____, which give the cells buoyancy and are composed of many small, hollow, cylindrical structures called _____ _____.

13. The enzyme _____ hydrolyzes the bond connecting the sugars in peptidoglycan and therefore can cause bacterial cells to lyse.

14. A _____ is a cell with its cell wall completely removed. A _____ is a cell having a weakened or partially removed cell wall. The latter can be formed when gram-negative bacteria are treated with penicillin. The penicillin inhibits _____ synthesis, but the outer membrane remains intact.

15. The _____ protein secretion pathway involves transport machinery called the _____, which transfers proteins from the cytoplasm across or into the plasma membrane. This pathway is also called the _____ secretion pathway. Proteins secreted by this pathway have a short segment at the amino terminal called the _____. It is removed as the protein is secreted from the cytoplasm.

16. If a bacterium is deflagellated, it will regenerate a flagellar filament. It is believed that _____ subunits, the protein subunits that make up the filament, are transported through the filament's hollow core. When they reach the tip, the subunits spontaneously aggregate. This is an example of _____.

17. Chains of the polysaccharide _____, which is observed in most bacterial cell walls, are joined by cross-links between the peptides of each chain. In many cases the carboxyl group of the terminal D-alanine in one peptide is connected directly to the amino group of diaminopimelic acid in the other chain. However, in some bacteria a _____ _____ connects the two peptides.

MULTIPLE CHOICE

For each of the questions below select the *one best* answer.

1. Which of the following is NOT a function of the plasma membrane?
   a. It retains the cytoplasm.
   b. It acts as a selectively permeable barrier, allowing some molecules to pass while preventing the movement of others.
   c. It maintains the various shapes of the bacteria.
   d. It is the location of a variety of metabolic processes, including respiration and photosynthesis.

2. Which of the following is NOT true of procaryotic plasmids?
   a. They can exist and replicate independently of the chromosome.
   b. They are required for host growth and/or reproduction.
   c. They may carry genes for drug resistance.
   d. They may carry genes that give the bacterium new useful metabolic activities.

3. Lipopolysaccharide (LPS), which is found in the outer membrane of gram-negative bacteria, is also known as
   a. teichoic acid.
   b. exotoxin.
   c. endotoxin.
   d. murein.

4. Why is it thought that gram-positive cells retain the primary stain of the Gram stain while gram-negative cells do not?
   a. Because the stain is bound to the thicker peptidoglycan layer.
   b. Because the alcohol removes the lipids of the outer membrane, and thus the trapped stain, from gram-negative bacteria.
   c. Because the alcohol shrinks the pores of the thick peptidoglycan layer of gram-positive cells.
   d. Both (b) and (c) are correct.

5. Penicillin inhibits cell wall synthesis, but cells will continue to grow normally in the presence of penicillin if they are maintained in a(n) _____ environment.
   a. hypotonic
   b. isotonic
   c. hypertonic
   d. nonpolar

6. Which of the following refers to a network of polysaccharides that extends from the surface of a bacterium and aids in attachment of the bacterium to surfaces?
   a. lipoteichoic acid
   b. lipopolysaccharide
   c. outer membrane
   d. glycocalyx

7. Which of the following is true of capsules?
   a. They help bacteria resist phagocytosis by host phagocytic cells.
   b. They protect bacteria from desiccation.
   c. They prevent the entry of bacterial viruses.
   d. All of the above are true of capsules.

8. What will motile bacteria do when they are in the presence of both attractants and repellents?
   a. move toward the attractant
   b. compare both signals and respond to the chemical with the most effective concentration
   c. move away from the repellent
   d. move in a random fashion

9. Why are endospores of great practical importance in industrial and medical microbiology?
   a. Because they are resistant to harsh environments and thus increase the survival of spore-forming bacteria as compared to that of bacteria that do not form spores.
   b. Because many endospore-formers are dangerous pathogens.
   c. Both (a) and (b) are correct.
   d. Neither (a) nor (b) is correct.

10. Where are extensive invaginations of the plasma membrane usually observed?
    a. in cyanobacteria and other photosynthetic bacteria
    b. in bacteria with high respiratory activity
    c. Both (a) and (b) are correct.
    d. Neither (a) nor (b) is correct.

11. Which of the following are examples of inclusion bodies or materials found in inclusion bodies?
    a. volutin granules
    b. poly-β-hydroxybutyrate (PHB)
    c. polyphosphate
    d. magnetosomes
    e. All of the above.

12. Which protein secretion pathway is used by some gram-negative pathogens to move virulence factors from the cytoplasm, through both the plasma membrane and the outer membrane, and into the host cells being attacked by the pathogen?
    a. type I protein secretion pathway
    b. type II protein secretion pathway
    c. type III protein secretion pathway
    d. sec-dependent protein secretion pathway

13. Which protein secretion pathway that moves proteins from the cytoplasm is also known as the ABC protein secretion pathway because it involves an ATP-binding cassette?
    a. type I protein secretion pathway
    b. type II protein secretion pathway
    c. type III protein secretion pathway
    d. sec-dependent protein secretion pathway

14. Which of the following is the primary lipid component of the membranes of extreme thermophiles?
    a. C_{20} diethers
    b. C_{40} tetraethers
    c. sulfolipids
    d. cholesterol

15. Which of the following strengthens the cell membranes of cell wall-less archaea?
    a. diglycerol tetraethers
    b. lipopolysaccharides
    c. glycoproteins
    d. All of the above are correct.

16. Populations can be cured of their plasmids by treatments that inhibit plasmid replication but that do not affect host DNA replication. Which of the following has been used as a curing agent?
    a. UV and ionizing radiation
    b. acridine mutagens
    c. thymine starvation
    d. All of the above have been used as curing agents.

17. Which of the following is NOT encoded within a virulence plasmid?
    a. proteins that destroy other bacteria
    b. genes that render the bacterium less susceptible to host defense mechanisms
    c. genes for the production of toxigenic substances
    d. All of the above may be encoded within virulence plasmids.

TRUE/FALSE

_____ 1. The cell membrane is a rigid and relatively static structure.

_____ 2. Gas vacuoles are membranous structures that regulate buoyancy in cyanobacteria and other photosynthetic bacteria. However, even though they are membranous, their vesicle walls contain no lipids.

_____ 3. Sedimentation coefficients (measured in Svedberg units) are directly proportional to the molecular weight of a particle and are unaffected by volume or shape.

_____ 4. Gram-positive cell walls have a thick layer of peptidoglycan but a rather simple overall structure, while gram-negative cell walls have a thinner peptidoglycan layer but a more complex overall structure.

_____ 5. Lipoteichoic acids are teichoic acids that are connected to lipids in the plasma membrane.

_____ 6. Porin proteins are found in the outer membrane of gram-negative bacteria and function in the transport of molecules into the cell.

_____ 7. Mycoplasmas must be maintained in an isotonic environment because they lack a cell wall.

_____ 8. In the construction of flagella, the protein flagellin is added at the base of the flagellum so that first the tip is constructed and then pushed outward by the addition of new material at the base.

_____ 9. Bacterial flagellar rotation appears to be powered by proton gradients or sodium gradients, but not by ATP hydrolysis.

_____ 10. Procaryotic organisms are so remarkably uniform that nearly all genera contain all of the structures described in this chapter.

_____ 11. Pleomorphic bacteria are uniformly club-shaped.

_____ 12. Chemotactic receptors are directly coupled to the basal body of the flagella.

_____ 13. Archaeal membranes often have a monolayer rather than a bilayer structure.

_____ 14. The cytoplasmic matrix of procaryotic cells contains a cytoskeleton.

_____ 15. Gliding motility is a means by which bacteria coast along solid surfaces.

_____ 16. Endospores have a high content of dipicolinic acid.

_____ 17. The glycocalyx is also called the cell envelope.

_____ 18. A plasmid that can exist independent of the host chromosome but that cannot be integrated into the host chromosome is called an episome.

_____ 19. For a bacterium to be resistant to five different antibiotics, it must carry five different R factor plasmids simultaneously.

CRITICAL THINKING

1. Discuss why the plasma membrane is considered the external boundary of the cell, even though other structures outside of the plasma membrane are considered part of the cellular anatomy.

2. Discuss the nature of flagellar-mediated movement of bacteria. In particular, discuss how the direction of rotation affects the direction of movement, and speculate how this movement is altered in the presence of chemical attractants and repellents for which the bacterium has the appropriate chemoreceptors. Give plausible mechanisms of the response to chemical attractants and/or repellents.

ANSWER KEY

Terms and Definitions

1. g, 2. c, 3. ii, 4. qq, 5. nn, 6. x, 7. ll, 8. mm, 9. a, 10. ff, 11. w, 12. q, 13. r, 14. aa, 15. t, 16. gg, 17. pp, 18. y, 19. cc, 20. bb, 21. a, 22. l, 23. z, 24. u, 25. ee, 26. d, 27. kk, 28. jj, 29. m, 30. b, 31. f, 32. j, 33. oo, 34. n, 35. s, 36. o, 37. e. 38. i. 39. p, 40. dd, 41. k, 42. h, 43. hh, 44. rr, 45. v

Flagellar Arrangements

1. monotrichous, 2. amphitrichous, 3. lophotrichous, 4. peritrichous

Bacterial Shapes

1. coccus, 2. coccobacillus, 3. bacillus, 4. vibrio, 5. spirilla, 6. spirochete

Bacterial Arrangements

1. diplococcus, 2. streptococcus, 3. staphylococcus, 4. tetrad, 5. sarcinae

Fill in the Blank

1. plasma membrane; environment 2. bilayer; hydrophilic; hydrophobic 3. fluid mosaic; integral; peripheral 4. lipid A; core polysaccharide; O 5. hypertonic; plasmolysis 6. periplasmic space; exoenzymes 7. hypotonic; inward; cell wall 8. flagella; filament; basal body; hook 9. counterclockwise; clockwise; tumble (twiddle) 10. run; tumbles (twiddles); run; less; normal 11. exosporium; spore coat; cortex; spore cell wall; core 12. gas vacuoles; gas vesicles 13. lysozyme 14. protoplast; spheroplast; peptidoglycan 15. sec-dependent; translocon; general; signal peptide 16. flagellin; self-assembly 17. peptidoglycan; peptide interbridge

Multiple Choice

1. c, 2. b, 3. c, 4. d, 5. b, 6. d, 7. d, 8. b, 9. c, 10. c, 11. e, 12. c, 13.a, 14. b, 15. d, 16. d, 17. a

True/False

1. F, 2. T, 3. F, 4. T, 5. T, 6. T, 7. F, 8. F, 9. T, 10. F, 11. F, 12. F, 13. T, 14. F, 15. T, 16. T, 17. F, 18. F, 19. F

4 Eucaryotic Cell Structure and Function

CHAPTER OVERVIEW

This chapter focuses on eucaryotic cell structure and function. Although procaryotic organisms are immensely important in microbiology, eucaryotic microorganisms—such as fungi, algae, and protozoa—are also prominent members of many ecosystems, and some have medical significance as etiological agents of disease as well. The chapter concludes with a comparison of eucaryotic and procaryotic cells.

CHAPTER OBJECTIVES

After reading this chapter you should be able to:

- identify major eucaryotic cell structures in a drawing or photomicrograph
- discuss the various elements of the cytoskeleton (microfilaments, intermediate filaments, microtubules) with regard to their structure and various functions within the cell
- discuss the composition, structure, and function of each of the internal organelles, such as the endoplasmic reticulum, Golgi apparatus, lysosomes, ribosomes, mitochondria, chloroplasts, nucleus, and nucleolus
- discuss the mechanism of endocytosis and the difference between phagocytosis and pinocytosis
- compare mitosis and meiosis
- compare and contrast procaryotes and eucaryotes

CHAPTER OUTLINE

I. An Overview of Eucaryotic Cell Structure
    A. Eucaryotic cells have membrane-delimited nuclei
    B. In addition to the nucleus, eucaryotic cells have other membrane-bound organelles that perform specific functions within the cells; this allows simultaneous independent control of cellular processes
    C. Eucaryotic cells also contain a large intracytoplasmic membrane complex, which provides a large surface area allowing greater respiratory and photosynthetic activity; this complex of membranes also serves as a transport system to move materials to different cell locations
II. The Plasma Membrane and Membrane Structure
    A. Eucaryotic membranes are phospholipid bilayers with sphingolipids and cholesterol, but the lipids in the outer layer have been shown to differ from those in the inner layer
    B. Certain regions in the membranes, called lipid rafts, span the bilayer and allow lipids in the adjacent layers to interact
III. The Cytoplasmic Matrix, Microfilaments, Intermediate Filaments, and Microtubules
    A. The cytoplasmic matrix, although superficially featureless, provides the complex environment required for many cellular activities
    B. The cytoskeleton is a vast network of interconnected filaments important for motion and a scaffold for maintaining cell organization
    C. Microfilaments (4 to 7 nm) composed of actin may be scattered throughout the matrix or organized into networks and parallel arrays; they play a major role in cell motion and changes in cell shape
    D. Microtubules are hollow cylinders (25 nm) composed of tubulin that help maintain cell shape, are involved (with microfilaments) in cellular movement, participate in intracellular transport of substances, and participate in organelle movements; they also form the mitotic spindle during cell division and are present in cilia and flagella

E. Intermediate filaments (8 to 10 nm), along with microfilaments and microtubules, are major components of the cytoskeleton, and are particularly prominent in nuclear lamina

IV. Organelles of the Biosynthetic-Secretory and Endocytic Pathways
   A. The endoplasmic reticulum
      1. The endoplasmic reticulum (ER) is a complex set of internal membranes that may have ribosomes attached (rough endoplasmic reticulum; RER) or may be devoid of ribosomes (smooth endoplasmic reticulum; SER)
      2. The ER has many important functions:
         a. It transports proteins, lipids, and other materials within the cell
         b. ER-associated enzymes and ribosomes synthesize lipids and many proteins
         c. It is a major site of cell membrane synthesis
   B. The Golgi apparatus
      1. The Golgi apparatus is a set of membrane sacs (cisternae) that is involved in the modification, packaging, and secretion of materials; the cisternae exist in stacks called dictyosomes
      2. The Golgi apparatus is present in many eucaryotic cells, but many fungi and ciliated protozoa lack it
   C. Lysosomes are membrane-bound vesicles that contain hydrolase enzymes needed for intracellular digestion of all types of macromolecules
   D. The biosynthetic-secretory pathway
      1. Proteins destined for the cell membrane, lysosomes, or secretion are transported in vesicles that bud off the ER and join the cis face of the Golgi apparatus; the proteins are modified and transported in vesicles that bud off of the trans face of the Golgi apparatus
      2. Transport vesicles move the material to the cell membrane or lysosome; secretory vesicles hold their contents until signaled to release them through fusion with the plasma membrane
      3. Proteasomes are a nonlysosomal protein degradation system that has been recently discovered in eucaryotic cells, a few bacteria, and many archaea
   E. The endocytic pathway
      1. Endocytosis is the process in which the cell takes up solutes or particles by enclosing them in vesicles (endosomes) pinched off from the plasma membrane
         a. Phagocytosis—endocytosis of large particles by engulfing them into a phagocytic vacuole (phagosome)
         b. Pinocytosis—endocytosis of small amounts of liquid with its solute molecules; there are three types of pinocytosis: fluid-phase endocytosis; receptor-mediated endocytosis using clathrin-coated pits and vesicles; and a type of endocytosis that forms special vesicles (caveolae), whose contents are not degraded
      2. Most endosomes fuse with early lysosomes (newly formed lysosomes) to form late lysosomes; late lysosomes are important to cell functioning
         a. Some function as food vacuoles, digesting and releasing nutrients into the cytoplasm
         b. Some function to destroy invading bacteria
         c. Autophagosomes fuse lysosomes to selectively digest portions of the cell's own cytoplasm as part of the normal turnover of cellular components
      3. Undigested materials accumulate in lysosomes called residual bodies

V. Eucaryotic Ribosomes
   A. Eucaryotic ribosomes (80S with 60S and 40S subunits) are generally larger than procaryotic ribosomes (70S)
   B. Eucaryotic ribosomes, like their procaryotic counterparts, are responsible for synthesis of cellular proteins; they can either be attached to the ER or are free in the cytoplasm
      1. ER-associated ribosomes synthesize integral membrane proteins or proteins that are secreted out of the cell
      2. Free ribosomes synthesize nonsecretory, nonmembrane proteins

VI. Mitochondria
   A. Mitochondria are the site of energy-generating tricarboxylic acid (TCA) cycle activity and the formation of ATP by electron transport and oxidative phosphorylation
   B. Mitochondria have both an inner membrane and an outer membrane enclosing a fluid matrix

      1.   The inner and outer membranes have different lipids and enzymes

      2.   The enzymes of the TCA cycle and the β-oxidation pathway for fatty acids are located within the matrix

      3.   Electron transport and oxidative phosphorylation occur only in the inner mitochondrial membrane

    C.   Mitochondria use their own DNA and their own ribosomes to synthesize some of their proteins; mitochondrial DNA and mitochondrial ribosomes are similar to bacterial DNA and ribosomes in terms of size and structure; mitochondria reproduce by binary fission

VII. Chloroplasts

    A.   Chloroplasts are the site of photosynthesis in protists and higher plants

    B.   Chloroplasts are surrounded by two membranes; the inner membrane encloses a fluid matrix called the stroma; within the stroma is a system of flattened sacs called thylakoids which often form stacks known as grana

      1.   The formation of carbohydrates from carbon dioxide and water (dark reactions) occurs in the stroma

      2.   The trapping of light energy to generate ATP, NADPH, and oxygen (light reaction) occurs in the thylakoid membranes of the grana

VIII. The Nucleus and Cell Division

    A.   Nuclei are membrane-bound structures that house most of the genetic material of the cell

    B.   Nuclear structure

      1.   Chromatin is the dense fibrous material seen within the nucleoplasm of the nucleus; the nucleoplasm is the DNA-containing part of the nucleus; when the cell is dividing, chromatin condenses into visible chromosomes

      2.   The nucleus is bounded by the nuclear envelope, a double-membrane structure, which is penetrated by nuclear pores; the nuclear pores allow materials to be transported into or out of the nucleus

      3.   The nucleolus is a very noticeable structure within the nucleus; it is involved in the synthesis of ribosomes

    C.   Mitosis and meiosis

      1.   The cell cycle is the total sequence of events in the growth-division cycle of a cell

          a.   Cell growth takes place in interphase, the portion of the cycle between periods of mitosis

          b.   Mitosis is a process of nuclear division in which duplicated genetic material is distributed equally to two daughter nuclei so that each has a full set of chromosomes and genes

      2.   Meiosis is a complex, two-stage process of nuclear division in which the number of chromosomes in the resulting daughter cells is reduced from the normal (diploid) number to one-half of that number (haploid)

IX. External Cell Coverings

    A.   Some cells have a rigid cell wall but many do not; cell walls of eucaryotic microbes vary in composition, but are generally chemically simpler than peptidoglycan

    B.   Other cells, such as some protists, have a pellicle, which is a rigid layer of components just beneath the plasma membrane

X. Cilia and Flagella

    A.   Cilia and flagella are locomotor structures that differ in length and how they propel the cell

    B.   Cilia and flagella are structurally very similar; both are membrane-bound cylinders composed of microtubules, in a 9+2 arrangement, embedded in a matrix

XI. Comparison of Procaryotic and Eucaryotic Cells

    A.   Eucaryotes have a membrane-delimited nucleus and many complex membrane-bound organelles, each of which performs a separate function for the cell

    B.   Procaryotes lack a membrane-delimited nucleus and internal membrane-bound organelles; they do not undergo mitosis, meiosis, or endocytosis performed by many eucaryotes

    C.   Despite the significant differences between procaryotes and eucaryotes, they have remarkable biochemical similarities: the same basic chemical composition, the same genetic code, and some of the same basic metabolic processes

TERMS AND DEFINITIONS

Place the letter of each term in the space next to the definition or description that best matches it.

_____ 1. Intracellular structures that perform specific functions

_____ 2. Minute protein fibers (4 to 7 nm diameter), either scattered within the cytoplasmic matrix or organized into networks and parallel arrays, that may play a role in cell movement

_____ 3. Hollow cylinders (25 nm diameter) that help maintain cell shape and that participate in cell movement

_____ 4. Protein fibers (8 to 10 nm diameter) that help maintain cell shape and that participate in cell movement

_____ 5. Combination of microtubules and filaments that support cell shape and movement

_____ 6. Arrangement of microtubules along the length of flagella and cilia

_____ 7. Endoplasmic reticulum to which are attached many ribosomes

_____ 8. Endoplasmic reticulum that is mostly devoid of attached ribosomes

_____ 9. The general process of importing solutes or particles by enclosing them in vesicles pinched off from the plasma membrane

_____ 10. The specific process of importing large particles by enclosing them in vesicles pinched off from the plasma membrane

_____ 11. The specific process of importing solutes but not large particles by enclosing them in vesicles pinched off from the plasma membrane

_____ 12. Spherical complexes of proteins that are observed on the inner membrane of mitochondria and are responsible for the synthesis of ATP during cellular respiration

_____ 13. Organelles of algae and higher plants that often possess pigments and are the sites of synthesis and storage of food reserves

_____ 14. Process of nuclear division that maintains the ploidy (number of sets of chromosomes) of the parent cell

_____ 15. Infoldings of the mitochondrial inner membrane

_____ 16. Process of nuclear division in which the number of chromosomes is reduced to one-half of the original number

_____ 17. A dense region found in the chloroplast that is composed of protein surrounded by polysaccharides

_____ 18. Vesicles formed by endocytosis

_____ 19. The total sequence of events in the growth-division cycle between the end of one division and the end of the next

_____ 20. A protein that functions in the movement of cilia and eucaryotic flagella

a. axoneme
b. cell cycle
c. cristae
d. cytoskeleton
e. dynein
f. endocytosis
g. endosomes
h. F_1 particles
i. intermediate filaments
j. meiosis
k. microfilaments
l. microtubules
m. mitosis
n. organelles
o. phagocytosis
p. pinocytosis
q. plastids
r. polysomes (polyribosomes)
s. pyrenoid
t. rough endoplasmic reticulum
u. smooth endoplasmic reticulum

IDENTIFICATION OF EUCARYOTIC STRUCTURES

Label the appropriate structures indicated on the accompanying figure using the following terms. In the space beside each term provide a *brief* description of its structure and/or function.

1. Chromatin:

2. Golgi apparatus:

3. Lysosome:

4. Mitochondrion:

5. Nucleolus:

6. Nucleus:

7. Plasma membrane:

8. Ribosomes:

9. RER:

10. SER:

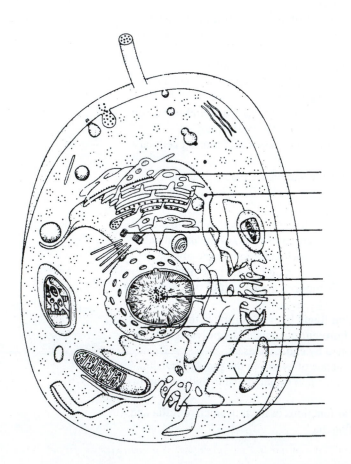

COMPARISON OF PROCARYOTIC AND EUCARYOTIC CELLS

Complete the following table by providing a *brief* description of the properties/structures listed for both procaryotes and eucaryotes.

Property/Structure	Procaryote	Eucaryote
Organization of Genetic Material		
1. Membrane-bound nucleus		
2. Chromosome structure		
3. Number of chromosomes		
4. Nucleolus		
5. Plasmids		
Organelles		
6. Mitochondria		
7. Chloroplasts		
8. Plasma membrane		
9. Flagella		
10. Cilia		
11. Endoplasmic reticulum		
12. Golgi apparatus		
13. Lysosomes		
14. Ribosomes		
Functions		
15. Photosynthesis		
16. Mitosis		
17. Meiosis		
18. Differentiation		

MATCHING

Match the following structures with the appropriate functions. Answers may be used once, more than once, or not at all.

_____ 1.	Major route by which proteins, lipids, and other materials are transported through the cell	a. cell wall
_____ 2.	Involved in the modification, packaging, and secretion of materials	b. chloroplasts
		c. cilia
_____ 3.	Contain the enzymes needed to digest all types of macromolecules	d. endoplasmic reticulum
_____ 4.	Structures responsible for the synthesis of proteins	e. flagella
_____ 5.	Responsible for ATP synthesis by electron transport and oxidative phosphorylation	f. Golgi apparatus
		g. lysosomes
_____ 6.	Major site of cell membrane synthesis	h. mitochondria
_____ 7.	Responsible for the synthesis of ATP and carbohydrate using light as the energy source	i. nucleolus
		j. nucleus
_____ 8.	Responsible for the production of ribosomes	k. pellicle
_____ 9.	Repository for the cell's genetic information	l. proteasome
_____ 10.	A complex structure or set of structures lying underneath the plasma membrane that gives some eucaryotic cells their characteristic shape	m. ribosomes
_____ 11.	A rigid structure outside the plasma membrane that gives some cells their characteristic shape	
_____ 12.	Short fibers containing microtubules that are responsible for locomotion for some eucaryotes	
_____ 13.	Long fibers containing microtubules that are responsible for locomotion for some eucaryotes	
_____ 14.	A huge cylindrical complex of proteins that plays a major role in the degradation and recycling of proteins, as well as the production of peptides for antigen presentation during immunological responses	

FILL IN THE BLANK

1. Organelles are located in an apparently featureless, homogeneous substance called the _____ _____, which is also the location of many important biochemical processes.
2. The Golgi apparatus consists of _____ in stacks, often called _____, which can be clustered in one region or scattered throughout the cell.
3. The _____ theory proposes that mitochondria, chloroplasts, and perhaps other organelles were once free-living procaryotes, which were engulfed by an ancestral phagocytic cell, survived within the host cell, eventually lost their ability to live independently of the host, and became an organelle.
4. Newly formed lysosomes, or _____ lysosomes, fuse with phagocytic vesicles to yield _____ lysosomes; _____ _____ are lysosomes that have accumulated large quantities of indigestible material.
5. Enzymes and electron carriers involved in electron transport and oxidative phosphorylation are located only in the _____ membrane of the mitochondrion, while the enzymes of the tricarboxylic acid cycle are located in the _____.
6. Photosynthetic reactions occur in separate compartments of the chloroplast. The formation of carbohydrate from carbon dioxide and water (the dark reaction) takes place in the _____ or fluid matrix, while the trapping of light energy to form ATP, NADPH, and oxygen (the light reaction) takes place in the _____ of the chloroplast.

42

7. The nucleus is bounded by the _____ _____, which consists of an inner and outer membrane. It is penetrated by many _____ _____, each of which is surrounded by a ring of granular and fibrous material called the annulus.

8. The period between mitotic divisions, during which most cell growth takes place, is known as _____.

9. Many protozoa and some algae have a structure consisting of a relatively rigid layer just beneath the plasma membrane and including the plasma membrane. This structure is called the _____, and although it is not as rigid as a _____ _____, it does provide a characteristic shape to a cell that possesses it.

10. Cilia beats have two distinctive phases. In the _____ stroke, the cilium moves like an oar to propel the organism through the water. The cilium then bends along its length while it is pulled forward during the _____ stroke.

11. Material is transported from the ER to the _____ face of the Golgi apparatus. It moves through the Golgi apparatus by budding off the cisternal edges and moving to the next sac. After it has completed its journey through the Golgi apparatus and has been appropriately modified, it is secreted from the _____ face for transport to its final destination.

12. Several kinds of pinocytosis have been identified. In _____ _____, extracellular fluid is taken in nonselectively. This may be a method for recycling the plasma membrane. A second type of pinocytosis, called _____ _____, results in the formation of _____ _____. This type of pinocytosis occurs at specialized regions of the membrane coated on the cytoplasmic side with _____. A third type of pinocytosis forms _____ (little caves), which may be involved in signal transduction and active import of certain molecules (e.g., folic acid). Unlike other _____, these vesicles do not fuse with _____ and their contents are not degraded.

13. Within the _____ of a chloroplast is a complex internal membrane system called _____. In some algae, several disk-like _____ are stacked on each other like coins to form _____.

MULTIPLE CHOICE

For each of the questions below select the *one best* answer.

1. What are the flattened sacs of the endoplasmic reticulum called?
   a. thylakoids
   b. cristae
   c. cisternae
   d. vacuomes

2. Which of the following is one of the more important functions of the Golgi apparatus?
   a. synthesis of ribosomes
   b. synthesis of lysosomes
   c. synthesis of nucleosomes
   d. synthesis of mesosomes

3. Which of the following do cells use to selectively digest portions of their own cytoplasm?
   a. autophagic vacuole
   b. turnover lysosome
   c. suicide vacuole
   d. recycling vacuole

4. Which of the following is NOT a function of the mitochondrion?
   a. tricarboxylic acid cycle enzyme reactions
   b. electron transport
   c. oxidative phosphorylation
   d. All of the above are functions of the mitochondrion.

5. A cell with a diploid number of 8 chromosomes (2n=8) undergoes mitosis. How many chromosomes does each of the daughter cells have?
   a. 16
   b. 4
   c. 8
   d. 12

6. If the same cell described in question #5 undergoes meiosis instead of mitosis, the resulting daughter cells will each have _____ chromosomes.
   a. 16
   b. 4
   c. 8
   d. 12
7. Which of the following directs the construction of flagella and/or cilia?
   a. axoneme
   b. tubulin
   c. centriole
   d. basal body

8. Which term refers to stacks of cisternae in the Golgi apparatus?
   a. stigmata
   b. golgisomes
   c. dictyosomes
   d. lamellosomes

TRUE/FALSE

_____ 1. In cells that produce large quantities of lipid, the endoplasmic reticulum is mostly devoid of ribosomes and is referred to as smooth endoplasmic reticulum.

_____ 2. The Golgi apparatus is the major site of cell membrane synthesis.

_____ 3. Both eucaryotes and procaryotes have ribosomes, but the eucaryotic ribosome is generally larger and more complex than the procaryotic ribosome.

_____ 4. The majority of mitochondrial proteins are manufactured under the direction of mitochondrial DNA by mitochondrial ribosomes.

_____ 5. DNA is replicated during the S period of interphase.

_____ 6. The cell walls of eucaryotic cells are constructed of polysaccharides (e.g., cellulose) that are simpler than the peptidoglycan found in bacterial cell walls.

_____ 7. Unlike procaryotes, many eucaryotes lack an external cell wall.

_____ 8. Proteins to be secreted out of the cell are synthesized by free ribosomes (those that are not attached to the ER).

_____ 9. Lysosomes join with phagosomes for defensive purposes as well as for acquisition of nutrients.

_____ 10. Although different structurally, procaryotes and eucaryotes are metabolically quite similar.

CRITICAL THINKING

1. Describe how the "division of labor" associated with internal, membrane-delimited organelles enables eucaryotic cells to be more efficient and thereby allows them to grow much larger than procaryotes grow.

2. Both procaryotes and eucaryotes are capable of using large particles as nutrient sources. Eucaryotes bring them into the cell by endocytosis. Procaryotes cannot do this; they export digestive enzymes to digest the particles externally and then transport the resulting small nutrient molecules into the cell. Why is the procaryotic process much less efficient? Use diagrams to support your answer.

ANSWER KEY

Terms and Definitions

1. n, 2. k, 3. l, 4. i, 5. d, 6. a, 7. t, 8. u, 9. f, 10. o, 11. p, 12. h, 13. q, 14. m, 15. c, 16. j, 17. s, 18. g, 19. b, 20. e

Matching

1. d, 2. f, 3. g, 4. m, 5. h, 6. d, 7. b, 8. i, 9. j, 10. k 11. a 12. c 13. e, 14. l

Fill in the Blank

1. cytoplasmic matrix 2. cisternae; dictyosomes 3. endosymbiotic 4. early; late; residual bodies
5. inner; matrix 6. stroma; thylakoids 7. nuclear envelope; nuclear pores 8. interphase 9. pellicle; cell wall
10. effective; recovery 11. cis (forming); trans (maturing) 12. fluid-phase endocytosis; receptor-mediated endocytosis; coated vesicles; clathrin; caveolae; endosomes; lysosomes 13. stroma; thylakoids, thylakoids; grana

Multiple Choice

1. c, 2. b, 3. a, 4. d, 5. c, 6. b, 7. d, 8. c

True/False

1. T, 2. F, 3. T, 4. F, 5. T, 6. T, 7. T, 8. F, 9. T, 10. T

5 Microbial Nutrition

CHAPTER OVERVIEW

This chapter describes the basic nutritional requirements of microorganisms. Cells must have a supply of raw materials and energy in order to construct new cellular components. This chapter also describes the processes by which microorganisms acquire nutrients and provides information about the cultivation of microorganisms.

CHAPTER OBJECTIVES

After reading this chapter you should be able to:

- list the ten elements that microorganisms require in large amounts (macronutrients/macroelements) and the six elements that they require in trace amounts (micronutrients/trace elements)
- list the major nutritional categories and give the source of carbon, energy, and electrons for each
- compare and contrast the various processes (passive diffusion, facilitated diffusion, active transport, group translocation) by which cells can obtain nutrients from the environment
- describe the various types of culture media for microorganisms (synthetic, complex, selective, differential) and tell how each is normally used in the study of microorganisms
- describe the techniques used to obtain pure cultures (spread plate, streak plate, pour plate)

CHAPTER OUTLINE

I.   The Common Nutrient Requirements
 A.   Macroelements, also known as macronutrients (C, O, H, N, S, P, K, Ca, Mg, Fe), are required by microorganisms in relatively large amounts
 B.   Trace elements or micronutrients (Mn, Zn, Co, Mo, Ni, Cu) are required in trace amounts by most cells and are often adequately supplied in the water used to prepare media or in the regular media components
 C.   Other elements may be needed by particular types of microorganisms
II. Requirements for Carbon, Hydrogen, and Oxygen
 A.   The requirements for carbon, hydrogen, and oxygen are often satisfied together by the same molecule
 B.   Heterotrophs use reduced, preformed organic molecules (usually from other organisms) as carbon sources
 C.   Autotrophs use carbon dioxide as their sole or principal carbon source
III. Nutritional Types of Microorganisms
 A.   All organisms need a source of energy and electrons
   1.   Energy
     a.   Phototrophs use light as their energy source

     b. Chemotrophs obtain energy from the oxidation of organic or inorganic compounds
    2. Electrons
     a. Lithotrophs use reduced inorganic compounds as their electron source
     b. Organotrophs use reduced organic compounds as their electron source
  B. Most microorganisms can be categorized as belonging to one of four major nutritional types depending on their sources of carbon, energy, and electrons:
    1. Photolithotrophic autotrophs
    2. Chemoorganotrophic heterotrophs
    3. Photoorganotrophic heterotrophs
    4. Chemolithotrophic autotrophs
  C. Some organisms show great metabolic flexibility and alter their metabolic patterns in response to environmental changes; mixotrophic organisms combine lithotrophic and heterotrophic metabolic processes, relying on inorganic energy sources and organic carbon sources

IV. Requirements for Nitrogen, Phosphorus, and Sulfur
  A. Nitrogen is needed for the synthesis of amino acids, purines, pyrimidines and other molecules; depending on the organism, nitrogen can be supplied by organic molecules, by assimilatory nitrate reduction, or by nitrogen fixation
  B. Phosphorus is present in nucleic acids, phospholipids, nucleotides, and other molecules; most microorganisms use inorganic phosphate to meet their phosphorus needs
  C. Sulfur is needed for the synthesis of certain amino acids and other molecules; most microorganisms meet their sulfur needs by assimilatory sulfate reduction

V. Growth Factors
  A. Growth factors are organic compounds required by the cell because they are essential cell components (or precursors of these components) that the cell cannot synthesize; there are three major classes:
    1. Amino acids—needed for protein synthesis
    2. Purines and pyrimidines—needed for nucleic acid synthesis
    3. Vitamins—function as enzyme cofactors
  B. Knowledge of specific growth factor requirements makes possible quantitative growth-response assays

VI. Uptake of Nutrients by the Cell
  A. Passive diffusion
    1. Passive diffusion is a process by which molecules move from an area of high concentration to an area of low concentration because of random thermal agitation
    2. Passive diffusion requires a large concentration gradient for significant levels of uptake
    3. Only a few small molecules (e.g., glycerol, H_2O, O_2, and CO_2) can be taken up by this mechanism given the amphipathic character of cell membranes
  B. Facilitated diffusion
    1. Facilitated diffusion is a process that involves a carrier molecule (permease) to increase the rate of diffusion; it only moves molecules from an area of higher concentration to an area of lower concentration
    2. It requires a smaller concentration gradient than passive diffusion
    3. The rate of diffusion plateaus when the carrier becomes saturated (i.e., when it is binding and transporting molecules as rapidly as possible)
    4. It is generally more important in eucaryotes than in procaryotes
  C. Active transport
    1. Active transport is a process that uses carrier proteins and metabolic energy to move molecules to the cell interior where the solute concentration is already higher than outside the cell (i.e., it runs against the concentration gradient)
    2. ATP-binding cassette transporters (ABC transporters) use ATP to drive transport against a concentration gradient; they are observed in bacteria, archaea, and eucaryotes
    3. Proton motive forces (i.e., proton gradients and sodium gradients) also can be used to power active transport

4. The active transport of two different substances can be linked: symport is the linked transport of two substances in the same direction; antiport is the linked transport of two substances in opposite directions

    D. Group translocation
        1. In this process molecules are modified as they are transported across the membrane
        2. The best known group translocation system is the phosphoenolpyruvate:sugar phosphotransferase system (PTS), which transports sugars into procaryotic cells while phosphorylating them
        3. Eucaryotes apparently do not use this type of transport
    E. Iron uptake
        1. Iron uptake is difficult because of the insolubility of ferric iron
        2. Many bacteria and fungi secrete siderophores to transport ferric ion into the cell

VII. Culture Media
    A. A culture medium is a solid or liquid preparation used to grow, transport, and store microorganisms; solidified media are usually made with the addition of agar
    B. Chemical and physical types of culture media
        1. Synthetic or defined media—these are media in which all components and their concentrations are known
        2. Complex media—these are media that contain some ingredients of unknown composition and/or concentration
    C. Functional types of media
        a. Supportive media (e.g., tryptic soy broth) are general-purpose media for the growth of many microorganisms
        b. Enriched media are supplemented by blood or other special nutrients to encourage the growth of fastidious heterotrophs
        c. Selective media favor the growth of particular microorganisms and inhibit the growth of others
        d. Differential media distinguish between different groups of bacteria on the basis of their biological characteristics

VIII. Isolation of Pure Cultures
    A. A pure culture is a clonal population of cells arising from a single cell
    B. The spread plate and streak plate are methods to separate cells on an agar surface such that each cell grows into a completely isolated colony (a macroscopically visible growth or cluster of microorganisms on a solid medium)
    C. The pour plate is a method that dilutes a culture sample to decrease the number of microorganisms, then mixes the dilution with molten agar before pouring into a petri dish
    D. Microbial growth on agar surfaces helps microbiologists identify bacteria because individual species often form colonies of characteristic size, shape, color, and appearance (morphology) on a particular growth medium; most growth occurs near the edges of the colony

TERMS AND DEFINITIONS

Place the letter of each term in the space next to the definition or description that best matches it.

_____ 1. Elements required by microorganisms in relatively large amounts

_____ 2. Elements required in trace amounts sufficiently supplied in water and regular media components

_____ 3. Organisms that can use carbon dioxide as their sole or principal source of carbon

_____ 4. Organisms that use reduced, preformed organic molecules as carbon sources

_____ 5. Organisms that obtain energy from the oxidation of organic or inorganic compounds

_____ 6. Organisms that obtain electrons from the oxidation of inorganic compounds

_____ 7. Organisms that obtain electrons from the oxidation of organic compounds

_____ 8. Organisms that obtain energy from light

_____ 9. Required organic compounds that are essential cell components (or precursors of such components) and that cannot be synthesized by the organism

_____ 10. Small organic molecules that make up all or part of enzyme cofactors

_____ 11. The process in which molecules move from a region of higher concentration to one of lower concentration as a result of random thermal agitation

_____ 12. Carrier proteins embedded in the plasma membrane that increase the rate of diffusion of specific molecules across selectively permeable membranes

_____ 13. The diffusion process that is aided by action of a carrier protein

_____ 14. Transport of molecules to areas of higher concentration (i.e., against a concentration gradient) with the input of metabolic energy

_____ 15. Linked transport of two substances in the same direction

_____ 16. Linked transport of two substances in opposite directions

_____ 17. A process in which molecules are chemically modified and simultaneously transported into the cell

_____ 18. Low molecular weight molecules that complex with ferric ion and supply it to the cell

_____ 19. A growth medium in which all components and their specific concentrations are known

_____ 20. A growth medium that contains some ingredients of unknown composition and/or concentration

_____ 21. A growth medium that favors the growth of some microorganisms and inhibits the growth of other microorganisms

_____ 22. A growth medium that distinguishes between different groups of bacteria on the basis of their biological characteristics

a. active transport
b. antiport
c. autotrophs
d. chemotrophs
e. complex medium
f. differential medium
g. facilitated diffusion
h. group translocation
i. growth factors
j. heterotrophs
k. lithotrophs
l. macronutrients (macroelements)
m. micronutrients (trace elements)
n. organotrophs
o. passive diffusion
p. permeases
q. phototrophs
r. selective medium
s. siderophores
t. symport
u. synthetic (defined) medium
v. vitamins

MACRONUTRIENTS/MICRONUTRIENTS

Indicate whether each of the following is a macronutrient or a micronutrient and briefly describe how cells use it.

Element	Type of Nutrient	Common Uses
1. Magnesium (Mg)		
2. Zinc (Zn)		
3. Oxygen (O)		
4. Sulfur (S)		
5. Iron (Fe)		
6. Manganese (Mn)		
7. Nitrogen (N)		
8. Cobalt (Co)		
9. Phosphorus (P)		
10. Carbon (C)		
11. Hydrogen (H)		
12. Potassium (K)		
13. Molybdenum (Mo)		
14. Nickel (Ni)		
15. Calcium (Ca)		
16. Copper (Cu)		

FILL IN THE BLANK

1. _____ _____ are special culture dishes that consist of two round halves. The top half overlaps the bottom half.

2. _____ media contain undefined components such as _____, which are protein hydrolysates prepared by partial proteolytic digestion of meat casein, soy meal, gelatin, and other protein sources.

3. ATP-binding cassette transporters (also known as _____ _____) are found in bacteria, archaea, and eucaryotes. They bind and hydrolyze ATP to drive uptake of sugars or amino acids. Thus, they are an example of _____ _____.

4. Iron uptake is made difficult by the great insolubility of ferric ion. Many bacteria overcome this by secreting _____, which are compounds of low molecular weight that can complex with ferric ion and supply it to the cell.

5. Media that contain some ingredients of unknown chemical composition are called _____ media. This type of medium is often used because the nutritional requirements of a particular microbe are unknown and thus a _____ medium cannot be constructed.

6. If a solid medium is needed for cultivation of microorganisms, liquid media can be solidified by the addition of _____, a sulfated polymer extracted from algae, which can be melted in boiling water and can be cooled to about ___°C before hardening.

7. A macroscopically visible growth or cluster of microorganisms on a solid medium is called a _____. These growths can be obtained by using _____ _____, _____ _____, or _____ _____ methods.

8. In many procaryotes, sugars are transported into the cell while being phosphorylated. The phosphate donor for the process is phosphoenolpyruvate and the transport system is called _____ _____ system. It is an example of _____ _____.

9. Mixotrophic organisms combine _____ and _____ metabolic processes, relying on inorganic energy sources and organic carbon sources.

10. Microorganisms of the nutritional type _____ contribute greatly to the chemical transformations of elements that continually occur in ecosystems.

NUTRITIONAL TYPES OF MICROORGANISMS

Complete the table below by supplying the missing information.

Nutritional Type	Source of Energy	Source of Electrons	Source of Carbon
1. Photolithotrophic autotrophy		Inorganic compounds	
2. Photoorganotrophic heterotrophy			Organic compounds
3. Chemolithotrophic autotrophy	Inorganic compounds		
4.	Organic compounds		Organic compounds
5.	Inorganic compounds	Inorganic compounds	Organic compounds

MULTIPLE CHOICE

For each of the questions below select the *one best* answer.

1. Which of the following is NOT a major class of growth factors?
   a. amino acids
   b. purines and pyrimidines
   c. vitamins
   d. All of the above are major classes of growth factors.

2. Which of the following processes can be used to concentrate nutrients from dilute nutrient sources?
   a. active transport
   b. group translocation
   c. Both (a) and (b) are correct.
   d. None of the above are correct.

3. When there are several transport systems for the same substance, in what way do the systems differ?
   a. in the energy source they use
   b. in their affinity for the transported solute
   c. in the nature of their regulation
   d. All of the above are correct.

4. Which of the following is NOT a characteristic of active transport?
   a. saturable rate of uptake
   b. requires an expenditure of metabolic energy
   c. can transport materials against a concentration gradient
   d. All of the above are characteristics of active transport.

5. Which of the following can be used as an energy source by bacteria to drive active transport?
   a. ATP hydrolysis
   b. proton motive force
   c. Both (a) and (b) are correct.
   d. Neither (a) nor (b) is correct.

6. Which of the following is a good method for obtaining isolated pure cultures of a microorganism?
   a. spread plate
   b. pour plate
   c. streak plate
   d. All of the above are good methods for obtaining isolated pure cultures.

SELECTIVE AND DIFFERENTIAL MEDIA

Mannitol salt agar is a culture medium that contains a high salt (NaCl) concentration, mannitol (a fermentable sugar), and a chemical pH indicator that is yellow at acidic conditions and red at alkaline conditions. (Acids are released when microorganisms ferment mannitol.) This medium also contains other carbohydrates that allow growth of nonfermenting, halophilic organisms (i.e., nonfermenting organisms that tolerate high salt concentrations). Nonhalophilic organisms will not grow on mannitol salt agar.

For each of the following situations, place the letter of the term that *best* describes the way the medium is being used. Then list the organism(s) that will grow, the organisms that will not grow, and the color of the pH indicator.

_____ 1. Onto mannitol salt agar you inoculate a halophilic mannitol fermenter, a halophilic mannitol nonfermenter, and a nonhalophilic mannitol fermenter.

_____ 2. Onto mannitol salt agar you inoculate a halophilic mannitol fermenter and a halophilic mannitol nonfermenter.

_____ 3. Onto mannitol salt agar you inoculate a halophilic mannitol nonfermenter that is pigmented yellow, and a halophilic mannitol nonfermenter that is pigmented red. These two organisms show the same pigmentation (yellow and red, respectively) on a general purpose medium such as nutrient agar.

_____ 4. Onto mannitol salt agar you inoculate a halophilic mannitol nonfermenter and a nonhalophilic mannitol fermenter.

a. selective medium
b. differential medium
c. both selective and differential mediums
d. neither selective nor differential medium

TRUE/FALSE

_____ 1. Micronutrients are normally a part of enzymes and cofactors where they aid in the catalysis of reactions and the maintenance of protein structure.

_____ 2. Most microorganisms require large amounts of sodium (Na).

_____ 3. Transport of materials against a concentration gradient that requires expenditure of metabolic energy is called facilitated diffusion.

_____ 4. Permease proteins resemble enzymes in their specificity for the substance to be transported; each carrier is selective and will transport only closely related solutes.

_____ 5. Microorganisms usually have only one transport system for each nutrient.

_____ 6. Agar is an excellent hardening agent because it is not usually degraded by microorganisms.

_____ 7. Media can be selective or differential, but cannot be both selective and differential.

_____ 8. Heterotrophs usually obtain preformed, partially reduced organic molecules from other organisms.

_____ 9. Facilitated diffusion is generally more important in procaryotes than in eucaryotes.

_____ 10. Quantitative growth-response assays use the growth of microorganisms as a way of measuring the amount of a specific, limited growth nutrient in a particular growth medium.

_____ 11. Nitrogen, phosphorous, and sulfur may be obtained from organic molecules and by reduction and assimilation of oxidized inorganic molecules.

_____ 12. Generally, the most rapid growth occurs at the center of a colony, where nutrients and other resources are more plentiful.

_____ 13. Most nutrients needed by cells are brought into the cell with the aid of carrier proteins.

_____ 14. Colony size and appearance can help microbiologist identify bacteria.

CRITICAL THINKING

1. In this figure, which line corresponds to the situation for passive diffusion? Which line corresponds to the situation for facilitated diffusion? Explain.

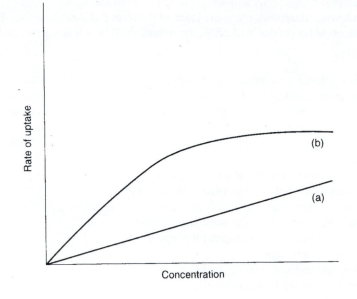

2. A great majority of microorganisms studied thus far are either photolithotrophic autotrophs or chemoorganotrophic heterotrophs. Why? Where would you expect to find photolithotrophic heterotrophs? Chemolithotrophic autotrophs?

3. Many bacteria have several different mechanisms for transport of a single substance. Compare the advantages of multiple mechanisms with those of a single transport mechanism for any given substance.

ANSWER KEY

Terms and Definitions

1. l, 2. m, 3. c, 4. j, 5. d, 6. k, 7. n, 8. q, 9. i, 10. v, 11. o, 12. p, 13. g, 14. a, 15. t, 16. b, 17. h, 18. s, 19. u, 20. e, 21. r, 22. f

Macronutrients/Micronutrients

1. macro, 2. micro, 3. macro, 4. macro, 5. macro, 6. micro, 7. macro, 8. micro, 9. macro, 10. macro, 11. macro, 12. macro, 13. micro, 14. micro, 15. macro, 16. micro

Fill in the Blank

1. Petri dishes 2. complex; peptones 3. ABC transporters; active transport 4. siderophores 5. complex; defined 6. agar; 40 7. colony; spread plate; streak plate; pour plate 8. phosphoenolpyruvate: sugar phosphotransferase; group translocation 9. lithotrophic; heterotrophic10. chemolithotroph

Nutritional Types of Microorganisms

1. light; CO_2 2. light; organic 3. inorganic; CO_2 4. chemoorganotrophic heterotroph; organic 5. mixotroph

Multiple Choice

1. d, 2. c, 3. d, 4. d, 5. c, 6. d

Selective and Differential Media

1. c, 2. b, 3. d, 4. a

True/False

1. T, 2. F, 3. F, 4. T, 5. F, 6. T, 7. F, 8. T, 9. F, 10. T, 11. T, 12. F, 13. T, 14. T

6 Microbial Growth

CHAPTER OVERVIEW

This chapter describes the basic nature of microbial growth. Several methods for the measurement of microbial growth are described, as are different systems for studying microbial growth. The influence of various environmental factors on the growth of microorganisms is also discussed and the chapter ends with a consideration of microbial growth in natural environments. This latter section sets the stage for more detailed discussions of microbial ecology in chapters 28–30.

CHAPTER OBJECTIVES

After reading this chapter you should be able to:

- name the various phases of growth that occur in closed culture systems and describe what is occurring in each phase
- determine from experimental data the various parameters (number of generations, specific growth rate constant, mean generation time) that describe microbial growth in mathematical terms
- compare and contrast the various methods for measuring microbial growth
- describe the various types of continuous culture systems and explain the differences in their function
- describe the influence of various environmental factors (water availability, pH, temperature, oxygen concentration, pressure, radiation) on the growth of microorganisms
- categorize microorganisms according to the environmental factors that are conducive to optimal growth of the organism
- discuss three growth-related phenomena observed in natural environments (growth limitations by environmental factors, viable but nonculturable cells, and quorum sensing)

CHAPTER OUTLINE

I. Growth
   A. Growth is an increase in cellular constituents that may result in an increase in cell size, an increase in cell number, or both
   B. Because observing growth of single cells is difficult, microbiologists usually study growth of a population of microbes
II. The Prokaryotic Cell Cycle
   A. Most procaryotes reproduce by binary fission; other reproduce by budding or fragmentation; during binary fission, a parent cell grows, elongates, replicates its genome, and then divides into two equal daughter cells
   B. Chromosome replication and partitioning
      1. Replication of the circular chromosome through a replisome begins at the origin of replication and ends at the terminus
      2. Partitioning of the chromosome copies into the daughter cells is performed with the help of MreB protein that is similar to eucaryotic actin
   C. Cytokinesis is the process by which the parent cell is divided into two daughter cells; a tubulinlike protein (FtsZ) forms a dynamic Z ring that constricts to cause cell division

III. The Growth Curve
   A. Population growth is usually analyzed in a closed system called a batch culture; it is usually plotted as the logarithm of cell number versus the incubation time
   B. Lag phase is the period of apparent inactivity during which the cells are adapting to a new environment and preparing for reproductive growth, usually by synthesizing new cell components; it varies considerably in length depending upon the condition of the microorganisms and the nature of the medium
   C. Exponential (log) phase is the period during which the organisms are growing at the maximal rate possible given their genetic potential, the nature of the medium, and the conditions under which they are growing; the population is most uniform in terms of chemical and physical properties during this period
   D. Stationary phase
      1. In this period the number of viable microorganisms remains constant either because metabolically active cells stop reproducing or because the reproductive rate is balanced by the rate of cell death
      2. Microbial populations enter stationary phase for several reasons including nutrient limitation, toxic waste accumulation, and possibly cell density
   E. Senescence and death
      1. There are two alternative hypothesis
      2. Starved cells may enter a state called viable but nonculturable (VBNC), a dormant state relieved by an appropriate change in environment
      3. Alternatively, some cells in the population may undergo programmed cell death and their corpses are used to feed the survivors
   F. Phase of prolonged decline is observed in some long-term cultures where successive waves of genetically distinct variants extend the viability of the culture for months or years
   G. The mathematics of growth
      1. Mean growth rate constant is the number of generations per unit time, often expressed as generations per hour
      2. Mean generation (doubling) time is the time required for the number of cells in a population to double
      3. Generation times vary markedly with the species of microorganism and environmental conditions; they can range from 10 minutes for some bacteria to several days for some eucaryotic microorganisms
IV. Measurement of Microbial Growth
   A. Measurement of cell numbers
      1. Direct counting of cells can be done by direct microscopic observation on specially etched slides (such as Petroff-Hausser chambers or hemacytometers) or by using electronic counters (such as Coulter Counters, which count microorganisms as they flow through a small hole or orifice); this does not distinguish between live and dead cells
      2. Membrane filters can be used to determine viable cell counts by staining with specific dyes or for direct cell counts of aquatic samples (where cell concentrations are too low for direct counting)
      3. Viable cell counts require plating of diluted samples (using a pour plate or spread plate) onto suitable growth media and monitoring colony formation; this method counts only reproductively active viable cells; because it is not possible to be certain that each colony arose from a single cell, results are usually expressed as colony-forming units (CFUs)
   B. Measurement of cell mass may be used to approximate the number of microorganisms if a suitable parameter proportional to the number of microorganisms present is used; suitable parameters may be dry weight, light scattering (turbidity) in liquid solutions, or biochemical determinations of specific cellular constituents such as protein, DNA, or ATP

V.   The Continuous Culture of Microorganisms
     A.   Continuous culture methods are used to maintain cells in the exponential growth phase at a constant biomass concentration for extended periods of time; these conditions are met by continual provision of nutrients and removal of wastes and biomass
     B.   The chemostat is a continuous culture device that maintains a constant growth rate by supplying a medium containing a limited amount of an essential nutrient at a fixed rate and by removing medium that contains microorganisms at the same rate
     C.   The turbidostat is a continuous culture device that regulates the flow rate of a medium through the vessel in order to maintain a predetermined turbidity or cell density; there is no limiting nutrient
VI. The Influence of Environmental Factors on Growth
     A.   Microorganisms grow in a variety of environmental conditions; certain microorganisms, referred to as extremophiles, grow under harsh conditions that would kill most other organisms
     B.   Solutes and water activity
          1.   Microorganisms use a variety of mechanisms to protect themselves from damage or death when in environments having solute concentrations higher or lower than their cytoplasm; these protective mechanisms include inclusion bodies, mechanosensitive channels in the plasma membrane, rigid cell walls, compatible solutes, and contractile vacuoles
          2.   Microbes vary in their ability to adapt to habitats with low water activity; osmotolerant organisms can grow in solutions of both high and low water activity; halophiles require environments containing NaCl or other salts above 0.2 M; extreme halophiles require more than 2 M NaCl for growth
          3.   Water activity (a_w), the amount of water available to microorganism, is reduced by the interaction of water with solute molecules
     C.   pH
          1.   pH is the negative logarithm of the hydrogen ion concentration
          2.   Each species has a pH growth range and pH growth optimum: acidophiles grow best between pH 0 and 5.5; neutrophiles grow best between pH 5.5 and 8.0; alkalophiles grow best between pH 8.5 and 11.5; and extreme alkalophiles grow best at pH 10.0 or higher
          3.   Microorganisms can usually adjust to changes in environmental pH by maintaining an internal pH that is near neutrality; some bacteria also synthesize protective proteins (acid shock proteins) in response to pH
          4.   Microorganisms often change the pH of their environment with waste products requiring pH buffers be added to some microbial media
     D.   Temperature
          1.   As the temperature rises, there is an increase in the growth rate of microorganisms due to increasing rates of enzyme reactions; eventually a temperature becomes too high and microorganisms are damaged by protein denaturation, membrane disruption, and other phenomena
          2.   Organisms exhibit distinct cardinal temperatures (minimal, maximal, and optimal growth temperatures); microbes are adapted to their preferred temperature range by a variety of mechanisms (e.g., lipid content of membranes and heat-stable enzymes)
               a.   Psychrophiles can grow well at 0°C, have optimal growth at 15°C or lower, and usually will not grow above 20°C
               b.   Psychrotrophs (facultative psychrophiles) also can grow at 0°C, but have growth optima between 20°C and 30°C, and growth maxima at about 35°C
               c.   Mesophiles have growth optima of 20 to 45°C, minima of 15 to 20°C, and maxima of about 45°C or lower
               d.   Thermophiles have growth optima of 55 to 65°C, and minima around 45°C
               e.   Hyperthermophiles have growth optima of 80 to 110°C and minima around 55°C

E. Oxygen concentration
    1. An organism able to grow in the presence of O_2 is an aerobe; one that grows in its absence is an anaerobe
        a. Obligate (strict) aerobes are completely dependent on atmospheric O_2 for growth
        b. Facultative anaerobes do not require O_2 for growth, but grow better in its presence
        c. Aerotolerant anaerobes ignore O_2 and grow equally well whether it is present or not
        d. Obligate (strict) anaerobes do not tolerate O_2 and die in its presence
        e. Microaerophiles require lower levels (2 to 10%) for growth because normal atmospheric levels of O_2 (21%) are damaging to the cell
    2. The different relationships with O_2 are due to several factors including inactivation of proteins and the effect of toxic O_2 derivatives (superoxide radical, hydrogen peroxide, and hydroxyl radical), which oxidize and destroy cellular constituents; many microorganisms possess enzymes that protect against toxic O_2 derivatives (superoxide dismutase and catalase)

F. Pressure
    1. Barotolerant organisms are adversely affected by increased pressure, but not as severely as are nontolerant organisms
    2. Barophilic organisms require, or grow more rapidly in the presence of, increased pressure

G. Radiation
    1. There are many types of electromagnetic radiation, including visible light, ultraviolet (UV) light, infrared rays, radio waves, and ionizing radiation; some of these can be harmful to organisms
    2. Ionizing radiation such as Xrays or gamma rays is very harmful to microorganisms due to their high energies; low levels produce mutations and may indirectly result in death, whereas high levels are directly lethal by direct damage to cellular macromolecules or through the production of oxygen free radicals
    3. Ultraviolet radiation damages cells by causing the formation of thymine dimers in DNA; this damage can be repaired by photoreactivation (repairs thymine dimers by direct splitting when the cells are exposed to blue light) or by dark reactivation (repairs thymine dimers by excision and replacement in the absence of light)
    4. Although visible light is generally beneficial, when too intense, it can cause formation of singlet oxygen, a highly reactive oxidizing agent; many microorganisms that are airborne or live on exposed surfaces use carotenoid pigments for protection against photooxidation

VII. Microbial Growth in Natural Environments
  A. Growth limitation by environmental factors
    1. Microbial environments are complex, constantly changing, often oligotrophic (low in nutrients) and may expose a microorganism to overlapping gradients of nutrients and environmental factors
        a. Liebig's law of the minimum states that the total biomass of an organism will be determined by the nutrient present in the lowest concentration relative to the organism's requirements
        b. Shelford's law of tolerance states that there is a range of environmental factors, below and above which microorganisms cannot survive and grow, regardless of the nutrient supply
    2. In response to oligotrophic environments and intense competition, many microorganisms change their morphology or physiology or both
  B. Counting and identifying microorganisms in natural environments
    1. Identifying and counting microorganisms is hampered by the inability of many microbes to be grown under laboratory conditions; only about 1% of environmental microorganisms have been cultured
    2. When microorganisms are stressed they can remain viable but lose the ability to grow on media normally used for their cultivation (viable but nonculturable cells)
    3. Numerous microscopic and molecular procedures to identify and count viable but nonculturable cells have been developed

C. Biofilms
   1. Biofilms are complex, slime-encased communities of microbes that form on surfaces; this is an ancient lifestyle and most environmental microbes are sessile (attached) rather than planktonic (free-floating)
   2. Biofilm formation involves cell deposition, attachment, replication and growth, excretion of extracellular polysaccharide matrix, maturation, and colonization of the biofilm
   3. Microorganisms in biofilms are less sensitive to antibiotics and pose problems for bone implants and other medical devices
D. Quorum sensing and microbial populations
   1. Quorum sensing (autoinduction) is a process by which bacteria can communicate and behave cooperatively; it is thought that autoinducers may serve to sense population densities, to determine diffusion rates of molecules secreted into the environment, and to regulate the expression of virulence factors
   2. Bacteria secrete chemical signals and use them to communicate with each other; gram-negative bacteria use acyl homoserine lactones (AHLs) as signals; gram-positive bacteria often use oligopeptide signals

TERMS AND DEFINITIONS

Place the letter of each term in the space next to the definition or description that best matches it.

_____ 1. An increase in cellular constituents that may or may not be accompanied by an increase in cell number

_____ 2. A culture in a closed vessel with a single batch of medium to which no fresh medium is added and from which no waste products are removed

_____ 3. The length of time it takes for a population of microorganisms to double in number

_____ 4. The number of generations per unit time, usually expressed as the number of generations per hour

_____ 5. A process by which bacteria communicate and carry out certain activities in a density-dependent fashion

_____ 6. The number of cells per unit volume that are able to grow and reproduce

_____ 7. An approximation of the number of viable microorganisms based on the number of colonies that form on solid media after plating dilute solutions such that each colony probably arises from a single viable microorganism

_____ 8. The energy required by the cell to maintain itself

_____ 9. The amount of water available to a microorganism

_____ 10. A culture system with constant environmental conditions maintained through continual nutrient provision and waste removal

_____ 11. An open system in which sterile medium containing an essential nutrient in limiting quantities is fed into the culture vessel at the same rate as medium containing microorganisms is removed

_____ 12. An open system in which the flow rate of a medium through the vessel is automatically regulated to maintain a predetermined turbidity or cell density

_____ 13. Organisms that grow over a fairly wide range of water activity

_____ 14. Organisms that require high levels of sodium chloride in order to grow

_____ 15. Organisms that have their growth optimum between pH 0 and 5.5

_____ 16. Organisms that have their growth optimum between pH 5.5 and 8.0

_____ 17. Organisms that have their growth optimum between pH 8.5 and 11.5

_____ 18. Minimum, maximum, and optimum growth temperatures

_____ 19. Organisms that grow well at 0°C and have optimum growth temperatures of 15°C or lower

_____ 20. Organisms that can grow well at 0°C but have optimum growth temperatures between 20°C and 30°C

a. acidophiles
b. aerotolerant anaerobes
c. alkalophiles
d. barophiles
e. barotolerant
f. batch culture
g. biofilm
h. cardinal temperatures
i. chemostat
j. colony forming units
k. continuous (open) culture system
l. extremophiles
m. facultative anaerobes
n. growth
o. halophiles
p. hyperthermophiles
q. Liebig's law
r. maintenance energy
s. mean generation time (g)
t. mean growth rate constant (k)
u. mesophiles
v. microaerophiles
w. neutrophils
x. obligate aerobes
y. obligate anaerobes
z. origin of replication
aa. osmotolerant
bb. psychrophiles
cc. psychrotrophs (facultative psychrophiles)
dd. quorum sensing
ee. replisome
ff. Shelford's law
gg. thermophiles
hh. turbidostat
ii. viable but nonculturable
jj. viable cell count
kk. water activity

_____ 21. Organisms with growth temperature minima of 15 to 20°C, optima around 20 to 45°C, and maxima at 45°C or lower

_____ 22. Organisms that have growth temperature minima around 45°C and optima between 55 to 65°C

_____ 23. Organisms that are completely dependent upon atmospheric oxygen for growth

_____ 24. Organisms that do not require oxygen for growth but grow better in its presence

_____ 25. Organisms that ignore oxygen and grow equally well in its presence or in its absence

_____ 26. Organisms that do not tolerate oxygen and die in its presence

_____ 27. Organisms that require oxygen at levels in the range of 2 to 10% for growth and are damaged by the normal atmospheric levels of oxygen (20%)

_____ 28. Organisms that are not adversely affected by increased barometric pressure

_____ 29. Organisms that grow more rapidly at increased barometric pressure

_____ 30. Organisms that can grow in harsh environments that would kill most living organisms

_____ 31. Organisms with growth optima of 80 to 100°C

_____ 32. A law stating that the total biomass of an organism will be determined by the nutrient present in the lowest concentration relative to the organism's requirements

_____ 33. A law stating that there are limits to environmental factors below and above which a microorganism cannot survive and grow, regardless of the nutrient supply

_____ 34. The complex involved in replication of the procaryotic chromosome

_____ 35. Replication of the procaryotic chromosome begins at this structure

_____ 36. Starved cells may enter this dormant state

_____ 37. A complex, slime-encased microbial community

BACTERIAL GROWTH CURVE

Look at the accompanying bacterial growth curve. Label each of the four phases and describe what is occurring in each phase.

(*a*) _____
_____
_____

(*b*) _____
_____
_____

(*c*) _____
_____

(*d*) _____
_____
_____

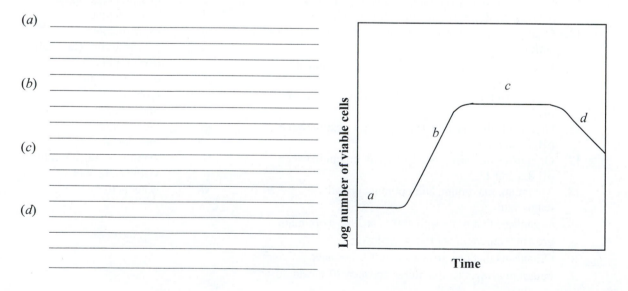

FILL IN THE BLANK

1. An increase in cellular constituents is termed _____. If a microorganism is _____, an increase in cell size will result, but not an increase in the number of cells. If cells reproduce by processes such as _____ _____ or _____, an increase in cell number could result.

2. Oxygen is readily reduced, leading to the production of a variety of toxic oxygen-containing compounds. These include _____ _____, _____ _____, and _____ _____. Aerobes have enzymes that destroy the toxic compounds (e.g., superoxide dismutase destroys _____ _____, and catalase destroys _____ _____), whereas strict anaerobes do not.

3. In _____ culture systems (closed systems), exponential growth lasts for only a few generations, whereas in _____ culture systems (open systems), a microbial population can be maintained in the exponential growth phase for extended periods of time.

4. A _____ is a continuous culture system equipped with a photocell that measures the turbidity of a culture and automatically regulates the flow rate of media through the vessel in order to maintain a predetermined cell density.

5. Microorganisms that can grow in habitats of high and low water activity are said to be _____, while those that require low water activity due to high concentrations of NaCl or other salts are called _____.

6. When microbes are grown in a closed system (also called a _____ _____), the resulting growth curve has four phases: the _____, _____ or_____, _____, and _____phases.

7. In obligate _____, oxygen serves as the terminal electron acceptor during respiration.

8. For organisms that are _____ anaerobes, the efficiency of growth increases when the organism is shifted from an anaerobic environment to an aerobic environment, while _____ anaerobes show no change in growth efficiency when the same shift occurs.

9. Organisms that are not affected by increased pressure are _____, while those that grow more rapidly at increased pressure are _____.

10. _____ radiation damages organisms, primarily by inducing the formation of thymine dimers in DNA. These flaws can be repaired by blue light, which directly splits them—a process called _____. They can also be repaired by excision and replacement in the absence of light—a process called _____.

11. Organisms that have growth optima of 55 to 65°C are called _____ while those with growth optima of 80 to 100°C are called _____.

12. In _____ _____systems, nutrients are constantly provided and wastes removed.

13. Absorption of visible light by molecules commonly found in microorganisms can lead to the generation of _____ _____, a powerful oxidizing agent that can rapidly destroy cells.

14. When a culture is growing exponentially it is exhibiting _____ growth. However, if the culture is shifted to a more nutrient-rich (shift up) or more nutrient-poor (shift down) medium, _____ growth can occur.

15. _____ environments are those that have low nutrient concentrations.

16. Most procaryotes reproduce by _____, while others use _____ or _____. Their chromosomes are _____ and replicated by a complex called a _____. The division of the cell, called _____, is performed with a _____ composed of tubulinlike protein.

17. Biofilm formation starts with _____ and _____. The organisms grow and _____ and exude an extracellular _____ that is primarily composed of _____. Other microbes can then _____the mature biofilm.

MICROBIAL GROWTH TEMPERATURES

The following is a list of organisms and their cardinal temperatures. For each organism, indicate whether the organism is psychrophilic, psychrotrophic, mesophilic, thermophilic, or hyperthermophilic.

Organism	Cardinal Temperatures (°C) Min.	Opt.	Max.	Psychrophilic, Psychrotrophic, Mesophilic, Thermophilic or Hyperthermophilic?
1. *Micrococcus cryophilus*	-4	10	24	
2. *Staphylococcus aureus*	6.5	30 to 37	46	
3. *Sulfolobus acidocaldarius*	60	80	85	
4. *Thermoplasma acidophilum*	45	59	62	
5. *Bacillus psychrophilus*	-10	23 to 24	28 to 30	

MICROBIAL GROWTH PROBLEMS

1. At 4:00 P.M. you inoculate 1×10^3 cells into a closed flask of nutrient broth. The cells have a lag phase that lasts one hour. At 9:00 P.M. the culture enters the stationary phase. At that time there are 6.5×10^7 cells in the flask. Calculate the number of generations that have occurred (n), the mean generation time (g) and the mean growth rate constant (k).

2. Continuing with the same culture of question #1, you measure the number of viable cells at 10:00 P.M., at 10:30 P.M., and at 11:00 P.M. and obtain the following results:

10:00 P.M.	6.5×10^7
10:30 P.M.	3.3×10^7
11:00 P.M.	1.6×10^7

    Construct a growth curve for this culture on the semilogarithmic graph paper provided. Be sure that you correctly label each axis. Explain what is occurring at each phase.

MULTIPLE CHOICE

For each of the questions below select the *one best* answer.

1. Which of the following is NOT a reason for the occurrence of a lag phase in a bacterial growth curve?
   a. The cells may be old and depleted of ATP, essential cofactors, and ribosomes, which must be synthesized before growth can begin.
   b. The current medium may be different from the previous growth medium; therefore, the cells must synthesize new enzymes to utilize different nutrients.
   c. The organisms may have been injured and thereby may require time to recover.
   d. All of the above are potential reasons for the occurrence of a lag phase.

2. Which of the following does NOT affect the growth rate observed during the exponential (log) growth phase?
   a. the genetic potential of the organism
   b. the number of cells originally inoculated into the culture vessel
   c. the composition of the medium
   d. the conditions of the incubation

3. In the stationary phase, the total number of viable microorganisms remains constant. What might cause this?
   a. a balance between the rate of cell division and the rate of cell death
   b. a cessation of cell division even though the cells remain metabolically active
   c. Both (a) and (b) may occur.
   d. Neither (a) nor (b) is correct.

4. Which of the following might cause cells to enter stationary phase?
   a. depletion of an essential nutrient
   b. lack of available oxygen
   c. accumulation of toxic waste products
   d. All of the above are correct.

5. For what purpose have membrane filter techniques been used?
   a. viable cell counts
   b. direct cell counts
   c. both viable and direct cell counts
   d. neither viable nor direct cell counts

6. Which of the following is NOT true about the use of colony-forming units as a measurement of bacterial population numbers?
   a. For best results, the samples should yield between 30 and 300 colonies.
   b. Because each colony arises from a single viable cell, the number of colonies is an accurate measure of the number of viable cells in the sample.
   c. Hot agar used in the pour plate technique may lead to an underestimate of the viable cell number compared to the spread plate technique.
   d. All of the above are true about the use of colony-forming units as a measurement of bacterial population numbers.

7. Which term is used to describe an organism with a pH growth optimum of 10.5?
   a. acidophile
   b. neutrophile
   c. alkalophile
   d. extreme alkalophile

8. What are compatible solutes?
   a. solutes that allow otherwise incompatible microorganisms to grow in the same habitat
   b. solutes that can coexist in the same cell without forming complexes that are toxic to the cell
   c. solutes that are compatible with metabolism and growth when at high intracellular concentrations
   d. None of the above correctly describes compatible solutes.

9. Which of the following is NOT a mechanism by which ionizing radiation kills cells?
   a. indirectly by accumulation of harmful mutations
   b. directly by damaging cellular macromolecules
   c. directly by the production of oxygen free radicals
   d. All of the above are mechanisms by which ionizing radiation kills cells.

10. What percentage of environmental
    microorganisms can be cultured in the
    laboratory?
    a. 100%
    b. 45%
    c. 25%
    d. 1%

TRUE/FALSE

_____ 1. When a young, vigorously growing culture is transferred to fresh medium with the same composition as the original medium, the lag phase is usually short or absent.

_____ 2. When a young, vigorously growing culture is transferred to fresh medium with a different composition from the original medium, the lag phase is usually short or absent.

_____ 3. The rate of growth in the exponential (log) phase is independent of the composition of the medium.

_____ 4. Exponential (log) phase cultures are usually used in biochemical studies because populations in this phase are relatively uniform in terms of their chemical and physiological properties.

_____ 5. The progress of the death phase can usually be followed experimentally by measuring total cell number, because cells lyse (disintegrate) after dying.

_____ 6. As the dilution rate in a chemostat is increased, the nutrient concentration remains relatively constant, the microbial population density remains stable, and the generation time shortens. However, the dilution rate can only be increased to a certain level beyond which the cells are washed out at a rate higher than they can reproduced, and the cell density drops dramatically.

_____ 7. Continuous culture systems are very useful because they enable us to study microbial growth at very low nutrient levels that approximate nutrient levels of natural environments.

_____ 8. The water activity of a solution is directly proportional to the osmotic pressure.

_____ 9. Most microorganisms maintain their internal pH at approximately the same level as their optimum growth pH.

_____ 10. Psychrotrophic and psychrophilic bacteria can both grow at 0°C, but they differ in their other cardinal temperatures: the psychrotrophs' optimum and maximum growth temperatures are higher than those of the psychrophiles.

_____ 11. Obligate anaerobes are usually poisoned by oxygen, but may grow in aerobic habitats if they are associated with facultative anaerobes that use up all of the available oxygen.

_____ 12. The mean generation time is the reciprocal of the mean growth rate constant.

_____ 13. Microbial cells are of roughly constant size; therefore, the amount of light scattering is proportional to the concentration of cells present in a culture.

_____ 14. In a chemostat, both the microbial population level and the generation time are related to the dilution rate.

_____ 15. The cell membranes of psychrophilic microorganisms have high levels of unsaturated fatty acids and remain semi-fluid when cold.

_____ 16. During balanced growth, cell components are synthesized at constant rates relative to one another.

_____ 17. Microorganisms rarely alter the pH of their environment; therefore, it is not necessary to buffer growth media.

_____ 18. Some bacteria produce starvation proteins that protect against the damaging effects of starvation.

_____ 19. During the phase of prolonged decline, genetically distinct variant can arise.

_____ 20. Microorganisms must adapt to the pH of their environment because they cannot change it.

CRITICAL THINKING

1. In this chapter, several methods used to determine microbial growth were presented. These included total cell counts, viable cell counts, and turbidity measurements and other methods of biomass determination. What are the advantages and disadvantages of these various methods, and under what circumstances might one method be preferred to another?

2. Thermophilic bacteria that can grow in temperatures above 100°C have been found. However, this is above the boiling point of water. In what natural habitats might you find these conditions, and how is an aqueous environment maintained at that high a temperature? What types of adaptations would you envision for organisms living in those habitats?

ANSWER KEY

Terms and Definitions

1. n, 2. f, 3. s, 4. t, 5. dd, 6. jj, 7. j, 8. r, 9. kk, 10. k, 11. i, 12. hh, 13. aa, 14. o, 15. a, 16. w, 17. c, 18. h, 19. bb, 20. cc, 21. u, 22. gg, 23. x, 24. m, 25. b, 26. y, 27. v, 28. e, 29. d, 30. l, 31. p, 32. q, 33. ff

Fill in the Blank

1. growth; coenocytic; binary fission; budding 2. superoxide radical; hydrogen peroxide; hydroxyl radical; superoxide radical; hydrogen peroxide; 3. batch; continuous 4. turbidostat 5. osmotolerant; halophiles 6. batch culture; lag; log; exponential; stationary; death 7. aerobes 8. facultative; aerotolerant 9. barotolerant; barophilic 10. ultraviolet; photoreactivation; dark reactivation 11. thermophiles; hyperthermophiles 12. continuous (open) 13. singlet oxygen 14. balanced; unbalanced 15. oligotrophic 16. binary fission; budding; fragmentation; circular; replisome; cytokinesis; Z ring 17. cell deposition; attachment; replicate; matrix; polysaccharide; colonize

Microbial Growth Temperatures

1. psychrophilic 2. mesophilic 3. hyperthermophilic 4. thermophilic 5. psychrotrophic

Microbial Growth Problems

1. n = 16 generations; g = 15 min (0.25 hr); k = 4 generations/hr

Multiple Choice

1. d, 2. b, 3. c, 4. d, 5. c, 6. b, 7. d, 8. c, 9. d, 10. d

True/False

1. T, 2. F, 3. F, 4. T, 5. F, 6. T, 7. T, 8. F, 9. F, 10. T, 11. T, 12. T, 13. T, 14. T, 15. T. 16. T, 17. F, 18. T, 19. T, 20. F

7 Control of Microorganisms by Physical and Chemical Agents

CHAPTER OVERVIEW

This chapter focuses on the control and the destruction of microorganisms by physical and chemical agents. This is a topic of great importance, because microorganisms may have deleterious effects, such as food spoilage and disease. It is therefore essential to be able to kill or remove microorganisms from certain environments in order to minimize their harmful effects.

CHAPTER OBJECTIVES

After reading this chapter you should be able to:

- compare and contrast the processes of disinfection, sanitization, antisepsis, and sterilization
- compare the difficulties encountered when trying to kill endospores with those encountered when trying to kill vegetative cells
- discuss the exponential pattern of microbial death
- discuss the influence of environmental factors on the effectiveness of various agents used to control microbial populations
- discuss the uses and limitations of various physical and chemical agents used to control microbial populations
- describe the procedures used to evaluate the effectiveness of various antimicrobial agents

CHAPTER OUTLINE

I. Definition of Frequently Used Terms
    A. Sterilization is the destruction or removal of *all* viable organisms from an object or from a particular environment
    B. Disinfection is the killing, inhibition, or removal of pathogenic microorganisms (usually on inanimate objects)
    C. Sanitization is the reduction of microbial populations to a safe level as determined by public health standards
    D. Antisepsis is the prevention of infection of living tissue by microorganisms
    E. Antimicrobial agents fall into one of two broad categories denoted by suffixes indicating effect
        1. -cide is a suffix indicating that the agent will kill the kind of organism in question (e.g., viricide, fungicide)
        2. -static is a suffix indicating that the agent will prevent the growth of the type of organism in question (e.g., bacteriostatic, fungistatic)
II. The Pattern of Microbial Death
    A. Microorganisms are not killed instantly when exposed to a lethal agent; rather, the population generally decreases by a constant fraction at constant intervals (exponential killing)
    B. A microorganism is usually considered dead when it is unable to grow in conditions that would normally support its growth and reproduction; however, the issue is confounded by organisms that are in a viable but nonculturable (VBNC) state
III. Conditions Influencing the Effectiveness of Antimicrobial Agent Activity
    A. Population size—larger populations take longer to kill than smaller populations

B. Population composition—populations consisting of different species or of cells at different developmental stages (e.g., endospores versus vegetative cells or young cells versus old cells) differ markedly in their sensitivity to various agents

C. Concentration or intensity of the antimicrobial agent—higher concentrations or intensities are generally more efficient, but the relationship is not linear

D. Duration of exposure—the longer the exposure, the greater the number of organisms killed

E. Temperature—a higher temperature will usually (but not always) increase the effectiveness of killing

F. Local environment—environmental factors (e.g., pH, viscosity, and concentration of organic matter) can profoundly influence the effectiveness of a particular antimicrobial agent

IV. The Use of Physical Methods in Control

    A. Heat

        1. Killing with moist heat

            a. Boiling water is effective against vegetative cells and eucaryotic spores

            b. Autoclaving (steam under pressure) at 121°C is effective against most vegetative cells and bacterial endospores

            c. Pasteurization, a process involving brief exposure to temperatures below the boiling point of water, reduces the total microbial population and thereby increases the shelf life of the treated material; it is often used for heat-sensitive materials that cannot withstand prolonged exposure to high temperatures

        2. Dry heat can be used to sterilize moisture-sensitive materials such as powders, oils, and similar items; it is less efficient than moist heat because it usually requires higher temperatures (160 to 170°C) and longer exposure times (2 to 3 hrs)

        3. Measuring heat killing efficiency

            a. The thermal death time (TDT) is the shortest time necessary to kill all microorganisms in a suspension at a specific temperature and under defined conditions

            b. The decimal reduction time (D, or D value) is the time required to kill 90% of the microorganisms or spores in a sample at a specific temperature

            c. The Z value is the increase in temperature required to reduce D to 1/10 of its previous value

            d. The F value is the time in minutes at a specific temperature (usually 250°F or 121°C) necessary to kill a population of cells or spores

    B. Low temperatures slow (refrigeration) or prevent (freezing) microbial growth and reproduction, but do not necessarily kill microorganisms; refrigeration and freezing are particularly important in food microbiology

    C. Filtration sterilizes heat-sensitive liquids and gases by removing microorganisms rather than destroying them

        1. Depth filters are thick fibrous or granular filters that remove microorganisms by physical screening, entrapment, and/or adsorption

        2. Membrane filters are thin filters with defined pore sizes that remove microorganisms, primarily by physical screening

        3. Air can be sterilized by passage through surgical masks, cotton plugs in culture vessels, and high-efficiency particulate air (HEPA) filters; HEPA filters are used in laminar flow biological safety cabinets to sterilize the air circulating in the enclosure

    D. Radiation

        1. Ultraviolet (UV) radiation is effective, but its use is limited to surface sterilization because UV radiation does not penetrate glass, dirt films, water, and other substances

        2. Ionizing radiation (Xrays, gamma rays, etc.) is effective and penetrates the material

V. The Use of Chemical Agents in Control

    A. Phenolics—laboratory and hospital disinfectants; act by denaturing proteins and disrupting cell membranes; widely used agents include Lysol and hexachlorophene

    B. Alcohols—widely used disinfectants and antiseptics; will not kill endospores; act by denaturing proteins and possibly by dissolving membrane lipids; ethanol and isopropanol are most widely used

C. Halogens—widely used antiseptics and disinfectants; iodine acts by oxidizing cell constituents and iodinating cell proteins; chlorine acts primarily by oxidizing cell constituents

D. Heavy metals—effective but usually toxic; act by combining with proteins and inactivating them

E. Quaternary ammonium compounds—cationic detergents used as disinfectants for food utensils and small instruments, and because of low toxicity, as antiseptics for skin; act by disrupting biological membranes and possibly by denaturing proteins; benzalkonium chloride and cetylpyridinium chloride are widely used

F. Aldehydes—reactive molecules that can be used as chemical sterilants; may irritate the skin; act by combining with nucleic acids and proteins and inactivating them; formaldehyde and glutaraldehyde are widely used

G. Sterilizing gases (e.g., ethylene oxide, betapropiolactone)—can be used to sterilize heat-sensitive materials such as plastic petri dishes and disposable syringes; act by combining with proteins and inactivating them; recently, vapor-phase hydrogen peroxide has been used to decontaminate large facilities

H. Chemotherapeutic agents—can be used internally to kill of inhibit microbial growth within host tissues; these selectively toxic agents are generally antibiotics

VI. Evaluation of Antimicrobial Agent Effectiveness

A. The phenol coefficient test is a useful initial screening test in which the potency of a disinfectant is compared to that of phenol

B. A more realistic test is the use dilution test; in this test, stainless steel cylinders are contaminated with specific bacterial species under carefully controlled conditions and then exposed to the disinfectant

TERMS AND DEFINITIONS

Place the letter of each term in the space next to the definition or description that best matches it.

_____ 1. Destruction or removal of all viable microorganisms from an object or a particular environment

_____ 2. An antimicrobial agent that kills algae

_____ 3. The killing, inhibition, or removal of microorganisms that may cause disease

_____ 4. Agents used to carry out disinfection of inanimate objects

_____ 5. Reduction of the microbial population to safe levels as determined by public health standards

_____ 6. The prevention of infection by controlling microorganisms on living tissue

_____ 7. Chemical agents applied to living tissue to prevent infection

_____ 8. The time in minutes at a specific temperature (usually 121.1°C) needed to kill a population of cells or spores

_____ 9. The shortest period of time needed to kill all the organisms in a microbial suspension at a specific temperature and under defined conditions

_____ 10. The time required to kill 90% of the microorganisms or spores in a sample at a specific temperature

_____ 11. The increase in temperature required to reduce D to 1/10 its value

a. algicide
b. antisepsis
c. antiseptics
d. bacteriocide
e. bacteriostatic
f. decimal reduction time (D)
g. detergents
h. disinfectant
i. disinfection
j. F value
k. fungistatic
l. germicide
m. pasteurization
n. phenol coefficient test
o. sanitization
p. sterilization
q. thermal death time (TDT)
r. viricide
s. Z value

_____ 12. The process in which a microbial population is reduced by raising the temperature of heat-sensitive materials to something less than the boiling point of water for relatively short periods of time

_____ 13. An antimicrobial agent that inhibits the growth of fungi

_____ 14. Organic molecules that serve as wetting agents and emulsifiers because they have both polar hydrophilic ends and nonpolar hydrophobic ends; such molecules that are cationic are useful as disinfectants

_____ 15. A test that measures the efficiency of a disinfectant by comparing it to phenol

_____ 16. An antimicrobial agent that kills bacteria

_____ 17. An antimicrobial agent that inhibits the growth of bacteria

_____ 18. An antimicrobial agent that kills pathogens (and many nonpathogens but not necessarily endospores)

_____ 19. An antimicrobial agent that destroys viruses

FILL IN THE BLANK

1. A _____ will not necessarily sterilize an object because viable spores may still remain.

2. _____ heat is thought to kill effectively by degrading nucleic acids, denaturing proteins, and disrupting cell membranes.

3. Pasteurization of milk involves heating it to 63°C for 30 minutes or to 72°C for 15 seconds. The latter protocol is called _____ pasteurization. Pasteurization procedures do not _____ milk but do kill _____ and reduce the total microbial population in milk, thereby drastically slowing spoilage and increasing shelf life. Some milk products (e.g., individual serving coffee creamers) are made by heating milk to 140 to 150°C for 1 to 3 seconds. This process is called _____ sterilization.

4. Gamma radiation, a type of _____ radiation has been used to pasteurize meats, fish, and other foods. However, this process has not been widely employed because of expense and because of concerns about the long-term effects on the irradiated foods.

5. _____ are germicidal because they denature proteins and disrupt cell membranes. However, their use is limited because they have a disagreeable odor and may cause irritation of the skin.

6. The two most popular _____ germicides are ethanol and isopropanol.

7. Iodine is an effective antiseptic that kills by _____ cell constituents. However, there are some problems associated with its use. Recently, iodine has been complexed with an organic carrier to form an _____. This preparation is water soluble, stable, and nonstaining, and releases _____ slowly, thus minimizing skin burns and irritation.

8. _____ is the disinfectant of choice for municipal water supplies and swimming pools, and is also employed in the dairy and food industries. Campers frequently use it in the form of halazone tablets to purify small volumes of drinking water.

9. Heavy metals are effective antimicrobial agents, but they are often toxic. Therefore, only a few are regularly used. For instance, _____ _____ is applied to the eyes of infants to prevent ophthalmic gonorrhea, and _____ _____ is used as an algicide in lakes and swimming pools.

10. _____ _____ compounds such as benzalkonium chloride kill most bacteria, but do not kill _Mycobacterium tuberculosis_ or spores. They have the advantage of being stable, nontoxic, and bland, but they are inactivated by hard water and soaps.

11. The two commonly used _____ are formaldehyde and glutaraldehyde. They are highly reactive and are _____ and can therefore be used as chemical sterilants.

12. Treatment with boiling water kills vegetative forms, but _____ must be used to destroy endospores. These instruments heat materials in a chamber that reaches _____°C and _____ pounds of pressure.

13. _____ and _____ do not kill or remove microorganisms, but they do slow or stop microbial growth.

14. One of the most important air filtration systems is _____ _____ _____ filters, which are often used in _____ _____ _____ safety cabinets.

15. In the _____ _____ test, stainless steel cylinders are contaminated with specific bacterial species under controlled conditions. After being briefly dried, the cylinders are immersed in the _____ being tested for 10 minutes, transferred to culture media, and incubated for 2 days.

MULTIPLE CHOICE

For each of the questions below select the *one best* answer.

1. Which of the following is NOT a reason for studying methods of destroying microorganisms?
   a. It makes microbial research possible.
   b. preservation of food
   c. prevention of disease
   d. All of the above are reasons for studying methods of destroying microorganisms.

2. What is an agent that specifically kills fungi but not other kinds of microorganisms?
   a. germicide
   b. fungicide
   c. germistatic agent
   d. fungistatic agent

3. Which will require a longer time to kill?
   a. a large population of microorganisms
   b. a small population of microorganisms
   c. Neither; killing will be equally rapid in either a large or a small microbial population.
   d. There is no way to predict which will require a longer time to kill.

4. Which of the following may contribute to resistance to killing?
   a. formation/existence of endospores
   b. increased age of the culture
   c. inherent/genetic differences among organisms
   d. All of the above may contribute to resistance to killing.

5. Which of the following is NOT a method by which depth filters normally sterilize materials passing through them?
   a. exclusion of materials too large to penetrate the filter
   b. entrapment of material in the channels of the filter
   c. destruction of cells by interaction with the filter's harmful chemical composition
   d. adsorption of cells to the surface of the filter material

6. Which of the following contributes to the effectiveness of a disinfectant or sterilizing agent?
   a. concentration of the agent
   b. duration of exposure to the agent
   c. temperature
   d. All of the above contribute.

7. Which of the following is normally used to remove microorganisms from air?
   a. depth filters
   b. membrane filters
   c. HEPA filters
   d. None of the above is correct.

8. Which of the following is usually used to define microbial death?
   a. The organism will not grow on minimal medium.
   b. The organism will not grow on a medium that normally supports its growth.
   c. The organism no longer retains its original shape and structures.
   d. None of the above adequately define microbial death.

9. The process of heating milk products to 72°C for 15 seconds in order to reduce the microbial population is called _____ pasteurization.
   a. flash
   b. fast
   c. ultrafast
   d. None of the above is correct.

10. Which of the following is a property of an ideal disinfectant?
    a. effective against a wide variety of infectious agents
    b. nontoxic to people
    c. noncorrosive for common materials
    d. All of the above are properties of an ideal disinfectant.

TRUE/FALSE

_____ 1. In sanitization, the inanimate object is usually cleaned as well as partially disinfected.

_____ 2. Antiseptics are generally not as toxic as disinfectants.

_____ 3. Although many agents kill or control the growth of nonpathogenic organisms as well as pathogenic organisms, only their effects on pathogens are important in evaluating their usefulness.

_____ 4. Although agents normally kill a constant fraction of the microbial population in a constant period of time, the rate of killing may decrease when the population has been greatly reduced. One reason for this may be the survival of a resistant strain of the organism.

_____ 5. Dry heat sterilization is generally faster than moist heat sterilization.

_____ 6. Since filtration removes rather than destroys microorganisms, it does not truly sterilize the materials passing through the filter.

_____ 7. Ultraviolet radiation is normally used to sterilize air and exposed surfaces, but it cannot be used to sterilize interior compartments because it does not penetrate into those compartments.

_____ 8. Only anionic detergents are useful as disinfectants.

_____ 9. Ethylene oxide is used to sterilize heat-sensitive materials such as plastic petri dishes, and it is particularly effective because it can penetrate even the plastic wrap used to package these plates. Therefore, they can be sealed before sterilization, and the chance of contamination after treatment can be eliminated.

_____ 10. The phenol coefficient is a direct indication of disinfectant potency during normal use.

_____ 11. Death of microorganisms is defined as the loss of the ability to grow on a medium that would normally support growth.

_____ 12. A sterile object is completely free of viable microorganisms.

_____ 13. Most chemical agents are effective sterilants.

_____ 14. Membrane filters are porous membranes made of cellulose acetate or other synthetic materials. They are used to remove most vegetative cells, but not viruses, from liquids.

CRITICAL THINKING

1. Pasteurization can take place by either of two methods. In one a lower temperature is used for a longer period of time, while in the other a higher temperature is used for a shorter period of time. Which method is more advantageous and why?

2. Consider the various chemicals that can be used as disinfectants and antiseptics. What criteria would you use to select the right chemical for a particular task? Give at least two specific examples of situations. Then give your selection for each situation and defend your choice.

ANSWER KEY

Terms and Definitions

1. p, 2. a, 3. i, 4. h, 5. o, 6. b, 7. c, 8. j, 9. q, 10. f, 11. s, 12. m, 13. k, 14. g, 15. n, 16. d, 17. e, 18. l, 19. r

Fill in the Blank

1. disinfectant 2. Moist 3. flash; sterilize; pathogens; ultrahigh-temperature 4. ionizing 5. Phenolics 6. alcohol 7. oxidizing; iodophore; iodine 8. Chlorine 9. silver nitrate; copper sulfate 10. Quaternary ammonium 11. aldehydes; sporicidal 12. autoclaves; 121; 15 13. Refrigeration; freezing 14. high-efficiency particulate air; laminar flow biological 15. use dilution; disinfectants

Multiple Choice

1. d, 2. b, 3. a, 4. d, 5. c, 6. d, 7. c, 8. b, 9. a, 10. d

True/False

1. T, 2. T, 3. F, 4. T, 5. F, 6. F, 7. T, 8. F, 9. T, 10. F, 11. T, 12. T, 13. F, 14. T

8 Metabolism: Energy, Enzymes, and Regulation

CHAPTER OVERVIEW

This chapter discusses energy and the laws of thermodynamics. The participation of energy in cellular metabolic processes and the role of adenosine-5′-triphosphate (ATP) as the energy currency of cells are examined. The chapter concludes with a discussion of enzymes as biological catalysts: how they work, how they are affected by their environment, and how they are regulated.

CHAPTER OBJECTIVES

After reading this chapter you should be able to:

- discuss the first and second laws of thermodynamics and show how they apply to biological systems
- discuss enthalpy, entropy, and free energy and their application to biological reactions
- discuss the use of ATP as the energy currency of the cell and show how it is used to couple energy-yielding exergonic reactions with energy-requiring endergonic reactions
- discuss reduction potential and its relationship to exergonic and endergonic processes
- describe the role of enzymes in the catalysis of biological reactions, and discuss the ways in which enzymes are influenced by their environment
- discuss the need for metabolic regulation
- describe metabolic channeling
- describe how enzyme activity can be controlled by allosteric regulation and covalent modification
- describe how feedback inhibition can be used to control the activity of a metabolic pathway

CHAPTER OUTLINE

I. An Overview of Metabolism
 A. Metabolism is the total of all chemical reactions in the cell, including both energy-conserving reactions (catabolism or breakdown) and energy-requiring reactions (anabolism or synthesis)
 B. Interactions among the five major nutritional types of microorganisms are critical for the functioning of the biosphere; for instance, photoautotrophs and chemolithoautotrophs trap energy and use some of it to transform carbon dioxide into organic molecules; the organic molecules then serve as sources of carbon and energy for chemoheterotrophs, which in turn oxidize the organic molecules by processes such as aerobic respiration, releasing carbon dioxide
II. Energy and Work
 A. Energy is the capacity to do work or to cause particular changes
 B. Living cells carry out three major types of work
 1. Chemical work—synthesis of complex molecules
 2. Transport work—nutrient uptake, waste elimination, ion balance
 3. Mechanical work—internal and external movement

III. The Laws of Thermodynamics
  A. The science of thermodynamics analyzes energy changes in a collection of matter called a system; all other matter in the universe is called the surroundings
  B. First law—energy can be neither created nor destroyed
    1. The total energy in the universe remains constant
    2. Energy may be redistributed either within a system or between the system and its surroundings
  C. Second law—physical and chemical processes proceed in such a way that the disorder of the universe (entropy) increases to the maximum possible
  D. Energy is measured in calories where 1 calorie is the amount of heat energy needed to raise the temperature of 1 gram of water from 14.5 to 15.5°C; one calorie of heat equals about 4.2 joules
IV. Free Energy and Reactions
  A. The changes in energy that can occur in chemical reactions are expressed by the equation for free energy change ($\Delta G = \Delta H - T \cdot \Delta S$); free energy change ($\Delta G$) is the amount of energy in a system that is available to do work
  B. The change in free energy of a chemical reaction is directly related to the equilibrium constant of the reaction
    1. The standard free energy change (ΔG^{0}) is the change in free energy under standard conditions of concentration, pH, pressure, and temperature
    2. When ΔG^{0} is negative, the equilibrium constant is greater than one and the reaction goes to completion as written; the reaction is said to be exergonic and releases energy
    3. When ΔG^{0} is positive, the equilibrium constant is less than one and little product will be formed at equilibrium; the reaction is said to be endergonic and requires energy
V. The Role of ATP in Metabolism
  A. ATP is a high-energy molecule used to capture, store, and provide chemical energy; removal of the terminal phosphate by hydrolysis goes almost to completion with a large negative free energy change (i.e., the reaction is strongly exergonic); ATP also has high phosphate group transfer potential
  B. These characteristics make ATP well suited for its role as an energy currency; ATP is formed from ADP and P_i by energy-trapping processes; exergonic breakdown of ATP can be coupled with various endergonic reactions to facilitate their completion
VI. Oxidation-Reduction Reactions, Electron Carriers, and Electron Transport Systems
  A. The release of energy during metabolic processes normally involves oxidation-reduction reactions
    1. Oxidation-reduction (redox) reactions involve the transfer of electrons from an electron donor to an electron acceptor
    2. The equilibrium constant for an oxidation-reduction reaction is called the standard reduction potential (E_0) and is a measure of the tendency of the electron donor to lose electrons; the more negative the reduction potential, the better it is as an electron donor
  B. When electrons are transferred from an electron donor to an electron acceptor with a more positive reduction potential, free energy is released and can be used to form ATP; an electron transport system (ETS) is a series of electron carriers
  C. Electron transport is important in a variety of metabolic processes (e.g., respiration and photosynthesis); cells move electrons by using a variety of electron carriers organized into a chain (an ETS); electron carriers include NAD^+, $NADP^+$, flavoproteins, coenzymes, and cytochromes; these carriers differ in terms of how they carry electrons, and this impacts how they function in electron transport chains
VII. Enzymes
  A. Structure and classification of enzymes
    1. Enzymes are protein catalysts with great specificity for the reaction catalyzed and the molecules acted upon
      a. A catalyst is a substance that increases the rate of a reaction without being permanently altered
      b. The reacting molecules are called substrates and the substances formed are products

2. An enzyme may be composed only of protein or it may be a holoenzyme, consisting of a protein component (apoenzyme) and a nonprotein component (cofactor)
    a. Prosthetic group—a cofactor that is firmly attached to the apoenzyme
    b. Coenzyme—a cofactor that is loosely attached to the apoenzyme; it may dissociate from the apoenzyme and carry one or more of the products of the reaction to another enzyme
B. The mechanism of enzyme reactions
    1. Enzymes increase the rate of a reaction but do not alter the equilibrium constant (or the standard free energy change) of the reaction
    2. Enzymes lower the activation energy required to bring the reacting molecules together correctly to form the transition-state complex; once the transition state has been reached the reaction can proceed rapidly
    3. Enzymes bring substrates together at the active site to form an enzyme-substrate complex; this can lower activation energy in several ways:
        a. Local concentrations of the substrates are increased at the active (catalytic) site of the enzyme
        b. Molecules at the active site are oriented properly for the reaction to take place
C. The effect of environment on enzyme activity
    1. The amount of substrate present affects the reaction rate, which increases as the substrate concentration increases until all available enzyme molecules are binding substrate and converting it to products as rapidly as possible
        a. When no further increase in reaction rate occurs with increases in substrate concentration, a reaction is said to be proceeding at maximal velocity (V_{max})
        b. The Michaelis constant (K_m) of an enzyme is the substrate concentration required for the reaction to reach half maximal velocity; it is used as a measure for the apparent affinity of an enzyme for its substrate
    2. Enzyme activity is also affected by alterations in pH and temperature; each enzyme has specific pH and temperature optima; extremes of pH, temperature, and other factors can cause denaturation (loss of activity due to disruption of enzyme structure)
D. Enzyme inhibition
    1. Competitive inhibition occurs when the inhibitor binds at the active site and thereby competes with the substrate (if the inhibitor binds, then the substrate cannot, and no reaction occurs); this type of inhibition can be overcome by adding excess substrate
    2. Noncompetitive inhibition occurs when the inhibitor binds to the enzyme at some location other than the active site and changes the enzyme's shape so that it is inactive or less active; this type of inhibition cannot be overcome by the addition of excess substrate
VIII. The Nature and Significance of Metabolic Regulation
    A. Regulation is essential for microorganisms to conserve energy and material and to maintain metabolic balance despite frequent changes in their environment
    B. Metabolic processes can be regulated in three major ways:
        1. Metabolic channeling—the localization of metabolites and enzymes in different parts of a cell
        2. Increasing or decreasing the number of enzyme molecules present (regulation of gene expression); this type of regulation is discussed in chapter 12
        3. Stimulation or inhibition of critical enzymes in a pathway (posttranslational regulation)
IX. Metabolic Channeling
    A. Compartmentation is a common mechanism for metabolic channeling; enzymes and metabolites are distributed in separate cell structures or organelles
    B. Channeling can also occur within a compartment
    C. Channeling can generate marked variations in metabolite concentrations and therefore directly affect enzyme activity
X. Control of Enzyme Activity
    A. Allosteric regulation—regulation of enzyme activity by an effector or modulator, which binds reversibly and noncovalently to a regulatory site on the enzyme; the regulatory site is distinct from the catalytic site; the effect can be positive or negative

B. Covalent modification of enzymes—regulation of enzyme activity by the reversible covalent addition or removal of a chemical group (e.g., phosphate, methyl group, adenylic acid); the effect can be positive or negative

C. Feedback Inhibition
1. Every pathway has at least one pacemaker enzyme that catalyzes the slowest (rate-limiting) reaction in the pathway; often this is the first reaction in a pathway
2. In feedback inhibition (end product inhibition), the end product of the pathway inhibits the pacemaker enzyme
3. In branched pathways, balance between end products is maintained through the use of regulatory enzymes at branch points; multiple branched pathways often use isoenzymes, each under separate and independent control

D. Chemotaxis
1. A phosphorelay system is used to regulate the chemotactic response of bacteria and determine directional motion of flagellum; the system includes a sensor kinase and a response regulator
2. Attractants (or repellants) are detected by chemoreceptors, leading to transfer of phosphoryl groups to proteins by sensor kinases, which interact with response regulators that control the flagellar motor

TERMS AND DEFINITIONS

Place the letter of each term in the space next to the definition or description that best matches it.

_____ 1. During this phenomenon, the breakdown of ATP to ADP and P_i releases energy to do work for the cell; other processes trap energy by reforming ATP from ADP and P_i

_____ 2. The science that analyzes energy changes in a collection of matter

_____ 3. A law stating that energy can be neither created nor destroyed

_____ 4. The unit of measurement that describes the amount of heat needed to raise the temperature of 1 gram of water from 14.5°C to 15.5°C

_____ 5. The unit of measurement for the amount of work capable of being done

_____ 6. A law stating that physical and chemical processes occur in such a way that randomness (disorder) increases to a maximum

_____ 7. Describes the randomness or disorder of a system

_____ 8. The heat content of a system; in cells it is about that same as the total energy of the system

_____ 9. A measure of the energy of a reaction that is available to do useful work

_____ 10. Occurs when the forward rate of a reaction equals the reverse rate

_____ 11. A reaction that releases energy (ΔG is negative)

_____ 12. A reaction that requires an input of energy (in addition to the activation energy) in order to proceed (ΔG is positive)

_____ 13. Reactions in which there is a transfer of electrons from an electron donor to an electron acceptor

_____ 14. Protein catalysts with great specificity for the reaction catalyzed and the molecules acted upon

_____ 15. A substance that increases the rate of a reaction without being permanently altered by the reaction

_____ 16. The reacting molecules in an enzyme-catalyzed reaction

_____ 17. The molecules formed by a chemical reaction

_____ 18. A complex formed during a reaction that is composed of the substrates; it resembles both the substrates and the products of the reaction

_____ 19. The energy required to bring reacting molecules together in the correct way to reach the transition state

_____ 20. A special place on the surface of an enzyme where the substrates are brought together in the proper orientation for a reaction to occur

a. activation energy
b. active site (catalytic site)
c. allosteric enzymes
d. ATP
e. calorie
f. catalyst
g. compartmentation
h. competitive inhibitor
i. denaturation
j. effectors (modulators)
k. endergonic reaction
l. energy cycle
m. enthalpy
n. entropy
o. enzymes
p. equilibrium
q. exergonic reaction
r. feedback (end product) inhibition
s. first law of thermodynamics
t. free energy change
u. isoenzymes
v. joule
w. maximal velocity (V_{max})
x. metabolic channeling
y. Michaelis constant (K_m)
z. noncompetitive inhibitor
aa. oxidation-reduction (redox) reactions
bb. pacemaker enzyme
cc. products
dd. regulatory enzymes
ee. second law of thermodynamics
ff. substrates
gg. thermodynamics
hh. transition state complex

_____ 21. The high-energy molecule used by cells as their energy currency

_____ 22. Term that describes the velocity of a reaction when all available enzyme molecules are binding substrate and converting it to product as rapidly as possible

_____ 23. Constant that is equal to the substrate concentration at which an enzyme-catalyzed reaction reaches half maximal velocity

____ 24. An enzyme inhibitor that binds to an enzyme at the active site and thereby prevents the substrate from binding and reacting

____ 25. An enzyme inhibitor that binds to an enzyme at some location other than the active site and alters the enzyme's shape so that it is inactive or less active

____ 26. Disruption of an enzyme's structure with loss of activity caused by extremes of pH, temperature, or other factors

____ 27. The phenomenon in which metabolic pathways are regulated by controlling the intracellular location of the metabolites and enzymes involved in the pathway

____ 28. Differential distribution of enzymes and metabolites among separate cell structures or organelles

____ 29. Enzymes whose activity and shape are altered by noncovalent binding of a small molecule

____ 30. Small molecules that alter the activity of allosteric enzymes

____ 31. The enzyme that catalyzes the slowest or rate-limiting reaction in a pathway

____ 32. The process by which the end product of a metabolic pathway inhibits the first enzyme in the pathway

____ 33. Different enzymes that catalyze the same reaction, but that may be regulated independently of one another

FILL IN THE BLANK

1. The flows of carbon and energy in an ecosystem are intimately related. Light energy is trapped by _____ organisms when they use carbon dioxide and sunlight during _____ to make complex organic molecules. Some of this trapped energy is obtained by _____ organisms when they use the former as food and degrade the organic molecules. These molecules are often degraded by a process called _____ _____, which releases carbon dioxide.

2. The science of _____ analyzes energy changes in a collection of matter called a _____. All other matter in the universe is called the _____.

3. For any reaction, when $\Delta G^{0'}$ is negative, the equilibrium constant is _____ than one, the reaction is said to be _____, and the reaction goes to completion in the way it is written. However, in an _____ reaction, $G^{0'}$ is positive, the equilibrium constant is _____ than one, and the reaction is unfavorable; therefore, little product will be formed at equilibrium under standard conditions.

4. ATP is ideally suited for its role as energy currency. It is formed in energy-trapping and energy-generating processes such as _____ _____, fermentation, and _____. In the cell's economy, ATP breakdown, an _____ reaction, can be coupled with various _____ reactions to facilitate their completion.

5. In a redox reaction, electrons are transferred from an _____ _____ to an _____ _____. The two molecules are referred to as a redox couple.

6. The equilibrium constant for a redox reaction is called the _____ _____ _____, and it is a measure of the tendency of the electron donor to ____ _____ electrons. The more _____ this value, the better it is as an electron donor.

7. A number of enzymes are pure proteins. However, some enzymes consist of a protein component, the _____, plus a nonprotein component called a _____. The two together constitute the _____. When the nonprotein component is firmly attached to the protein it is referred to as a _____ _____; when it is loosely attached, it is referred to as a _____.

8. The _____ _____ (K_m) is equal to the substrate concentration at which an enzyme-catalyzed reaction reaches half maximal velocity. It is used as a measure of the apparent _____ of an enzyme for its substrate. The lower the value of K_m, the _____ the substrate concentration at which the enzyme catalyzes the reaction.

9. An inhibitor that binds at the active site and thereby prevents the binding of the substrate is called a _____ inhibitor; while an inhibitor that binds at a location other than the active site, thus altering the enzyme's shape so that it is inactive or less active, is called a _____ inhibitor.

10. _____ is the capacity to do work. Living cells carry out three major types of work. The synthesis of complex molecules is _____ work; nutrient uptake, waste elimination, and the

maintenance of ion balances is _____ work; and internal and external movement is _____ work.

11. Enzymes bind substrates at their _____ _____, forming an _____ complex.

12. Enzymes speed reactions by _____ the activation energy of the reaction.

13. ATP is a _____ molecule, and it has a high _____ _____ _____ _____, which means that it readily transfers phosphate to water. ATP is made when a third phosphate is added to _____ during processes such as photosynthesis, fermentation, or aerobic respiration.

14. Flavoproteins are proteins bearing the electron carrier _____ or _____.

15. The phenomenon known as _____ _____ localizes metabolites and enzymes in different parts of a cell and in doing so influences the activity of metabolic pathways. In some cases, metabolites and enzymes are distributed among separate cell structures or organelles. This is called_____.

16. Enzyme activity can be regulated by small molecules known as _____ or _____. Such enzymes are called _____ enzymes. The small molecules bind by noncovalent forces to a _____ site that is different from the catalytic site.

17. Every metabolic pathway has at least one _____ enzyme that catalyzes the slowest or rate-limiting reaction in the pathway. Since other reactions proceed more rapidly than this reaction, changes in the activity of the _____ enzyme directly alter the speed with which a pathway operates.

18. In_____ _____when the end product of a pathway becomes concentrated, it inhibits the _____ enzyme and slows its own synthesis. As the end product concentration _____, pathway activity once again _____ and more end product is formed.

19. The regulation of multiply branched pathways often involves _____ to catalyze the same step. In this situation, excess of a single end product _____ but does not completely block pathway activity because some _____ are still active.

20. Aspartate carbamoyl transferase is an _____ enzyme that is inhibited by noncovalent binding of CTP and activated by noncovalent binding ATP.

21. Glycogen phosphorylase can be regulated by reversible attachment of phosphate to the enzyme. This is an example of_____ _____ _____ of an enzyme to control its activity.

22. Energy is made available when electrons are transferred from redox couples with more _____ reduction potentials to those with more _____ reduction potentials.

23. The sporulation and chemotaxis regulatory systems of bacteria are examples of _____ _____ systems, which have two major components: a sensor kinase and a response regulator. In the chemotaxis regulatory system, _____ _____ proteins (MCPs) contribute to the regulation of flagellar rotation.

MULTIPLE CHOICE

For each of the questions below select the *one best* answer.

1. What is the amount of heat energy needed to raise the temperature of 1.0 gram of water from 14.5°C to 15.5°C called?
   a. joule
   b. calorie
   c. erg
   d. thermal unit

2. What is the equilibrium constant (K_{eq}) for the reaction $A + B \rightarrow C + D$?
   a. $\dfrac{[A][B]}{[C][D]}$
   b. $\dfrac{[C][D]}{[A][B]}$
   c. $\dfrac{[A][D]}{[B][C]}$
   d. $\dfrac{[B][C]}{[A][D]}$

3. Living organisms use a variety of electron carriers to aid in the cycle of energy flow. Which of the following is used as an electron carrier?
   a. NAD^+
   b. $NADP^+$
   c. ubiquinone
   d. All of the above are used as electron carriers.

4. Which of the following is true about enzymes?
   a. Enzymes are catalysts, and therefore, they increase the rate of a reaction without being permanently altered by the reaction.
   b. Enzymes are proteins that can be denatured by changes in pH or temperature.
   c. Enzymes are highly specific for the substrates they bind.
   d. All of the above are true about enzymes.

5. For a reaction to occur, the reacting molecules must be brought together in the correct way to form the transition-state complex. This requires an input of energy. What is this energy called?
   a. activation energy
   b. free energy
   c. entropy
   d. enthalpy

6. Which of the following is NOT a way in which enzymes lower the activation energy required for a reaction?
   a. bringing the substrates together at the active site; in effect, concentrating them
   b. binding the substrates so that they are correctly oriented to form the transition-state complex
   c. increasing molecular motion, thereby providing kinetic energy to drive the reaction
   d. All of the above are ways in which enzymes lower the activation energy required for a reaction.

7. Which of the following is NOT a function of the transport work done by a cell?
   a. uptake of nutrients
   b. elimination of waste products
   c. maintenance of internal/external ion balances
   d. All of the above are functions of cellular transport work.

8. Which of the following is a reason for metabolic regulation?
   a. conservation of material
   b. conservation of energy
   c. maintaining metabolic balance
   d. All of the above are reasons for metabolic regulation.

9. A small molecule binds to an allosteric enzyme and thereby increases the activity of the enzyme. What is the small molecule called?
   a. positive effector
   b. negative effector
   c. prosthetic group
   d. cofactor

10. Which of the following is NOT true about the regulation of branched metabolic pathways?
    a. There are usually separate regulatory enzymes for each branch, as well as a regulatory enzyme that controls the flow of carbon into the entire set of possible branches.
    b. An excess of one end product will usually completely inhibit the activity of the branch responsible for the synthesis of that particular end product.
    c. An excess of one end product will usually slow the flow of carbon into the entire set of branched pathways.
    d. All of the above are true about the regulation of branched metabolic pathways.

11. Which of the following is an expression of the amount of energy made available for useful work?
    a. $\Delta G = \Delta H + T \cdot \Delta S$
    b. $\Delta G = T \cdot \Delta S - \Delta H$
    c. $\Delta G = \Delta H - T \cdot \Delta S$
    d. none of the above

TRUE/FALSE

_____ 1. A reaction will occur spontaneously if the free energy of the system decreases during the reaction (i.e., if ΔG is negative).

_____ 2. Since ΔS is a measure of disorder, a decrease in ΔS will lead to a decrease in ΔG, and therefore, the reaction will proceed spontaneously.

_____ 3. The value of $\Delta G^{0'}$ indicates how fast a reaction will reach equilibrium.

_____ 4. Redox couples that have greater negative reduction potentials will donate electrons to couples that have higher positive potentials. This is the basis for the functioning of electron transport chains.

_____ 5. Ferredoxin is a nonheme iron protein that is active in photosynthetic electron transport.

_____ 6. Enzymes increase the rate of a reaction but do not alter the equilibrium constants for the reactions they catalyze.

_____ 7. When the amount of enzyme present is held constant, the rate of a reaction will continue to increase as long as the substrate concentration increases.

_____ 8. Enzyme activity can be greatly affected by the pH and temperature of the environment in which the enzyme must function.

_____ 9. The ultimate source of most biological energy is visible sunlight through the process of photosynthesis.

_____ 10. The standard free energy change is unrelated to the equilibrium constant.

_____ 11. Cytochromes contain iron atoms in heme groups or similar iron-porphyrin rings.

_____ 12. Covalent modification represents a reversible way of controlling enzyme activity because the modified form has an altered activity (either higher or lower) than the unmodified form.

_____ 13. Usually the last step in a pathway is catalyzed by a pacemaker enzyme.

_____ 14. When a biosynthetic pathway branches to form more than one end product, an excess of one of the end products will only inhibit the branch of the pathway involved in the synthesis of that particular product, while an excess of all the end products will usually inhibit the flow of carbon into the entire pathway.

CRITICAL THINKING

1. Consider the following diagram of the energy flow for a particular reaction. Is the reaction exergonic or endergonic? What does the diagram indicate? How would the use of an enzyme catalyst affect the energy flow? Indicate this on the diagram, and also indicate the energy of activation and the free energy change of both the catalyzed and uncatalyzed reactions.

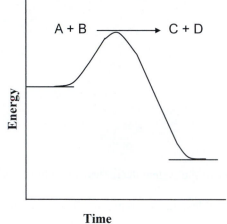

2. Each of the following two diagrams indicates the rate of a reaction as a function of substrate concentration. In each case an inhibitor is present (not the same one) and the substrate concentration is not saturating the enzyme. Explain the difference between the two situations.

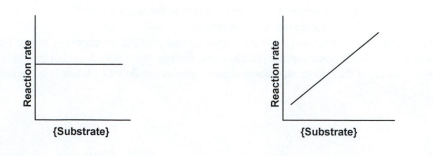

ANSWER KEY

Terms and Definitions

1. l, 2. gg, 3. s, 4. e, 5. v, 6. ee, 7. n, 8. m, 9. t, 10. p, 11. q, 12. k, 13. aa, 14. o, 15. f, 16. ff, 17. cc, 18. hh, 19. a, 20. b, 21. d, 22. w, 23. y, 24. h, 25. z, 26. i, 27. x, 28. g, 29. c, 30. j, 31.bb, 32. r, 33. u

Fill in the Blank

1. photoautotrophic; photosynthesis; chemoheterotrophic; aerobic respiration 2. thermodynamics; system; surroundings 3. greater; exergonic; endergonic; less 4. aerobic respiration; photosynthesis; exergonic; endergonic 5. electron donor; electron acceptor 6. standard reduction potential; give up; negative 7. apoenzyme; cofactor; holoenzyme; prosthetic group; coenzyme 8. Michaelis constant; affinity; lower 9. competitive; noncompetitive 10. Energy; chemical; transport; mechanical; 11. active site; enzyme-substrate 12. decreasing 13. high-energy; phosphate group transfer potential; ADP 14. FAD; FMN; 15. metabolic channeling; compartmentation 16. effectors; modulators; allosteric; regulatory 17. pacemaker; pacemaker 18. feedback inhibition; pacemaker; decreases; increases 19. isoenzymes; slows; isoenzymes 20. allosteric 21. reversible covalent modification 22. negative; positive 23. two-component phosphorelay; methyl-accepting chemotaxis

Multiple Choice

1. b, 2. b, 3. d, 4. d, 5. a, 6. c, 7. d, 8. d, 9. a, 10. d, 11. c

True/False

1. T, 2. F, 3. F, 4. T, 5. T, 6. T, 7. F, 8. T, 9. T, 10. F, 11. T, 12. T, 13. F, 14. T

9 Metabolism: Energy Release and Conservation

CHAPTER OVERVIEW

This chapter presents an overview of energy release and conservation mechanisms beginning with glucose degradation to pyruvate. Fermentation, aerobic respiration, and anaerobic respiration are then examined. The consideration of chemoorganoheterotrophic metabolism concludes with a discussion of the catabolism of lipids, proteins, and amino acids. Chemolithotrophic metabolism follows and the chapter concludes with a discussion of the trapping of energy by photosynthesis.

CHAPTER OBJECTIVES

After reading this chapter you should be able to:

- discuss the difference between catabolism and anabolism
- describe the various pathways for the catabolism of glucose to pyruvate
- list the various types of fermentations and give examples of their practical importance
- discuss the tricarboxylic acid (TCA) cycle and its central role in aerobic metabolism
- describe the electron transport process, and compare and contrast the electron transport systems of eucaryotes with those of procaryotes
- describe oxidative phosphorylation and the chemiosmotic hypothesis
- compare and contrast aerobic respiration, fermentation, and anaerobic respiration of organic molecules
- describe in general terms the catabolism of molecules other than carbohydrates
- list the common energy sources used by chemolithotrophs
- describe the mechanisms by which chemolithoautotrophs generate ATP and NADH
- discuss the photosynthetic light reactions
- compare and contrast the light reactions of eucaryotes (and cyanobacteria) with those of green (or purple) photosynthetic bacteria

CHAPTER OUTLINE

I.   Chemoorganotrophic Fueling Processes
    A.   Chemotrophic microorganisms vary in terms of their energy source (light, organic molecules, inorganic molecules), and also in terms of their electron acceptors
        1.   If the energy source is oxidized and degraded *with* the use of an exogenous electron acceptor, the process is called *respiration*; in aerobic respiration the final electron acceptor is oxygen, whereas in anaerobic respiration the final electron acceptor is a molecule other than oxygen
        2.   If an *organic* energy source is oxidized and degraded *without* the use of an exogenous electron acceptor, the process is called *fermentation*
II.   Aerobic Respiration
    A.   For aerobic chemoorganoheterotrophic organisms, catabolism is often a three-stage process during which nutrients are fed into common degradative pathways
        1.   breakdown of polymers and other large molecules into their constituent parts
        2.   initial degradation of the constituents' parts
        3.   completion of degradation accompanied by the generation of many ATP molecules
    B.   These common pathways function both catabolically and anabolically and are said to be amphibolic, including glycolysis and the tricarboxylic acid (TCA) cycle
III.   The Breakdown of Glucose to Pyruvate

A. The Embden-Meyerhof Pathway
   1. Known as the glycolytic pathway or glycolysis, it is the most common pathway for glucose degradation and it is found in all major groups of microorganisms
   2. Functions in the presence or absence of oxygen and is divided into two parts:
      a. In the 6-carbon sugar stage, glucose is phosphorylated twice to yield fructose 1,6-bisphosphate; this requires the expenditure of two molecules of ATP
      b. The 3-carbon sugar stage cleaves fructose 1,6-bisphosphate into two 3-carbon molecules, which are each processed to pyruvate; two molecules of ATP are produced by substrate-level phosphorylation from each of the 3-carbon molecules for a net yield of two molecules of ATP; two molecules of NADH are also produced per glucose molecule
B. The pentose phosphate pathway
   1. Also known as the hexose monophosphate pathway; produces a variety of 3-, 4-, 5-, 6-, and 7-carbon sugar phosphates
   2. Has several catabolic and anabolic functions
      a. Production of NADPH, which serves as a source of electrons for biosynthetic processes
      b. Production of 4- and 5-carbon skeletons that can be used for the synthesis of amino acids, nucleic acids, and other macromolecules
      c. Complete catabolism of hexoses and pentoses, yielding ATP and NADH (made by converting NADPH to NADH)
C. The Entner-Doudoroff pathway is an alternative glycolytic pathway used by some microbes; it produces pyruvate, ATP, NADPH, and NADH

IV. The Tricarboxylic Acid Cycle
A. Pyruvate can be oxidized to carbon dioxide by the tricarboxylic acid (TCA) cycle (or Krebs cycle) after first being converted to acetyl-CoA; this preliminary reaction is accompanied by the loss of one carbon atom as carbon dioxide
B. Acetyl-CoA reacts with oxaloacetate (a 4-carbon molecule) to produce a 6-carbon molecule, which is subsequently broken down to release two molecules of carbon dioxide, regenerating the oxaloacetate; during this process, the following occur:
   1. ATP is produced by substrate-level phosphorylation
   2. Three molecules of NADH and one molecule of $FADH_2$ are produced
C. Organisms that lack the complete TCA cycle usually have most of the TCA cycle enzymes, because one of the TCA cycle's major functions is to provide carbon skeletons for use in biosynthesis

V. Electron Transport and Oxidative Phosphorylation
A. The electron transport chain
   1. The mitochondrial electron transport chain uses a series of electron carriers to transfer electrons from NADH and $FADH_2$ to O_2
      a. Electron carriers are located within the inner membrane of the mitochondrion
      b. During oxidative phosphorylation, three ATP molecules may be synthesized when a pair of electrons passes from NADH to O_2; two ATP molecules may be synthesized when electrons from $FADH_2$ pass to O_2
   2. Although they operate according to the same fundamental principles, bacterial electron transport chains usually differ in structure: they may be branched, be composed of different electron carriers, or may be shorter than mitochondrial electron transport chains; bacterial electron transport chains are located in the plasma membrane
B. Oxidative phosphorylation
   1. The chemiosmotic hypothesis postulates that the energy released during electron transport is used to establish a proton gradient that produces a proton motive force (potential energy due to the difference in proton concentration and charge on either side of the membrane), which can be used to drive ATP synthesis, flagellar rotation, and transport of molecules across the membrane
   2. ATP synthesis is catalyzed by the ATP synthase complex, which is thought to behave like a small rotary motor where the energy released by the movement of protons across the membrane, down the concentration gradient, is used to make high-energy bonds in ATP
   3. Inhibitors of ATP synthesis fall into two main categories:

    a. Blockers that inhibit the flow of electrons through the system

    b. Uncouplers that allow electron flow, but disconnect it from oxidative phosphorylation

  C. ATP yield during aerobic respiration

    1. The net yield from the oxidation of a glucose molecule via glycolysis is 2 ATP

    2. The number of ATP generated by aerobic respiration of a glucose molecule depends on the precise nature of the electron transport system; mitochondrial respiration typically yields 36 ATP (plus 2 from glycolysis)

VI. Anaerobic Respiration

  A. Uses exogenous molecules other than oxygen as terminal electron acceptors; the most commonly used alternative electron acceptors are nitrate, sulfate, and CO_2

  B. Dissimilatory nitrate reduction occurs when nitrate is used as the terminal electron acceptor; if the nitrate is reduced to nitrogen gas, the process is called denitrification

  C. Anaerobic respiration is not as efficient in ATP synthesis as aerobic respiration because the alternative electron acceptors do not have as positive a reduction potential as O_2; despite this, anaerobic respiration is useful because it is more efficient than fermentation

  D. When several electron acceptors are present in the environment, microorganisms will use these acceptors in succession starting with oxygen, then moving to nitrate, manganese ion, ferric ion, sulfate, and finally carbon dioxide

VII. Fermentations

  A. Fermentation is a process in which an organism oxidizes the NADH produced by one of the glycolytic pathways by using pyruvate or one of its derivatives as an electron and hydrogen acceptor; thus the process involves the use of an endogenous electron acceptor and regenerates NAD^+ to act as the oxidant in glycolysis

  B. Fermentation takes place in the absence of oxidative phosphorylation and ATP is only generated through substrate-level phosphorylation

  C. Many different types of fermentations are known

    1. Alcoholic fermentations produce ethanol and CO_2

    2. Lactic acid fermentations produce lactic acid (lactate)

      a. Homolactic fermenters reduce almost all pyruvate to lactate

      b. Heterolactic fermenters form substantial amounts of products other than lactate

    3. Formic acid fermentation produces either mixed acids or butanediol

  D. Microorganisms can ferment substances other than sugars (e.g., amino acids); in the Strickland reaction, one amino acid is oxidized while a second acts as the electron acceptor

VIII. Catabolism of Carbohydrates and Intracellular Reserve Polymers

  A. Carbohydrates

    1. Most monosaccharides feed easily into the glycolytic pathway

    2. Disaccharides are cleaved into monosaccharides either by hydrolysis or phosphorolysis

    3. Polysaccharides are cleaved into smaller molecules either by hydrolysis or phosphorolysis; many however, are not easily degraded (e.g., cellulose, agar)

    4. Microorganisms also are capable of degrading xenobiotic molecules (foreign substances not formed by natural biosynthetic processes) such as pesticides

  B. Reserve polymers—when exogenous nutrients are absent, microorganisms catabolize internal carbon and energy stores (e.g., glycogen, starch); this is accomplished either by hydrolysis or phosphorolysis

IX. Lipid Catabolism

  A. Triglycerides are common energy sources; they are hydrolyzed to glycerol and fatty acids

  B. Fatty acids are catalyzed by the β-oxidation pathway, which successively shortens the chain by two carbons producing acetyl-CoA, NADH, and $FADH_2$; NADH and $FADH_2$ can be oxidized by an electron transport chain to produce ATP

X. Protein and Amino Acid Catabolism

  A. Proteins are degraded by proteases to their component amino acids

  B. Amino acids are first deaminated and then the remaining carbon skeletons are converted to pyruvate, acetyl-CoA, or a TCA-cycle intermediate

XI. Chemolithotrophy
   A. A metabolic process that uses inorganic molecules as a source of energy; the energy source is oxidized and the electrons are passed to an electron transport chain; the terminal electron acceptor is usually O_2, but sulfate and nitrate also are used; the most common electron donors (energy sources) are hydrogen, reduced nitrogen compounds, reduced sulfur compounds, and ferrous iron (Fe^{2+})
   B. Chemolithotrophs are usually autotrophs; they use the Calvin cycle to fix carbon dioxide; however, the oxidation of most inorganic molecules yields low levels of ATP; therefore, in order to fuel carbon dioxide fixation, chemolithotrophs must oxidize large amounts of inorganic material
   C. Some inorganic energy sources (e.g., ammonia and nitrite) cannot donate electrons directly to NAD^+; chemolithotrophs that use such energy sources must use their proton motive force to reverse electron flow in their electron transport chain in order to generate the NADH needed for biosynthetic processes
XII. Phototrophy
   A. During photosynthesis, energy from light is trapped and used to produce ATP and NADPH (light reactions), which are used to reduce carbon dioxide to form carbohydrates (dark reactions, discussed in chapter 10)
   B. The light reaction in oxygenic photosynthesis
       1. Oxygenic photosynthesis generates molecular oxygen when light energy is converted to chemical energy; this process is found in plants, algae, and cyanobacteria
       2. Chlorophyll molecules and a variety of accessory pigments (carotenoids and bilins) are used to form antennas; the antennas trap photons and transfer their energy to reaction-center chlorophylls; these special chlorophylls are directly involved in photosynthetic electron transport
       3. Eucaryotes and cyanobacteria have two photosystems; in each, electrons from the light-energized reaction-center chlorophylls are transferred to the associated electron transport chain
           a. Photosystem I can carry out cyclic photophosphorylation, producing ATP
           b. Photosystems I and II, working together, can carry out noncyclic photophosphorylation, producing ATP and NADPH; the electrons for noncyclic photophosphorylation are obtained from water, which is oxidized to O_2 (oxygenic photosynthesis)
       4. Photosynthetic electron transport takes place in membranes including the thylakoid membranes of chloroplasts; ATP can be generated through chemiosmosis
   C. The light reaction in anoxygenic photosynthesis
       1. Green and purple photosynthetic bacteria carry out anoxygenic photosynthesis (they do not use water as a source of electrons, so do not produce O_2), and they have different photosynthetic pigments called bacteriochlorophylls
       2. Many of the differences in the light reactions of the green and purple bacteria are due to the fact that they only have a single photosystem
       3. Green and purple bacteria are usually autotrophs that used NADH or NADPH for carbon dioxide fixation; three methods for making NADH are known:
           a. Reduction of NAD^+ directly by hydrogen gas
           b. Reverse electron flow
           c. A simplified form of noncyclic electron flow
   D. Rhodopsin-based phototrophy
       1. Some archaea use light as a source of energy but instead of using chlorophyll, these microbes use bacteriorhodopsin
       2. A retinal chromophore changes shape when excited by light and acts a proton pump; this creates a proton gradient that produces ATP through chemiosmosis

TERMS AND DEFINITIONS

Place the letter of each term in the space next to the definition or description that best matches it.

_____ 1. The total of all chemical reactions occurring in a cell

_____ 2. The breakdown of more complex molecules into simpler ones, accompanied by energy release and the trapping of some of that energy

_____ 3. The synthesis of complex molecules from simpler ones, accompanied by energy input

_____ 4. Metabolic pathways that function both catabolically and anabolically

_____ 5. Phosphorylation of ADP coupled to the exergonic breakdown of a high-energy substrate molecule

_____ 6. An alternative to the glycolytic pathway that may operate simultaneously with the glycolytic pathway

_____ 7. An alternative to glycolysis that is used by a few genera of bacteria

_____ 8. A cyclical pathway that is used by chemoorganotrophic organisms to generate energy and to provide carbon skeletons for use in biosynthesis

_____ 9. A series of electron carriers that transfer electrons to a terminal electron acceptor and liberate energy as they do so

_____ 10. Production of ATP during chemotrophic processes using energy liberated by an electron transport system

_____ 11. Potential energy that exists when there is a gradient of protons and an unequal distribution of charges across a membrane

_____ 12. The postulate that ATP synthesis is driven by proton diffusion down a gradient established during electron transport

_____ 13. Inhibitors that stop ATP synthesis without inhibiting electron transport

_____ 14. A type of chemoorganotrophic metabolism in which an endogenous organic molecule serves as the electron acceptor

_____ 15. A type of chemotrophic metabolism in which exogenous oxygen serves as the terminal electron acceptor

_____ 16. A type of chemotrophic metabolism in which an exogenous molecule other than oxygen serves as the terminal electron acceptor

_____ 17. A process in which excited electrons from a chlorophyll molecule return to that molecule; the electron flow drives the synthesis of ATP

_____ 18. A process in which excited electrons from a chlorophyll molecule are given to NADPH and are replaced with electrons from water; the electron flow drives the synthesis of ATP

_____ 19. Phycocyanin and phycoerythrin are examples of this type of accessory pigment

a. aerobic respiration
b. amphibolic pathways
c. anabolism
d. anaerobic respiration
e. catabolism
f. chemiosmotic hypothesis
g. cyclic photophosphorylation
h. electron transport chain
i. Entner-Doudoroff pathway
j. fermentation
k. metabolism
l. noncyclic photophosphorylation
m. oxidative phosphorylation
n. pentose phosphate (hexose monophosphate) pathway
o. phycobiliproteins
p. proton motive force
q. substrate level phosphorylation
r. tricarboxylic acid cycle (citric acid cycle; Krebs cycle)
s. uncouplers

METABOLIC ENERGY: GLYCOLYSIS AND THE TCA CYCLE

Consider the following metabolic process and answer the questions below. (Where numbers are asked for, give the number *per glucose molecule entering the pathway.*)

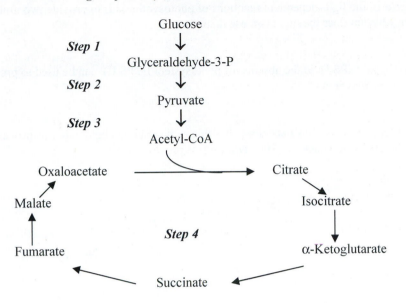

1. At which step(s) is there an investment of ATP? How many ATP molecules must be invested?

2. At which step(s) is there a production of ATP by substrate-level phosphorylation? How many are produced?

3. At which step(s) is there a production of NADH? How many are produced?

4. At which step(s) is there a production of $FADH_2$? How many are produced?

5. Considering that NADH and $FADH_2$ can be processed through the electron transport chain to produce ATP, what is the maximum net gain of ATP in eucaryotes?

6. In eucaryotes, not all NADH molecules give the same net yield of ATP. Which ones are different and why?

7. At which step(s) are CO_2 molecules released? How many are released at each step?

ENERGY METABOLISM: PHOTOSYNTHESIS

Consider the process of photosynthesis and answer the following questions:

1. The goal of the light-dependent reactions of photosynthesis is to provide two molecules for the cell. What are they? What does the organism use them for?

2. Photosystem I (PS I), in the absence of photosystem II (PS II), can be used to provide one of these molecules. Which one?

3. The process by which PS I accomplishes this is diagrammed below. In this process the electrons given up by P700 are cycled back to P700. What is this process called?

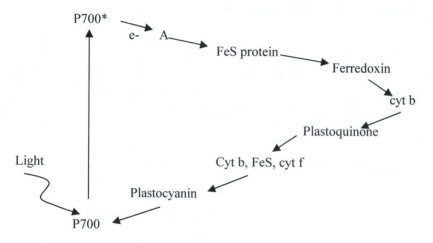

4. Why can't this process be used to provide the other molecule? (HINT: Consider the fate of the electrons given up by P700.)

5. How did the evolution of photosystem II (PS II) solve this problem? Diagram the overall process. (HINT: Use the diagram of PS I as your starting point.) On your diagram show where the two molecules referred to in question 1 are produced.

6. What is the fate of the electrons given up by P680 of PS II? How are they replaced?

ATP YIELD FROM THE AEROBIC OXIDATION OF GLUCOSE

Complete the table below by supplying the missing information. Answers should be expressed as molecules produced (yield) per molecule of glucose oxidized.

Metabolic Step	Number of Molecules Produced				Net Yield of
	ATP	GTP	NADH	FADH$_2$	ATP*
1. Glycolysis					
A. Glucose → Fructose 1,6-bisphosphate	___	___	___	___	___
B. Fructose 1,6-bisphosphate → Pyruvate	___	___	___	___	___
2. Pyruvate → Acetyl-CoA	___	___	___	___	___
3. Tricarboxylic Acid Cycle					
A. 6-Carbon Stage	___	___	___	___	___
B. 5-Carbon Stage	___	___	___	___	___
C. 4-Carbon Stage	___	___	___	___	___
Total Yield	___	___	___	___	___

*Assume maximum yield of ATP from NADH and FADH$_2$ processed through the electron transport chain

FILL IN THE BLANK

1. The breakdown of larger, more complex molecules into smaller, simpler ones with the release of energy is called _____. Some of this energy is trapped and made available for work in _____, which is the synthesis of complex molecules from simpler ones. The latter process takes energy to increase the _____ of a system.

2. _____ and _____ are not easily encompassed by the definition of catabolism, yet both provide ATP and reducing power (e.g., NADPH) for the organisms that use these processes.

3. In _____ _____, most reactions are freely reversible and can be used to both synthesize and degrade molecules. The few irreversible _____ steps are bypassed in biosynthesis with special enzymes that catalyze the reverse reaction. This allows independent _____ of catabolic and anabolic functions.

4. In the tricarboxylic acid (TCA) cycle, two carbons in the form of _____ are added to oxaloacetate at the start of the cycle. Subsequently, two carbons in the form of _____ _____ are removed, regenerating oxaloacetate. The reactions of the TCA cycle generate _____, _____, and _____, which can be used to make ATP.

5. In cyanobacteria and in photosynthetic eucaryotes, the most important pigments are _____, but other pigments are also important. These other pigments are called _____ pigments, and they include _____ and _____. All the photosynthetic pigments are assembled into arrays called _____. Within each array is a _____ _____ that is directly involved with photosynthetic electron transport.

6. A process often called _____ nitrate reduction occurs when certain bacteria use nitrate as the electron acceptor during _____ respiration. If the nitrate is reduced to nitrogen gas, the process is called _____.

7. The Pasteur effect is a drastic reduction in the rate of sugar catabolism that occurs in many microorganisms when they are changed from _____ to _____ conditions. This occurs because aerobic respiration is generally more efficient than _____.

8. The reduction of pyruvate to lactate is called _____ _____ fermentation. If organisms reduce almost all of their pyruvate to lactate, they are called _____ fermenters; organisms that form both lactate and substantial amounts of products other than lactate are called _____ fermenters.

9. Although electrons derived from sugars and other organic molecules are donated to endogenous electron acceptors during _____, or to exogenous molecular oxygen during _____ _____, some procaryotes have electron transport chains that can operate with exogenous electron acceptors other than oxygen, such as _____, _____, and _____ _____, in a process called _____ _____.

10. Disaccharides and polysaccharides can be processed as nutrients after first being cleaved to monosaccharides by either _____ or _____.

11. Fatty acids are metabolized by the _____ pathway in which 2-carbon units are cleaved off to form _____, which is channeled into the TCA cycle.

12. Proteins are catabolized by hydrolytic cleavage to amino acids by the action of enzymes called _____. The amino acids are then processed by first removing the amino group in a process called _____. In many cases the amino group is removed by transferring it to another carbon chain. This is called _____.

13. Almost all the energy of living organisms is ultimately derived from the process of _____. It provides the organisms that carry it out with the _____ and _____ needed to manufacture the organic material required for their growth. These organisms then serve as the base for most of the food chains in the biosphere, and in addition, these organisms are responsible for replenishing our supply of _____.

14. Photosynthesis as a whole is divided into two parts. In the _____ _____, light energy is trapped and converted to chemical energy, which is then used in the _____ _____ to reduce or fix _____ _____ and synthesize cellular constituents.

15. Green and purple photosynthetic bacteria do not use water as an electron source and do not produce _____. They are said to be _____, in contrast with cyanobacteria and eucaryotic photosynthesizers, which are almost always _____.

16. In chemoorganotrophic procaryotes, electron transport chains are located within the _____ _____, while in eucaryotes electron transport chains are located within the _____ _____.

17. Chemotrophic microorganisms use three types of electron acceptors during energy metabolism: endogenous acceptors are used during _____; exogenous _____ is used during aerobic respiration; and other exogenous acceptors are used during _____ _____.

18. The oxidation of carbohydrates, fatty acids, and other nutrients generates _____ and _____, which can in turn be oxidized by electron transport chains. The electrons flow through the chains from carriers with more _____ reduction potentials to those with more _____ reduction potentials.

19. Eucaryotic ATP synthesis takes place on small protein spheres, called _____ _____, which line the inner surface of the inner mitochondrial membrane. In procaryotes, these spheres are located on the inner surface of the _____ membrane.

20. Photosynthetic green and purple bacteria possess _____ rather than chlorophyll, and they only have a single _____, rather than two. Thus they are only able to carry out _____ photophosphorylation.

21. Bacteria and fungi carry out a variety of important fermentations. _____ fermentation is used to make numerous beverages, including beer and wine. _____ _____ fermentation results in the excretion of several acids including acetic, lactic, succinic, and formic acids. During _____ fermentation, acetoin is formed and subsequently reduced to _____. Both of these latter two fermentations are used to distinguish enteric bacteria and are the basis for diagnostic tests.

22. Nitrogen-oxidizing bacteria are _____ that are also called _____ bacteria. When ammonia is oxidized to nitrate by these bacteria, the process is called _____.

23. _____ usually incorporate carbon dioxide fixation through the Calvin cycle and produce NADH for that process by reversing electron flow.

MULTIPLE CHOICE

For each of the questions below select the *one best* answer.

1. Which of the following does NOT represent a major function of carbohydrates and other nutrients?
   a. use, without modification, as cellular constituents
   b. use, with modification, as cellular constituents
   c. oxidation to provide energy
   d. All of the above are major functions of carbohydrates and other nutrients.

2. Which of the following is the most common pathway for degradation of glucose to pyruvate?
   a. Entner-Doudoroff pathway
   b. pentose phosphate pathway
   c. phosphogluconate pathway
   d. Embden-Meyerhof pathway

3. The synthesis of ATP from ADP and P_i, when coupled with the exergonic enzymatic breakdown of a high-energy molecule, is called
   a. oxidative phosphorylation.
   b. chemiosmotic phosphorylation.
   c. substrate-level phosphorylation.
   d. conformational change phosphorylation.

4. Which of the following is NOT a useful outcome of the pentose phosphate pathway?
   a. production of ATP from the metabolism of glyceraldehyde 3-phosphate to pyruvate
   b. production of NADPH, which serves as a source of electrons during biosynthesis
   c. production of 4- and 5-carbon sugars for amino acid biosynthesis, nucleic acid synthesis, and photosynthesis
   d. All of the above are useful outcomes of the pentose phosphate pathway.

5. Which of the following statements is true about the similarities or differences between eucaryotic and procaryotic electron transport?
   a. They use the same electron carriers.
   b. In procaryotes electrons can enter at a number of points and leave through several terminal oxidases, while in eucaryotes electrons can enter at several points but can only leave at a single terminal oxidase.
   c. Procaryotic chains have the same or higher P/O ratios than eucaryotic mitochondrial electron transport chains.
   d. None of the above is true.

6. Which of the following is NOT true of fermentation?
   a. NADH is oxidized to NAD^+.
   b. The electron acceptor is pyruvate or a pyruvate derivative.
   c. Some fermentations generate additional ATP for the organism.
   d. All of the above are true of fermentation reactions.

7. What does it take to reduce one molecule of carbon dioxide into carbohydrate?
   a. three molecules of ATP
   b. two molecules of NADPH
   c. 10 to 12 quanta of light
   d. It takes all of the above.

8. Which of the following may serve as an energy source for chemolithotrophic organisms?
   a. hydrogen gas
   b. reduced nitrogen compound
   c. reduced sulfur compound
   d. All of the above are correct.

9. Which of the following is NOT true about fermentation?
   a. It can occur in both aerobic and anaerobic conditions.
   b. Electrons are transferred to the electron acceptor by an electron transport chain.
   c. It uses an endogenous electron acceptor.
   d. All of the above are true.

TRUE/FALSE

_____ 1. Pathways are either catabolic or anabolic, but cannot be both.

_____ 2. Glycolysis and the pentose phosphate pathway are mutually exclusive. An organism may use one or the other, but not both simultaneously.

_____ 3. Organisms that do not use the TCA cycle for energy production usually have most of the TCA cycle enzymes in order to provide carbon skeletons for use in biosynthesis.

_____ 4. The P/O ratio represents the number of ATP molecules generated when electrons are passed from a donor such as NADH or $FADH_2$ to oxygen.

_____ 5. Most of the ATP that is generated aerobically comes from the oxidation of NADH and $FADH_2$ in the electron transport chain.

_____ 6. In anaerobic respiration, ATP is formed primarily by electron transport chain activity.

_____ 7. Uncouplers stop ATP synthesis without inhibiting electron transport; they may even enhance the rate of electron flow.

_____ 8. In the absence of oxygen, the electron transport chain can't be used to oxidize NADH; therefore, NAD^+ is not regenerated.

_____ 9. Unlike mitochondrial electron transport, photosynthetic electron transport takes place in either the cytoplasm or the fluid matrix (stroma) of chloroplasts rather than in a membrane.

_____ 10. The processes of electron transport in procaryotes and eucaryotes are fundamentally different and use very different mechanisms to couple electron transport to ATP synthesis.

_____ 11. In fermentation, ATP is formed primarily by electron transport chain activity.

_____ 12. Chemolithotrophs incorporate carbon dioxide through the Calvin cycle; in some cases the NADH needed is obtained by reverse electron flow.

_____ 13. The Strickland reaction is an interesting fermentation in which one amino acid in a mixture serves as the energy source and is oxidized, while another amino acid acts as the electron acceptor.

CRITICAL THINKING

1. The authors make the statement, "The use of a few common catabolic pathways, each degrading many nutrients, greatly increases metabolic efficiency by avoiding the need for a large number of less metabolically flexible pathways." Explain the significance of this statement.

2. The overall yield of ATP in eucaryotes has a theoretical maximum of 38 ATP molecules per molecule of glucose. However, usually only 36 ATP molecules are produced. Explain this apparent discrepancy. Procaryotes often have lower P/O ratios than eucaryotic systems. Explain the significance of this relative to the aerobic yield of ATP.

ANSWER KEY

Terms and Definitions

1. k, 2. e, 3. c, 4. b, 5. q, 6. n, 7. i, 8. r, 9. h, 10. m, 11. p, 12. f, 13. s, 14. j, 15. a, 16. d, 17. g, 18. l, 19. o

Metabolic Energy: Glycolysis and the TCA Cycle

1. Step 1, 2 molecules 2. Step 2, 4 molecules; Step 4, 2 molecules 3. Step 2, 2 molecules; Step 3, 2 molecules; Step 4, 6 molecules 4. Step 4, 2 molecules 5. 36 molecules 6. Those from step 1 must be transported into mitochondria at the expense of 1 ATP per NADH so that the net yield is only 2 ATP per NADH. 7. Step 3, 2 molecules; Step 4, 4 molecules

Metabolic Energy: Photosynthesis

1. ATP, NADPH 2. ATP 3. cyclic photophosphorylation 4. The electrons given to NADPH cannot be recycled to P700, and the process would stop when P700 became electron deficient. 5. Electrons are replenished by P680. 6. They are replaced from water.

Fill in the Blank

1. catabolism; anabolism; order 2. Chemolithotrophy; photosynthesis 3. amphibolic pathways; catabolic; regulation; 4. acetyl-CoA; carbon dioxide; GTP; NADH; $FADH_2$ 5. chlorophylls; accessory; carotenoids; phycobiliproteins; antennas; reaction-center chlorophyll 6. anaerobic; dissimilatory; denitrification 7. anaerobic; aerobic; fermentation 8. lactic acid; homolactic; heterolactic 9. fermentation; aerobic respiration; nitrate; sulfate; carbon dioxide; anaerobic respiration 10. hydrolysis; phosphorolysis 11. β-oxidation; acetyl-CoA 12. proteases; deamination; transamination 13. photosynthesis; ATP; NADPH; oxygen 14. light reactions; dark reactions; carbon dioxide 15. oxygen; anoxygenic; oxygenic 16. plasma membrane; inner mitochondrial membrane 17. fermentation; oxygen; anaerobic respiration 18. NADH; $FADH_2$; negative; positive 19. ATP synthase; plasma 20. bacteriochlorophyll; photosystem; cyclic 21. Alcoholic; Mixed acid; butanediol; butanediol 22. chemolithotrophs; nitrifying bacteria; nitrification 23. Chemolithotrophs

Multiple Choice

1. a, 2. d, 3. c, 4. d, 5. b, 6. d, 7. d, 8. d, 9. b

True/False

1. F, 2. F, 3. T, 4. T, 5. T, 6. T, 7. T, 8. F, 9. F, 10. F, 11. F, 12. T, 13. T

10 Metabolism: The Use of Energy in Biosynthesis

CHAPTER OVERVIEW

This chapter presents an overview of anabolism starting with the fixation of carbon dioxide. It then focuses on the synthesis of carbohydrates; the assimilation of phosphorus, sulfur, and nitrogen; and the synthesis of amino acids, purines and pyrimidines, and lipids. The chapter concludes with a discussion of the synthesis of peptidoglycan and bacterial cell walls.

CHAPTER OBJECTIVES

After reading this chapter you should be able to:

- discuss the use of energy to construct more complex molecules and structures from smaller, simpler precursors
- discuss the way that biosynthetic pathways are organized to conserve genetic storage space, biosynthetic raw materials, and energy
- discuss the way that autotrophs use ATP and NADPH (or NADH) to reduce carbon dioxide and incorporate it into organic material
- describe the assimilation of phosphorus, sulfur, and nitrogen
- discuss the use of the TCA cycle as an amphibolic pathway and the need for anaplerotic reactions to maintain adequate levels of TCA cycle intermediates
- discuss the synthesis of glucose (gluconeogenesis) and other carbohydrates, fatty acids, triacylglycerols, purines and pyrimidines, and amino acids
- describe in general terms the synthesis of peptidoglycan and the construction of new cell walls

CHAPTER OUTLINE

I. Introduction
   A. Anabolism—the creation of order by the synthesis of complex molecules from simpler ones; it requires the input of energy
   B. Turnover—the continual degradation and resynthesis of cellular constituents
   C. The rate of biosynthesis is approximately balanced by that of catabolism, due to careful regulation of metabolic processes
II. Principles Governing Biosynthesis
   A. Biosynthetic metabolism follows a few general principles:
      1. The synthesis of large, complex molecules (macromolecules) from a limited number of simple structural units (monomers) saves much genetic storage capacity, biosynthetic raw material, and energy
      2. The use of many of the same enzymes for both catabolism and anabolism saves additional materials and energy
      3. Many biosynthetic pathways are reversals of catabolic pathways; many steps of the pathway are catalyzed by enzymes that participate in both catabolic and anabolic activities; however, some steps are catalyzed by two different enzymes: one that functions in the catabolic direction and a second that functions in the biosynthetic direction; this permits independent regulation of catabolism and anabolism
      4. Coupling some biosynthetic reactions with the breakdown of ATP (or other nucleoside triphosphates) drives anabolic pathways irreversibly in the direction of biosynthesis

5. In eucaryotic cells, anabolic and catabolic reactions involving the same constituents are frequently located in separate compartments for simultaneous but independent operation
6. Catabolic and anabolic pathways use different cofactors: catabolic oxidations produce NADH, which is a substrate for electron transport, while NADPH acts as an electron donor for anabolic pathways
    B. Once macromolecules have been made from simpler precursors, cell structures (e.g., ribosomes) form spontaneously from the macromolecules by a process known as self-assembly
III. The Precursor Metabolites
    A. Precursor metabolites are carbon skeletons used as building blocks for the synthesis of macromolecules; many are intermediates in glycolytic pathways and the TCA cycle
IV. The Fixation of CO_2 by Autotrophs
    A. Autotrophs use CO_2 as their sole or principal carbon source; carbon fixation requires much energy and reducing power
    B. The Calvin cycle
        1. The most widely used carbon fixation pathway is the Calvin cycle (reductive pentose phosphate cycle or Calvin Benson cycle) it consists of three phases, which occur in the chloroplast stroma of eucaryotes and possibly in the carboxysomes of certain bacteria
        2. The carboxylation phase—the enzyme ribulose 1,5-bisphosphate carboxylase catalyzes the addition of carbon dioxide to ribulose 1,5-bisphosphate, forming two molecules of 3-phosphoglycerate
        3. The reduction phase—3-phosphoglycerate is reduced to glyceraldehyde 3-phosphate
        4. The regeneration phase—a series of reactions is used to regenerate ribulose 1,5-bisphosphate and to produce carbohydrates such as fructose and glucose; this phase is similar to the pentose phosphate pathway and involves transketolase and transaldolase reactions
        5. The incorporation of one carbon dioxide takes three ATP molecules and two NADPH molecules; thus the formation of a single glucose molecule requires six turns through the cycle with an expenditure of 18 ATP molecules and 12 NADPH molecules; sugars formed in the Calvin cycle can then be used to synthesize other essential molecules
    C. Other CO_2-fixation pathways are used by some bacteria and archaea including the reductive TCA cycle, the 3-hydroxypropionate cycle, and the acetyl-CoA pathway
V. Synthesis of Sugars and Polysaccharides
    A. Synthesis of monosaccharides
        1. Heterotrophs synthesize glucose from noncarbohydrate precursors in a process called gluconeogenesis; this pathway is a functional reversal of glycolysis—it shares seven enzymes with the glycolytic pathway, reversing their catabolic direction, and uses several distinct enzymes or multi-enzyme systems to catalyze steps that cannot be directly reversed
        2. Once glucose and fructose are synthesized by gluconeogenesis, other sugars are manufactured; several of these other sugars are synthesized while attached to a nucleoside diphosphate
    B. Synthesis of polysaccharides also requires the use of nucleoside diphosphate sugars as precursors
    C. Synthesis of peptidoglycan
        1. A multistep process that involves two carriers: uridine diphosphate and bactoprenol; during the process a peptidoglycan repeat unit is formed and is attached to the growing peptidoglycan chain after being transported across the cytoplasmic membrane; cross-links are then formed by transpeptidation
        2. Autolysins carry out limited digestion of peptidoglycan, and provide acceptor ends for the addition of new peptidoglycan units
        3. Two general patterns of cell wall synthetic activity have been observed
            a. Many gram-positive cocci have only one or a few growth zones, usually at the site of septum formation
            b. Rod-shaped bacteria usually have growth sites scattered along the cylindrical portion of the cell as well as at the site of septum formation
        4. Peptidoglycan synthesis is very vulnerable to disruption by antimicrobial agents, including antibiotics such as penicillin; inhibition of any step in the process weakens the cell wall and can cause lysis

VI. The Synthesis of Amino Acids
    A.   Nitrogen assimilation
        1.   Ammonia incorporation
            a.   Many microorganisms use reductive amination to make alanine and glutamate, which are then used as sources of amino groups; the amino groups are transferred from alanine or glutamate to other carbon skeletons by transamination reactions
            b.   Other microorganisms use the enzymes glutamine synthetase and glutamate synthase to synthesize glutamate, which then acts as an amino group donor in transaminase reactions
        2.   Assimilatory nitrate reduction involves the reduction of nitrate to nitrite, then to hydroxylamine, and finally to ammonia, which can then be incorporated by the routes described above
        3.   Nitrogen fixation is the reduction of atmospheric nitrogen to ammonia; this is catalyzed by the enzyme nitrogenase, which is found in only a few species of bacteria and archaea; nitrogen fixation requires an expenditure of 16 ATP molecules and 8 electrons per N_2 reduced; the ammonia produced can be incorporated into organic molecules by the processes described above
    B.   Sulfur assimilation
        1.   Organic sulfur in the form of cysteine and methionine can be obtained from external sources
        2.   Assimilatory sulfate reduction is used to reduce inorganic sulfate before it is incorporated into cysteine
    C.   Amino acid biosynthetic pathways
        1.   Involves attachment of an amino group to a carbon skeleton
        2.   Carbon skeletons are derived from acetyl-CoA and from intermediates of the TCA cycle, glycolysis, and the pentose phosphate pathway
    D.   Anaplerotic Reactions
        1.   Biosynthetic functions of the TCA cycle are so important that many of its intermediates must be synthesized even when the TCA cycle is not functioning to catabolize pyruvate or to provide NADH for electron transport
        2.   Anaplerotic reactions replenish TCA cycle intermediates so that biosynthesis can occur; two major types of anaplerotic reactions have been observed
            a.   Anaplerotic carbon dioxide fixation (e.g., pyruvate carboxylase reaction)
            b.   Glyoxylate cycle—used by microorganisms that can grow on acetate as a sole carbon source; is a modified TCA cycle
VII. The Synthesis of Purines, Pyrimidines, and Nucleotides
    A.   These molecules are critical for all cells because they are used in the synthesis of ATP, several cofactors, RNA, and DNA; two types of bases are required: purines (adenine and guanine) and pyrimidines (uracil, cytosine, and thymine); a nucleoside includes the base and sugar, while a nucleotide also has the phosphate group
    B.   Phosphorus assimilation
        1.   Inorganic phosphates are incorporated through the formation of ATP by photophosphorylation, oxidative phosphorylation, and substrate-level phosphorylation
        2.   Organic phosphates obtained from the surroundings are hydrolyzed to release inorganic phosphates by enzymes called phosphatases
    C.   Purine biosynthesis is a very complex pathway in which seven different molecules (including folic acid) contribute parts to the final purine skeleton; the first purine product is the nucleotide inosinic acid, from which all other purine nucleotides can be made
    D.   Pyrimidine biosynthesis starts with aspartic acid and carbamoyl phosphate forming the initial pyrimidine product (orotic acid), which can then be converted to pyrimidine nucleotides
VIII. Lipid Synthesis
    A.   Fatty acid synthesis is catalyzed by fatty acid synthetase using the substrates acetyl-CoA and malonyl-CoA, the electron donor NADPH, and a small protein called acyl carrier protein (ACP), which carries the growing fatty acid chain; the fatty acid is lengthened by adding two carbons at a time to its carboxyl end

B. Triacylglycerols are formed from the reduction of dihydroxyacetone phosphate (a glycolytic pathway intermediate) to glycerol 3-phosphate, which then undergoes esterification with two fatty acids to form phosphatidic acid; this can then be used to produce triacylglycerol

C. Phospholipids also are produced from phosphatidic acid using a cytidine diphosphate (CDP) carrier

TERMS AND DEFINITIONS

_____ 1. The process by which cellular molecules and constituents are continually being degraded and resynthesized

_____ 2. Very large molecules that are polymers of smaller units

_____ 3. A pathway used by autotrophs to incorporate carbon dioxide into carbohydrate

_____ 4. The synthesis of glucose from noncarbohydrate precursors

_____ 5. The reduction of atmospheric gaseous nitrogen to ammonia

_____ 6. Reactions that replenish any TCA cycle intermediates that have been used in biosynthetic reactions

_____ 7. Enzymes that digest peptidoglycan just enough to provide acceptor ends for incorporation of new peptidoglycan units

_____ 8. Enzyme that releases inorganic phosphate from organic phosphate

_____ 9. An intermediate in assimilatory sulfate reduction

a. anaplerotic reactions
b. autolysins
c. Calvin cycle
d. gluconeogenesis
e. macromolecules
f. nitrogen fixation
g. phosphatase
h. phosphoadenosine 5′-phosphosulfate
i. turnover

FILL IN THE BLANK

1. Nitrogenous bases are either purines or pyrimidines. The purines are _____ and _____. The pyrimidines are _____, _____, and _____.

2. The _____ _____ _____ system is used to synthesize _____ _____ from acetyl-CoA, malonyl-CoA, and NADPH. During synthesis, the intermediates are attached to the _____ _____ protein.

3. The cell saves energy and materials by using many of the same enzymes for both _____ and _____. However, some steps are catalyzed by different enzymes to allow _____ _____ of the two processes.

4. After macromolecules have been constructed from simpler precursors, they are assembled into supra-molecular complexes or organelles by a process known as _____.

5. The three phases of the Calvin cycle are the _____ phase, the _____ phase, and the _____ phase. The initial phase is catalyzed by the enzyme _____ _____.

6. The photosynthetic production of one molecule of glucose requires _____ molecules of ATP and _____ molecules of NADPH, which are provided by the _____ reactions of photosynthesis.

7. The reduction of sulfate for use in the production of such compounds as the amino acid cysteine is called _____ sulfate reduction, while the reduction of sulfate as a terminal electron acceptor during anaerobic respiration is called _____ sulfate reduction.

8. Ammonia is assimilated by the activity of enzymes that catalyze reductive amination. Two important reductive amination reactions are catalyzed by alanine dehydrogenase and _____ _____, which synthesize alanine from pyruvate and _____ from α-ketoglutarate, respectively. Once these amino acids are made, their amino group can be transferred to other carbon skeletons by enzymes called _____. Another route for ammonia incorporation involves the sequential action of _____ _____ and _____ _____, followed by the transfer of amino groups to other carbon skeletons by _____. Nitrate and nitrogen gas can also serve as nitrogen sources. Nitrate is incorporated through _____ nitrate reduction, which is catalyzed by the enzymes _____ _____ and _____ _____. Nitrogen gas can be used by just those few microorganisms capable of carrying out _____ _____, which is catalyzed by the enzyme _____.

9. Biosynthesis of purines and pyrimidines is critical for all cells since these molecules are used in the synthesis of _____, _____, and _____, as well as several cofactors and other important cellular components.

10. A purine or pyrimidine base joined with a pentose sugar, either _____ or _____, is a _____. If one or more phosphate groups are attached to the sugar, it is called a _____.

11. Microorganisms synthesize various carbohydrates using the _____ pathway. Glucose, fructose, and mannose are either intermediates of the pathway or are made from its intermediates. Other sugars and polysaccharides are made while attached to nucleoside diphosphates. One of the most important nucleoside diphosphate sugars is _____ _____ _____ (UDPG).

12. Lipids called _____ are made from _____ _____ and glycerol phosphate by a pathway in which _____ _____ is an important intermediate.

13. Synthesis of _____ involves UDP derivatives and a lipid carrier called _____. This carrier transports N-acetyl muramic acid (NAM)–N-acetyl glucosamine (NAG)–pentapeptide units across the cell membrane. Cross-links are formed by _____.

14. The reduction of carbon dioxide to organic compounds is called ____ _____ _____. Heterotrophic organisms use _____ reactions to do this, for the purpose of replenishing TCA cycle intermediates. Only _____ do this as the principal method of fulfilling their carbon needs.

MULTIPLE CHOICE

For each of the questions below select the *one best* answer.

1. Polymers are large molecules composed of smaller units joined together. What are the smaller units called?
   a. monomers
   b. micromolecules
   c. multimers
   d. macromolecules

2. How do cells independently regulate anabolic and catabolic pathways?
   a. by using separate enzymes for reversal of key steps
   b. by compartmentation of anabolic and catabolic pathways
   c. by using different cofactors for anabolic and catabolic pathways
   d. All of the above are ways in which cells independently regulate anabolic and catabolic pathways.

3. Which of the following is NOT assimilated or incorporated in large quantities into organic molecules?
   a. nitrogen
   b. sodium
   c. phosphorus
   d. sulfur

4. Which of the following is NOT used to assimilate inorganic phosphate?
   a. oxidative phosphorylation
   b. substrate-level phosphorylation
   c. periplasmic phosphorylation
   d. photophosphorylation

5. Which of the following is NOT a source of the carbon skeletons used in the synthesis of amino acids?
   a. acetyl-CoA
   b. TCA cycle
   c. glycolysis
   d. All of the above are sources of carbon skeletons for the synthesis of amino acids.

6. Which reactions are used to replace TCA cycle intermediates so that the TCA cycle can continue to function when active biosynthesis is taking place?
   a. anaplerotic reactions
   b. amphibolic reactions
   c. anabolic reactions
   d. catabolic reactions

7. In which of the following ways are the anaplerotic CO_2 fixation reactions of heterotrophs different from the CO_2 fixation reactions found in autotrophs?
   a. They usually use pyruvate as the acceptor molecule rather than ribulose 1,5-bisphosphate.
   b. They are used to replenish TCA cycle intermediates and maintain metabolic balance rather than providing carbon for growth.
   c. Both (a) and (b) are correct.
   d. Neither (a) nor (b) is correct.

8. Cyanobacteria, some nitrifying bacteria, and thiobacilli have polyhedral inclusion bodies that contain the enzyme ribulose 1,5-bisphosphate carboxylase. These are thought to be the site of CO_2 fixation in these organisms. What are these inclusion bodies called?
   a. fixosomes
   b. carboxysomes
   c. plastosomes
   d. chloroplasts

9. Which of the following is true about the synthesis of macromolecules from monomeric subunits?
   a. It saves genetic storage capacity.
   b. It saves biosynthetic raw materials.
   c. It saves energy.
   d. All of the above are true.

TRUE/FALSE

_____ 1. The process of linking a few monomers together with a single type of covalent bond makes the synthesis of macromolecules an inefficient process.

_____ 2. In eucaryotic microorganisms, biosynthetic pathways are frequently located in cellular compartments that are different from those in which their corresponding catabolic pathways are located; this makes it easier for simultaneous but independent operation.

_____ 3. Most microorganisms have the ability to incorporate or fix carbon dioxide, but only autotrophs can use carbon dioxide as their sole or principal source of carbon.

_____ 4. Nitrogen fixation does not require the expenditure of much energy.

_____ 5. The enzyme nitrogenase is not highly specific and can reduce a number of compounds containing triple bonds, such as acetylene, cyanide, and azide.

_____ 6. Both purines and pyrimidine nucleotides are synthesized by first synthesizing the nitrogenous base and then adding ribose 5-phosphate to form the nucleotide.

_____ 7. Unsaturated fatty acids are those containing one or more carbon-carbon double bonds.

_____ 8. Membrane phospholipids are constructed from the products of glycolysis, fatty acid biosynthesis, and amino acid biosynthesis.

_____ 9. Because peptidoglycan lies outside of the cytoplasmic membrane, all of the steps in its synthesis must take place outside the membrane.

_____ 10. There are two general patterns of peptidoglycan synthesis. In cocci there are only one or a few sites of growth, with the principal growth zone located at the site of septum formation. In bacilli, active growth occurs at the site of septum formation, but also at multiple growth sites scattered along the cylindrical portion of the rod. Thus, growth is more diffusely distributed.

_____ 11. Nitrogen fixation can consume up to 20% of the ATP generated by the host plant.

_____ 12. Phosphorus can be obtained from both inorganic and organic phosphates.

_____ 13. The glyoxylate cycle is a modified TCA cycle that is used to synthesize TCA intermediates.

CRITICAL THINKING

1. What would be the consequences for a cell if anaplerotic reactions did not exist? Consider in your discussion the interrelationship between catabolic and anabolic reactions and the needs of an organism growing on limited nutritional sources.

2. There are a number of different ways to assimilate nitrogen. Only a relatively few species of bacteria, however, are capable of utilizing gaseous atmospheric nitrogen (nitrogen fixation). Why do you think such a variety of assimilatory pathways exist? Why do you think so few organisms are equipped to use gaseous atmospheric nitrogen, even though it is quite abundant?

3. Biosynthetic processes (e.g., gluconeogenesis) frequently are not direct reversals of the related catabolic processes (e.g., glycolysis). However, some of the steps involved in the overall pathway may be direct reversals. What is the advantage to the organism to have separate pathways for synthesis and degradation? Furthermore, what is the advantage to the organism for the substantial overlap (i.e., directly reversed steps) within these pathways?

ANSWER KEY

Terms and Definitions

1. i, 2. e, 3. c, 4. d, 5. f, 6. a, 7. b, 8. g, 9. h

Fill in the Blank

1. adenine; guanine; thymine; cytosine; uracil 2. fatty acid synthetase; fatty acids; acyl carrier 3. catabolism; anabolism; independent regulation 4. self-assembly 5. carboxylation; reduction; regeneration; ribulose 1,5-bisphosphate carboxylase 6. 18; 12; light 7. assimilatory; dissimilatory 8. glutamate dehydrogenase; glutamate; transaminases; glutamine synthetase; glutamate synthase; transaminases; assimilatory; nitrate reductase; nitrite reductase; nitrogen fixation; nitrogenase 9. ATP; RNA; DNA 10. ribose; deoxyribose; nucleoside; nucleotide 11. gluconeogenesis; uridine diphosphate glucose 12. triacylglycerols; fatty acids; phosphatidic acid 13. peptidoglycan; bactoprenol; transpeptidation 14. carbon dioxide fixation; anaplerotic; autotrophs

Multiple Choice

1. a, 2. d, 3. b, 4. c, 5. d, 6. a, 7. c, 8. b, 9. d

True/False

1. F, 2. T, 3. T, 4. F, 5. T, 6. F, 7. T, 8. T, 9. F, 10. T, 11. T, 12. T, 13. T

11   Microbial Genetics: Gene Structure, Replication, and Expression

CHAPTER OVERVIEW

This chapter presents the basic concepts of molecular genetics: storage and organization of genetic information in the DNA molecule. The chapter includes a description of the synthesis of RNA (transcription) and proteins (translation), the two processes involved in gene expression. Primary emphasis is given to the genetics of bacteria.

CHAPTER OBJECTIVES

After reading this chapter you should be able to:

* discuss the structural and compositional differences between DNA and RNA
* discuss the association of proteins with DNA and describe the differences in the types of proteins associated with procaryotic and eucaryotic DNA
* discuss the flow of genetic information from DNA to RNA to protein, and discuss the relationship between the nucleotide sequences of DNA and RNA and the amino acid sequences of proteins
* describe the replication of DNA and the processes used to minimize errors and correct those errors that do occur
* discuss the nature of the genetic code
* define a gene and discuss controlling elements, such as promoters and operators
* discuss the nature of a bacterial gene
* discuss the synthesis of RNA during transcription
* describe the similarities and differences between eucaryotic and procaryotic RNA transcription
* discuss the synthesis of proteins during translation and describe the role(s) of the various components required for this process

CHAPTER OUTLINE

I.   Genetics Vocabulary
    A.   Genome—all the genes present in a cell or virus; procaryotes normally have one set of genes (haploid) whereas eucaryotic microbes usually have two sets (diploid)
    B.   Genotype—the specific set of genes an organism possesses
    C.   Phenotype—the collection of characteristics of an organism that an investigator can observe
II.   DNA as Genetic Material
    A.   Griffith (1928) demonstrated the phenomenon of transformation: live, nonvirulent bacteria could become virulent when mixed with dead, virulent bacteria
    B.   Avery, MacLeod, and McCarty (1944) demonstrated that DNA was the transforming principle (the material responsible for transformation to virulence in Griffith's experiments)

C. Hershey and Chase (1952) showed that for the T2 bacteriophage, only DNA was needed for infectivity; therefore, they proved that DNA was the genetic material

III. The Flow of Genetic Information
    A. DNA is the genetic material of cells; DNA is precisely copied by a process called replication
    B. A gene is a DNA segment that encodes a polypeptide, an rRNA (ribosomal RNA), or a tRNA (transfer RNA)
    C. Genes are expressed when the information they encode is transcribed, forming an RNA molecule (mRNA) complementary to the original DNA template
    D. mRNA (messenger RNA) molecules direct the synthesis of proteins; the decoding of the mRNA information is done by ribosomes during a process called translation

IV. Nucleic Acid Structure
    A. DNA Structure
        1. DNA is composed of purine and pyrimidine nucleosides that contain the sugar 2′-deoxyribose and are joined by phosphodiester bridges
        2. DNA is usually a double helix consisting of two chains of nucleotides coiled around each other
        3. The purine adenine (A) on one strand of DNA is always paired (through hydrogen bonds) with the pyrimidine thymine (T) on the other strand, while the purine guanine (G) is always paired with the pyrimidine cytosine (C); thus, the two strands are said to be complementary
        4. The two strands are not positioned directly opposite one another; therefore, a major groove and a smaller minor groove are formed by the double helix backbone
        5. The two polynucleotide chains are antiparallel (i.e., their sugar-phosphate backbones are oriented in opposite directions)
    B. RNA structure
        1. RNA differs from DNA in that it is composed of the sugar ribose rather than 2′-deoxyribose
        2. RNA differs from DNA in that it contains the pyrimidine uracil (U) instead of thymine
        3. RNA differs from DNA in that it usually consists of a single strand that can coil back on itself, rather than two strands coiled around each other
        4. Three different kinds of RNA exist: ribosomal (rRNA), transfer (tRNA), and messenger (mRNA); they differ from one another in function, site of synthesis in eucaryotic cells, and structure
    C. The organization of DNA in cells
        1. In procaryotes, DNA exists as a closed circular, supercoiled molecule associated with basic (histonelike) proteins
        2. In eucaryotes, DNA is more highly organized; it is associated with basic proteins (histones) and is coiled into repeating units known as nucleosomes; archaea also use histonelike proteins for nucleoprotein complexes called archaeal nucleosomes

V. DNA Replication
    A. It is important that DNA is accurately replicated to prevent deleterious changes; although rapid (750–1000 base pairs per second), the process has a low frequency of errors (one in 10^9 or 10^{10} bases)
    B. DNA replication is semiconservative: each parental strand of DNA is conserved, but the two strands are separated from each other and serve as templates for the production of new complementary daughter strands
    C. Patterns of DNA synthesis
        1. The procaryotic chromosome is a replicon; that is, it consists of a single origin of replication and is replicated as a unit; two replication forks (the sites of DNA synthesis) move in opposite directions from the origin until they meet at termination sites on the opposite side; at that point the newly synthesized chromosome is released
        2. Some small, closed circular DNA molecules, such as plasmids and some virus genomes, replicate by means of a rolling-circle mechanism
        3. The large, linear DNA molecules of eucaryotes employ multiple replicons to efficiently replicate the DNA within a reasonable time span

D. The replication machinery
   1. DNA polymerases are enzymes that catalyze the synthesis of complementary DNA strands in the 5' to 3' direction by adding new nucleotide monophosphates (from triphosphate substrates) to the 3'-hydroxyl group of the growing chain; the enzyme needs a primer (forming a double-stranded region) with a free 3'-hydroxyl to begin replication
   2. The DNA to be replicated is bound by DnaA proteins forming the replisome, unwound by helicases, gyrases, and topoisomerases, held unwound by single-stranded DNA binding proteins (SSBs), primed with RNA by primase, and replicated by DNA polymerase III to create complementary daughter strands
E. Events at the replication fork
   1. In *E. coli* the process of replication proceeds in the following way:
      a. DNA replication is initiated when DnaA protein binds to the origin of replication
      b. Helicases unwind the two strands of DNA and as they do so topoisomerases (e.g., DNA gyrase) relieve the tension caused by the unwinding process; SSBs keep the single strands apart
      c. Primases synthesize a small RNA molecule (approximately 10 nucleotides) that acts as a primer for DNA synthesis
      d. DNA polymerase III, like all DNA polymerases, synthesizes DNA in a 5' to 3' direction as it creates the complementary strand of DNA according to the base-pairing rules
      e. The two strands are synthesized in a different manner: on one strand (the leading strand), synthesis is continuous, while on the other (the lagging strand), a series of fragments (Okazaki fragments) are generated by discontinuous synthesis due to the antiparallel nature of DNA strands and the 5' to 3' direction of polymerization; a multiprotein complex called a replisome organizes all of these processes
      f. DNA polymerase I removes the primers and fills the resulting gaps
      g. DNA ligases join all DNA fragments to form a complete strand of DNA
   2. DNA polymerase III (the product of the *dnaE* gene) can proofread nascent DNA chains to remove (and replace) mismatched base pairs immediately
F. Termination of replication
   1. DNA replication stops when the replisome reaches a termination site (*ter*); in some bacteria replication stops when the two replication forks meet
   2. The circular chromosome copies are interlocked as catenanes that are separated by topoisomerase
G. Replication of linear chromosomes
   1. Linear chromosomes cannot maintain a free 3'-hydroxyl on each end (telomeres) of both DNA strands and therefore do not provide the primer needed for DNA polymerase
   2. In eucaryotes, the enzyme telomerase contains a small RNA molecule that acts as a template for the synthesis of extensions to telomeres and allowing for DNA polymerase to work closer to the ends of the chromosome; it is not clear how this problem is dealt with in procaryotes that have linear chromosomes

VI. Gene Structure
   A. Gene
      1. A linear sequence of nucleotides that has a fixed start point and end point, and that encodes a polypeptide, a tRNA, or an rRNA; if it encodes a single polypeptide it is also called a cistron
      2. With some exceptions, genes are not overlapping; there is a single starting point with one reading frame in which the three-base codons are in frame with the start and stop codons
      3. In prokaryotes, coding information is normally continuous although some bacterial genes are interrupted; in eucaryotes, most genes have coding sequences (exons) that are interrupted by noncoding sequences (introns); the mRNA transcripts are eucaryotes spliced to remove introns and connect exons, and alternative splicing sites may be present
   B. Genes that code for proteins
      1. Template strand—the strand that contains coding information and directs RNA synthesis

2. Promoter—a sequence of bases that is usually situated upstream from the coding region; serves as a recognition/binding site for RNA polymerase; different genes have different promoters, and promoters from different species vary in sequence

3. Leader sequence—a sequence of nucleotides that is transcribed but is not translated; contains a consensus sequence known as the Shine-Dalgarno sequence, which serves as the recognition site for the ribosome

4. Coding region—the sequence that begins immediately downstream of the leader sequence; starts with the template sequence 3'TAC5', which gives rise to mRNA codon 5'AUG3', the first translated codon (specifies N-formylmethionine in bacteria, methionine in archaea and eucaryotes)

5. Trailer sequence—nontranslated sequence of nucleotides located immediately downstream of the coding region and before the transcription terminator

6. Regulatory sites—sites where DNA-recognizing regulatory proteins bind to either stimulate or inhibit gene expression (e.g., operator)

C. Genes that code for tRNA and rRNA

1. tRNA genes have promoters, leader sequences, coding regions, and trailer sequences; noncoding regions are removed after transcription; more than one tRNA may be made from a single transcript; the tRNAs are separated by a spacer region, which is removed after transcription by special ribonucleases that may contain catalytic RNA molecules (ribozymes)

2. rRNA genes have promoters, leader sequences, coding regions and trailer sequences; all rRNA molecules are transcribed as a single large transcript, which is cut up after transcription, yielding the final rRNA products

VII. Transcription

A. Transcription is the synthesis of RNA under the direction of DNA

1. The RNA product is complementary to the DNA template

2. An adenine nucleotide in the DNA template directs the incorporation of a uracil nucleotide in the RNA; otherwise, the base pair rules are the same as for DNA replication

3. Three types of RNA are produced by transcription

a. mRNA carries the message that directs the synthesis of proteins; if it carries more than one gene transcript, it is called polycistronic or polygenic

b. tRNA molecules carry amino acids during protein synthesis

c. rRNA molecules are components of the ribosomes

B. Transcription in bacteria

1. RNA polymerase (a large, multi-subunit enzyme) is responsible for the synthesis of RNA; in bacteria, the core enzyme, consisting of four subunit types, catalyzes RNA synthesis and the holoenzyme includes a sigma factor that is not catalytic, but helps the core enzyme bind DNA at the appropriate site

2. Transcription involves three separate processes: initiation, elongation, and termination

3. A promoter is the region of the DNA to which RNA polymerase binds in order to initiate transcription; consensus sequences centered at 35 and 10 base pairs before the transcription starting point (including the Pribnow box) are important in directing RNA polymerase to the promoter

4. During elongation, the RNA polymerase unwinds the DNA helix to create an open complex (transcription bubble) of 16–20 base pairs within which a strand of RNA complementary to the DNA template strand is formed in the 5' to 3' direction

5. Terminators are regions of the DNA that signal termination of the transcription process; this often involves hairpin loops in the DNA template or the binding of rho factor protein

C. Transcription in eucaryotes

1. There are three major RNA polymerases

a. RNA polymerase II catalyzes mRNA synthesis; it requires several initiation factors and recognizes promoters that have several important elements (rather than just two as seen in procaryotes)

b. RNA polymerase I catalyzes rRNA (5.8S, 18S, and 28S) synthesis

c. RNA polymerase III catalyzes tRNA and 5S rRNA synthesis

112

2. Transcription yields large, monogenic RNA precursors (heterogeneous nuclear RNA; hnRNA) that must be processed by posttranscriptional modification to produce mRNA

    a. Adenylic acid is added to the 3′ end to produce a polyA sequence about 200 nucleotides long (polyA tail)

    b. 7-methylguanosine is added to the 5′ end by a triphosphate linkage (5' cap)

    c. These two modifications are believed to protect the mRNA from exonuclease digestion

3. Eucaryotic genes are split or interrupted such that the expressed sequences (exons) are separated from one another by intervening sequences (introns); the introns are represented in the primary transcript but are subsequently removed by a process called RNA splicing in a large spliceosome complex; some mRNAs are ribozymes capable of self-splicing

D. Transcription in the archaea tends to be more like that of eucaryotes, requiring a complex RNA polymerase, TATA-binding proteins, several transcription factors, and splicing out of mRNA introns; however, like bacteria, the archaea have only one RNA polymerase and polycistronic mRNAs

VIII. The Genetic Code

A. For polypeptide-coding genes, the DNA base sequence corresponds to the amino acid sequence of the polypeptide (colinearity)

B. Establishment of the genetic code—each codon that specifies a particular amino acid must be three bases long for each of the 20 amino acids to have at least one codon; thus, the genetic code consists of 64 codons

C. Organization of the code

1. Code degeneracy—many amino acids are encoded by more than one codon

2. Sense codons—61 codons that specify amino acids

3. Stop (nonsense) codons—three codons (UGA, UAG, UAA) that do not specify an amino acid, and that are used as translation (protein synthesis) termination signals

4. Wobble—describes the somewhat loose base pairing of a tRNA anticodon to the mRNA codon; wobble eliminates the need for a unique tRNA for each codon because the first two positions are sufficient to establish hydrogen bonding between the mRNA and the aminoacyl-tRNAs

IX. Translation

A. Translation is the synthesis of a polypeptide chain directed by the nucleotide sequence in a mRNA molecule; synthesis begins at the N-terminal and moves in the C-terminal direction

1. Ribosomes are the sites of translation

2. Polyribosomes are complexes of an mRNA molecule with several ribosomes

B. Transfer RNA and amino acid activation

1. The first stage of protein synthesis is the attachment of amino acids to tRNA molecules (catalyzed by aminoacyl-tRNA synthetases); this process is referred to as amino acid activation

2. Each tRNA has an acceptor end and can only carry a specific amino acid; it also has an anticodon triplet that is complementary to the mRNA codon triplet and leads to this specificity

C. The ribosome is a complex organelle constructed from several rRNA molecules and many polypeptides; has two subunits (in procaryotes: 50S and 30S; in eucaryotes: 40S and 60S)

D. Initiation of protein synthesis

1. In bacteria, the small subunit of the ribosome binds fMet-tRNA (Met-tRNA in archaea and eucaryotes) and then binds the mRNA at a special initiator codon (AUG); then the large subunit of the ribosome binds

2. Three protein initiation factors also are required in procaryotes (eucaryotes require more initiation factors)

E. Elongation of the polypeptide chain

1. Elongation involves the sequential addition of amino acids to the growing polypeptide chain; several polypeptide elongation factors are required for this process

2. The ribosome has three sites for binding tRNA molecules: peptidyl site (P site), aminoacyl site (A site), and exit site (E site)

3. Each new amino acid is positioned in the A site by its tRNA, which has an anticodon that is complementary to the codon on the mRNA molecule

4. The ribosomal enzyme peptidyl transferase catalyzes the formation of the peptide bonds between the amino acid (or growing peptide chain) held by the tRNA in the P site and the amino acid held by the tRNA in the A site; in doing so, the growing peptide chain is transferred to the tRNA in the A site; the 23S rRNA is a major component of this enzyme

5. After each amino acid is added to the chain, translocation occurs, thereby moving the ribosome to position the next codon in the A site and the tRNA carrying the growing peptide chain in the P site

F. Termination of protein synthesis

1. Takes place at any one of three special codons (UAA, UAG, or UGA); three polypeptide release factors aid in the recognition of these codons; the ribosome hydrolyzes the bond between the completed protein and the final tRNA, and the protein is released from the ribosome, which then dissociates into its two component subunits

2. Protein synthesis is expensive, using five high-energy bonds to add one amino acid to the chain

G. Protein folding and molecular chaperones

1. Molecular chaperones are special helper proteins that aid the nascent polypeptide in folding to its proper shape; many have been identified and they include heat-shock proteins and stress proteins; in addition to helping polypeptides fold, chaperones are important in the transport of proteins across membranes

2. Protein conformation is a direct function of amino acid sequence; proteins have self-folding, structurally independent regions called domains; in eucaryotes, the domains fold immediately upon synthesis, whereas in procaryotes the domains do not fold until the complete protein is synthesized

H. Protein splicing—before folding, part of the polypeptide is removed; such splicing removes intervening sequences (inteins) from the sequences (exteins) that remain in the final product

TERMS AND DEFINITIONS

_____ 1. Small, basic proteins that help organize the structure of DNA in the eucaryotes

_____ 2. A structure formed by coiling DNA around the surface of a complex of histones

_____ 3. The process by which DNA is very precisely copied

_____ 4. The process by which the base sequence of a gene is used to direct the synthesis of an RNA molecule

_____ 5. The process by which the base sequence of a mRNA is used to direct the synthesis of a protein

_____ 6. The Y-shaped part of the DNA molecule where replication takes place

_____ 7. Small fragments (100 to 1,000 bases long) formed by discontinuous replication of the lagging strand

_____ 8. A portion of the gene that precedes the translation initiation codon; it is transcribed but not translated

_____ 9. A portion of the gene that is immediately downstream of the coding region; it is transcribed but not translated

_____ 10. A sequence of nucleotides that codes for a polypeptide, rRNA, or tRNA

_____ 11. The strand of DNA for a particular gene that serves as the template for RNA synthesis

_____ 12. The region of DNA to which RNA polymerase binds; it is required for the initiation of transcription but is not transcribed itself

_____ 13. An origin of replication and the DNA that is replicated as a unit from that origin

_____ 14. A sequence of nucleotides that signals RNA polymerase to stop transcription

_____ 15. A nucleotide sequence centered at 10 nucleotides before the transcription start signal; it is the binding site for RNA polymerase

_____ 16. A complex of proteins consisting of the primase and its accessory proteins

_____ 17. Genetic code words that specify particular amino acids; each code word is three nucleotides long

_____ 18. The concept that there is more than one codon that specifies a particular amino acid

_____ 19. Those codons in the genetic code that actually do specify amino acids

_____ 20. Those codons in the genetic code that do not specify an amino acid; instead they cause termination of translation

_____ 21. The portion of a polypeptide-encoding gene that is both transcribed and translated

a. amino acid activation
b. cistron
c. code degeneracy
d. coding region
e. codons
f. exons
g. exteins
h. gene
i. heterogeneous nuclear RNA (hnRNA)
j. histones
k. inteins
l. leader sequence
m. leader sequence
n. nonsense (stop) codons
o. nucleosome
p. Okazaki fragments
q. polyribosome (polysome)
r. Pribnow Box
s. Pribnow box
t. primosome
u. promoter
v. promoter
w. replication
x. replication fork
y. replicon
z. ribozymes
aa. RNA splicing
bb. sense codons
cc. small nuclear RNA molecules
dd. spliceosome
ee. structural genes
ff. template strand
gg. terminator sequence
hh. trailer sequence
ii. transcription
jj. translation
kk. transpeptidation reaction

115

_____ 22. A segment of DNA that encodes a single polypeptide

_____ 23. Large eucaryotic RNA molecules that must be processed to produce smaller mRNA molecules

_____ 24. The process by which intervening sequences are removed from an initial RNA transcript to produce mature mRNA molecules

_____ 25. A complex formed when several ribosomes bind and translate a single mRNA molecule

_____ 26. The phenomenon in which amino acids are attached to the appropriate tRNA molecules

_____ 27. The reaction by which amino acids are added sequentially to form a polypeptide chain

_____ 28. RNA molecules with catalytic activity

_____ 29. Portions of a protein that remain after protein splicing

_____ 30. Portions of a protein that are removed by protein splicing

_____ 31. Genes that code for polypeptides

_____ 32. The expressed (coding) sequences of split (interrupted) genes

_____ 33. The region at which RNA polymerase binds before initiating transcription of downstream sequences

_____ 34. An untranslated sequence of 25 to 150 bases at the 5′ end of mRNA

_____ 35. Small RNA molecules found in the nucleus; they complex with proteins to form particles that form the RNA splicing machinery

_____ 36. A large complex of proteins and RNA that is responsible for RNA splicing

REPLICATION

Complete the table by providing a brief description of the functions of each of the components listed below:

Component	Description of Function
DNA polymerase III	
Helicase	
Single-strand binding proteins	
Topoisomerase (DNA gyrase)	
Primase	
DNA polymerase I	
DNA ligase	

TRANSCRIPTION AND TRANSLATION

Complete the table by providing a brief description of the functions of each of the components listed:

TRANSCRIPTION—BACTERIA

Component	Description of Function
RNA polymerase core enzyme	
Gene	
Template strand	
Sigma factor	
Promoter	
Operator	
Terminator and Rho factor	

TRANSCRIPTION—EUCARYOTES

Component	Description of Function
RNA polymerase I, II, and III	
Gene	
Template strand	
Promoter	
PolyA tail	
5′ cap	
Introns	
Exons	
Small nuclear RNA molecules	
Spliceosome	

TRANSLATION

Component	Description of Function
mRNA	
Ribosome	
Aminoacyl-tRNA synthetase	
tRNA	
Initiator codon	
Initiation factors	
Anticodon triplet	
Codon	
Elongation factors	
Peptidyl transferase	
Nonsense codons	
Release factors	
Molecular chaperones	

FILL IN THE BLANK

1. The two strands making up the double helix of DNA are said to be _____ (i.e., the purine adenine on one strand always pairs with the pyrimidine _____ on the other strand, while the purine _____ always pairs with the pyrimidine cytosine).

2. The two strands of DNA are not positioned directly opposite one another in the helical cylinder. Therefore, when the strands are twisted about one another, a wide _____ _____ and a narrower _____ _____ are formed by the backbone.

3. The two backbones of DNA are antiparallel with respect to the orientation of their sugars. If one end of a double helix is examined, the _____ end of one strand and the _____ end of the other strand will be seen.

4. In procaryotes, DNA is associated with basic proteins different from the _____ found associated with eucaryotic DNA.

5. During replication, both parental DNA strands are used as templates for DNA synthesis. This results in the formation of two replicas, each containing a parental DNA strand and a newly synthesized strand. This process is referred to as _____ replication because although the parental molecule is not conserved intact, the individual parental strands are conserved.

6. The two strands of a replicating DNA molecule do not completely dissociate from one another; instead, replicating DNA molecules have a Y shape and the actual replication process takes place at the junction of the arms of the Y. This junction is called the _____ _____. Replication begins at a single point and moves outward in both directions until the whole _____ (that portion of the DNA molecule containing a replication origin that is replicated as a single unit) has been replicated.

7. During _____, one DNA strand is synthesized continuously and the other is synthesized discontinuously. The fragments produced by discontinuous synthesis are called _____ fragments.

8. Replication errors can be detected and corrected by certain DNA polymerases before replication is completed. This is called _____.

9. Avery, MacLeod, and McCarty provided the first evidence that _____ carried the genetic information when they exposed extracts from virulent strains of *Pneumococcus* to enzymes that hydrolyze RNA, DNA, or protein. Only extracts exposed to enzymes that hydrolyzed _____ prevented the extract-treated nonvirulent strains from becoming virulent. This acquisition of virulence was called _____.

10. Hershey and Chase demonstrated that when the bacteriophage T2 infected its host cell, the _____ was injected into the host while the _____ remained outside. Since progeny viruses were produced, it was clear that the _____ was carrying the genetic information.

11. The genetic code must be contained in some sequence of the four nucleotides found in the linear DNA sequence. If taken individually, only four amino acids could be specified. If taken two at a time only _____ amino acids could be specified. Therefore, a code word or _____ must consist of at least three nucleotides. However, this would allow _____ possible combinations, which is more than the minimum of _____ needed to specify all the amino acids usually found in proteins.

12. There are 61 codons that specify amino acids. These are called _____ codons. The three codons, UAG, UGA, and UAA, which do not specify amino acids, are called _____ codons and cause termination of _____.

13. Although DNA is double stranded, only one of the strands contains coded information and directs RNA synthesis. This strand is called the _____ strand.

14. In addition to the coding region, each gene also has three noncoding regions. These are the _____, the _____ sequence, and the _____ sequence.

15. The _____ is all of the genes present in a cell or virus.

16. The leader sequence of an mRNA is transcribed but not _____. It contains a recognition sequence known as the _____ sequence, which serves as a _____ binding site.

17. DNA polymerases synthesize DNA in the _____ to _____ direction while reading the template in the _____ to _____ direction.

18. DNA polymerase III holoenzyme requires a primer (a short chain of nucleotides having a free 3' hydroxyl) to which deoxyribonucleoside triphosphates are added. When the primer is an RNA molecule, it is synthesized by an enzyme called _____.

19. Sigma factors enable RNA polymerase core enzyme to recognize and bind to _____. Different sigma factors recognize different sets of _____ and an alteration in the sigma factors available alters the expression of many genes and operons. This type of regulation is an example of a _____ regulatory system.

20. In eucaryotes, large RNA precursors called _____ _____ RNA (hnRNA) undergo _____ modification. This involves addition of polyA sequences to the _____end and a cap to the _____ end. In addition, introns are removed by a process called _____ _____.

21. Amino acids are activated for protein synthesis through a reaction that is catalyzed by_____ _____. There are at least _____ of these enzymes, each specific for a single amino acid and all the tRNAs to which it may properly be attached.

22. The ribosome has three sites for binding tRNA molecules: the _____ site (donor or P site), the _____ site (acceptor or A site), and the _____ site (E site). If the fourth codon is in the donor site, then the _____ codon is in the acceptor site. Once the tRNAs are positioned properly, the transpeptidation reaction, which is catalyzed by _____ _____, transfers the growing peptide chain to the tRNA in the _____ site, lengthening the chain by one amino acid. Then the tRNA carrying the growing chain is moved into the _____ site, the empty tRNA is moved to the _____ site, and the sixth codon is moved into the _____ site. The movement is referred to as _____. All of the above events are part of the _____ cycle of translation.

23. Procaryotic mRNA has untranslated _____ and _____ sequences at its ends. In addition, many procaryotic mRNA molecules are _____ (polycistronic); these mRNAs also have _____ regions between each coding region.

MULTIPLE CHOICE

For each of the questions below select the *one best* answer.

1. Which of the following nitrogenous bases is found in RNA but not in DNA?
   a. adenine
   b. thymine
   c. uracil
   d. guanine

2. Which of the following is NOT true about the structure of DNA?
   a. Purine and pyrimidine bases are attached to the 1'-carbon of the deoxyribose sugars.
   b. The bases extend toward the middle of the cylinder formed by the two chains.
   c. The bases are stacked on top of each other in the center with the rings forming parallel planes.
   d. All of the above are true about the structure of DNA.

3. In which of the following ways do mRNA, rRNA, and tRNA differ from one another?
   a. function
   b. site of synthesis in eucaryotes
   c. structure
   d. All of the above are correct.

5. Most amino acids are encoded by more than one codon. Which term is used to describe this characteristic of the genetic code?
   a. ambiguous
   b. degeneracy
   c. multiplicative
   d. repetitious

6. Which of the following is a function of molecular chaperones?
   a. protection against cellular stresses (e.g., heat-shock proteins)
   b. protein folding
   c. transport of proteins across membranes
   d. all of the above

7. What are the RNA molecules that carry amino acids to the ribosome during translation called?
   a. mRNA
   b. tRNA
   c. rRNA
   d. hnRNA

8. Which of the following is true when comparing initiation of translation in procaryotes and initiation of translation in eucaryotes?
   a. They are identical processes.
   b. They are similar, but procaryotes require more initiation factors.
   c. They are similar, but eucaryotes require more initiation factors.
   d. They are very different.

9. Translation is an energy-expensive process. How many high-energy bonds are required to add a single amino acid to a growing polypeptide chain?
   a. two
   b. three
   c. four
   d. five

10. Which of the following are removed from some proteins by protein splicing?
    a. exteins
    b. inteins
    c. exons
    d. introns

TRUE/FALSE

_____ 1. The chemical differences between RNA and DNA reside in their sugar and pyrimidine bases: RNA has ribose and uracil, while DNA has deoxyribose and thymine.

_____ 2. In addition to differing chemically from DNA, RNA is usually single stranded rather than double stranded.

_____ 3. Complementary base pairing and helical structure are not observed in RNA molecules.

_____ 4. Many procaryotes have only one origin of replication per DNA molecule, while most eucaryotes have multiple origins in each DNA molecule.

_____ 5. DNA replication has a high frequency of error.

_____ 6. The rate of replication in eucaryotes is the same as that in procaryotes.

_____ 7. DNA gyrase is a topoisomerase that removes superhelical twists produced in DNA during DNA replication.

_____ 8. Some small, circular DNA molecules (e.g., plasmids) are replicated by the rolling-circle mechanism.

_____ 9. Since a nucleotide sequence can be read in any of three different reading frames, each nucleotide sequence usually encodes three different polypeptide products in an overlapping fashion.

_____ 10. In both procaryotes and eucaryotes, the coding information within a gene is normally continuous (i.e., it is not interrupted by noncoding sequences).

_____ 11. For any given gene, only one strand is the template strand for RNA synthesis. However, different genes may use different strands as their templates.

_____ 12. Several tRNA molecules may be produced from a single transcript by posttranscriptional processing. The segments coding for the tRNAs are separated from each other by short spacer sequences that are removed during the processing.

_____ 13. All of the rRNAs are transcribed as a single, large precursor molecule that is cleaved to yield the final rRNA products. The spacer regions between the rRNA sequences usually contain tRNAs that are produced as well.

_____ 14. The concept of wobble in the codon-anticodon interaction eliminates the necessity for having a tRNA with an anticodon region specific for each of the 61 sense codons.

_____ 15. Ribosomes are composed of two separate subunits that come together as part of the initiation process and then dissociate immediately after termination.

_____ 16. Bacteria start protein synthesis with formylmethionine. However, the formyl group is removed after translation has been completed.

_____ 17. The poly-A sequence at the 3′ end of eucaryotic mRNA is believed to protect the mRNA from rapid enzymatic degradation.

_____ 18. Molecular chaperones are more important in the folding of eucaryotic proteins than in the folding of procaryotic proteins.

_____ 19. The initial amino acids linked together during translation are at the N-terminus of the protein.

_____ 20. Procaryotic proteins have domains that fold as they leave the ribosome. In contrast, eucaryotic proteins do not fold until completely synthesized.

CRITICAL THINKING

1. Diagram and discuss the process of semiconservative replication. Starting with a single bacterium having only one chromosome, trace the parental DNA through three replicative cycles (resulting in eight cells). In how many of these cells will you find DNA that was in the original parent? If the replication process was fully conservative rather than semiconservative, how would this affect your results?

2. Compare and contrast the transcription processes in procaryotes and eucaryotes. Be sure to include in your discussion the role(s) of promoters, various RNA polymerases, posttranscriptional modification, coupled transcription/translation, and split (interrupted) genes.

ANSWER KEY

Terms and Definitions

1. j, 2. o, 3. w, 4. ii, 5. jj, 6. x, 7. p, 8. l, 9. hh, 10. h, 11. ff, 12. u, 13. y, 14. gg, 15. r, 16. t, 17. e, 18. c, 19. bb, 20. n, 21. d, 22. b, 23. i, 24. aa, 25. q, 26. a, 27. kk, 28. z, 29. g, 30. k, 31. ee, 32. f, 33. v, 34. m, 35. cc, 36. dd

Fill in the Blank

1. complementary; thymine; guanine 2. major groove; minor groove 3. 5′; 3′ 4. histones 5. semiconservative 6. replication fork; replicon 7. replication; Okazaki 8. proofreading 9. DNA; DNA; transformation 10. DNA; protein; DNA 11. 16; codon; 64; 20 12. sense; stop (nonsense); translation 13. template 14. promoter; leader; trailer 15. genome 16. translated; Shine-Dalgarno; ribosome 17. 5′; 3′;3′; 5′ 18. primase 19. promoters; promoters; global 20. posttranscriptional; heterogeneous nuclear; 3′; 5′; RNA splicing 21. aminoacyl-tRNA synthetases; 20 22. peptidyl; aminoacyl; exit; fifth; peptidyl transferase; aminoacyl; peptidyl; exit; aminoacyl; translocation; elongation 23. leader; trailer; polygenic; spacer

Multiple Choice

1. c, 2. d, 3. d, 4. c, 5. b, 6. d, 7. b, 8. c, 9. d, 10. b

True/False

1. T, 2. T, 3. F, 4. T, 5. F, 6. F, 7. T, 8. T, 9. F, 10. F, 11. T, 12. T, 13. T, 14. F, 15. T, 16. T, 17. T, 18. F, 19. T, 20. F

12 Microbial Genetics: Regulation of Gene Expression

CHAPTER OVERVIEW

This chapter considers a variety of mechanisms by which gene expression is regulated. The discussion begins by giving the levels of regulation of gene expression. The regulation of transcription initiation includes induction and repression using the *lac*, *trp*, and *ara* operons as examples, and includes two-component regulatory systems. Regulation at the level of transcription continues with attenuation and riboswitches. Regulation at the level of transcription is discussed. Global regulatory systems include discussions of sigma factors, catabolite repression, quorum sensing, and sporulation. The chapter concludes with information about gene regulation in eucaryotes and archaea.

CHAPTER OBJECTIVES

After reading this chapter you should be able to:

- discuss levels of regulation of gene expression
- describe induction and repression using example operons
- discuss regulation of enzyme synthesis by negative and positive regulatory proteins
- describe two-component regulatory systems
- describe attenuation and riboswitches
- discuss regulation at the level of translation
- discuss global regulation of gene expression
- describe quorum sensing and catabolite repression
- contrast regulation in eucaryotes, bacteria, and archaea

CHAPTER OUTLINE

I. Levels of Regulation of Gene Expression
    A. Regulation of gene expression (controlling enzyme synthesis) occurs at different levels, including control of transcription, translation, and posttranslation
    B. Although there are similarities in the regulation of gene expression in organisms from different domains, there are many differences in chromosome organization, mRNA transcripts, signaling, and cell structure
II. Regulation of Transcription Initiation
    A. Induction and repression of enzyme synthesis
        1. Enzymes central to metabolic processes, routinely needed by cells, are encoded by housekeeping genes; these are constitutive genes that are continuously expressed
        2. Synthesis of enzymes involved in catabolic pathways are often inducible and are only expressed when needed; the initial substrate of the pathway (or some derivative of it) is usually the inducer; induction increases the amount of mRNA encoding the enzymes

3. Synthesis of enzymes involved in anabolic pathways is often repressible and expressed when biosynthesis of the end product is needed; the end product of the pathway usually acts as a corepressor; repression decreases the amount of mRNA encoding the enzymes

B. Control of transcription initiation by regulatory proteins
   1. Negative transcriptional control occurs when a repressor protein inhibits initiation of transcription; positive transcriptional control occurs when activator protein promotes initiation of transcription
   2. Repressor proteins bind to the operator, a region of DNA overlapping or downstream of the promoter, and block RNA polymerase binding; activator proteins bind activator-binding sites, often upstream of the promoter
   3. In inducible systems, the repressor protein is active until bound to the inducer (binding of inducer inactivates the repressor) whereas in repressible systems, the repressor is inactive until bound to the corepressor (binding of corepressor activates the repressor)
   4. In bacteria, a set of structural genes controlled by a particular operator and promoter is called an operon

C. The lactose operon: negative transcriptional control of inducible genes
   1. Even in operons where expression is repressed, a low basal level of transcription occurs; genes are only expressed when needed (substrate present; product absent)
   2. The lactose (*lac*) operon, which encodes genes for the catabolism of lactose, is an excellent example of negative regulation; binding of the *lac* repressor to the *lac* operators bends the DNA and inhibits RNA polymerase binding or blocks the movement of RNA polymerase
   3. The *lac* repressor is inactivated by binding the inducer, allolactose, a derivative of lactose; the presence of lactose induces expression of the *lac* operon by inhibiting repressor binding
   4. The *lac* operon also is regulated by catabolite activator protein (CAP), part of a global regulatory system

D. The tryptophan operon: negative transcriptional control of repressible genes
   1. The tryptophan (*trp*) operon, which encodes genes for the synthesis of tryptophan, is an excellent example of a repressible operon; the *trp* operon is expressed unless the *trp* repressor binds its corepressor, tryptophan, the end product of the pathway
   2. The *trp* operon also is controlled at the level of transcription elongation through attenuation

E. The arabinose operon: transcriptional control by a protein that acts both positively and negatively
   1. The arabinose (*ara*) operon encodes genes for the catabolism of arabinose
      a. When arabinose is absent the *ara* operon is repressed by the interaction of two AraC molecules at the operators
      b. When arabinose is present, this interaction is prevented and the AraC molecules stimulate expression

F. Two-component regulatory systems and phosphorelay systems
   1. Signal transduction systems regulate gene expression in response to environmental conditions, rather than metabolite levels
   2. In two-component regulatory systems, a sensor kinase protein in the plasma membrane senses changes in the environment and can phosphorylate a response-regulator protein, a DNA-binding protein that regulates gene expression
   3. Phosphorelay systems are longer pathways that use transfer of phosphoryl groups to control gene transcription and protein activity

III. Regulation of transcription elongation
   A. Attenuation
      1. There are two decision points for regulating transcription of anabolic pathways: initiation of transcription and continuation of transcription; attenuation regulates continuation of transcription
      2. In systems where transcription and translation are tightly coupled, ribosome behavior in the leader region of the mRNA can control continuation of transcription
         a. If ribosomes actively translate the leader region (attenuator), which contains several codons for the amino acid product of the operon, a transcription terminator forms and transcription will not continue

       b.   If ribosomes stall during translation of the leader region because the appropriate charged aminoacyl-tRNA is absent, the terminator does not form and transcription will continue

  B. Riboswitches
     1. Riboswitches (sensory RNAs) are a form of attenuation that does not involve the ribosome; regulation is based on differential folding of the mRNA leader sequence
     2. Alternative folding, creating antitermination and termination loops, is controlled by the binding of an effector molecule

IV. Regulation at the Level of Translation
  A. Regulation of translation by riboswitches is similar to the regulation of transcription by riboswitches, except here the binding of effector molecules changes the folding of the mRNA in such a way as to inhibit ribosomal binding and initiation of translation
  B. Regulation of translation by small RNA molecules
     1. Small RNAs (sRNAs), also called noncoding RNAs (ncRNAs), are involved in regulating cellular processes by directly pairing with mRNAs
     2. One kind of sRNA is antisense RNA, which is complementary to the leader sequence of an mRNA molecule, and specifically binds to it, thereby blocking translation

V. Global Regulatory Systems
  A. Overview
     1. Global regulatory systems affect many genes and pathways simultaneously, allowing for both independent regulation of operons as well as cooperation of operons
     2. A regulon is a group of genes or operons controlled by a common regulatory protein; a modulon is more complex and has a common regulatory protein that controls an operon network, but individual operons are controlled separately as well; a stimulon is a regulatory system in which all operons respond together to an environmental stimulus
  B. Mechanisms used for global regulation
     1. Bacteria produce a number of different sigma factors; each enables RNA polymerase to recognize and bind to specific promoters
     2. Alternate sigma factors available to RNA polymerase change gene expression
  C. Catabolite repression
     1. Diauxic growth—a biphasic growth pattern observed when a bacterium is grown on two different sugars (e.g., glucose and lactose)
     2. For *E. coli*, availability of glucose (the preferred carbon and energy source) causes a drop in cAMP levels, resulting in the deactivation of CAP (a positive regulator of several catabolic pathways, including the *lac* operon); deactivation of CAP allows the bacterium to use glucose preferentially over another sugar when both are present in the environment
  D. Quorum sensing
     1. Intercellular communication (quorum sensing) in procaryotes is important in gene regulation; even microbes of different species can affect each other's gene expression
     2. Quorum sensing is typically mediated by the release to homoserine lactones as autoinducers; in some cases a two-component regulatory system is involved
  E. Sporulation in *Bacillus subtilis*
     1. Sporulation is a complex process that is controlled at several levels through phosphorelay, transcription factors, posttranslational modifications, and alternate sigma factors
     2. Under certain environmental stimuli, the response-regulator protein (Spo0A) alters the expression over 500 genes, including alternate sigma factors that differentially control gene expression in the forespore and mother cell

VI. Regulation of gene expression in *Eucarya* and *Archaea*
  A. In eucaryotes, numerous general transcription factors are required; there are regulatory transcription factors that are specific to one or more genes and alter the rate of transcription; activators bind regulatory sites called enhancers, while repressors bind sites called silencers
  B. sRNAs act in eucaryotes as antisense RNAs and function at the level of translation; the smallest types are called microRNAs (miRNAs); some sRNAs work with the spliceosome

C. Gene regulation in archaea is not understood; in most cases archaeal regulatory proteins function like bacterial activators and repressors, while in others they function like eucaryotic regulatory transcription factors

TERMS AND DEFINITIONS

Place the letter of each term in the space next to the definition or description that best matches it.

_____ 1. Small molecules that inactivate repressor proteins and thereby increase the synthesis of certain enzymes

_____ 2. Small molecules that activate repressor proteins and thereby decrease the synthesis of certain enzymes

_____ 3. The site on the DNA to which a repressor binds

_____ 4. A promoter, an operator, and the structural genes that they control

_____ 5. A transcription termination site found in the leader region of certain operons that controls continuation of transcription of that operon

_____ 6. An RNA molecule that specifically binds to a target RNA, thereby preventing the utilization of that target RNA

_____ 7. Regulate gene expression in response to environmental conditions

_____ 8. Plasma membrane protein that senses changes in the environment.

_____ 9. DNA-binding protein that regulates gene expression after phosphorylation by a sensor kinase

_____ 10. Affect many genes and pathways simultaneously

_____ 11. Common regulatory protein controls operon network, but individual operons also controlled separately

_____ 12. Regulatory system where operons respond together to environmental stimulus

_____ 13. Proteins that enable RNA polymerase to recognize and bind to specific promoters

_____ 14. Intercellular communication important in gene regulation

_____ 15. In eukaryotes, these bind regulatory sites called enhancers

_____ 16. The smallest sRNAs in eucaryotes that act at the level of translation

_____ 17. Sensory RNAs that participate in a form of attenuation that does not involve the ribosome.

_____ 18. A collection of genes or operons that is controlled by a common regulatory protein

a. activators
b. antisense RNA
c. attenuator
d. corepressors
e. global regulatory systems
f. inducers
g. microRNAs (miRNAs)
h. modulon
i. operator
j. operon
k. quorum sensing
l. regulon
m. response-regulation protein
n. riboswitches
o. sensor kinase
p. sigma factors
q. signal transduction systems
r. stimulon

FILL IN THE BLANK

1. If *E. coli* is grown in a medium that contains both glucose and lactose, it uses _____ preferentially until this sugar is exhausted. Then after a short lag, growth resumes using _____ as a carbon source. This biphasic growth pattern is called _____ growth.

2. For negatively regulated operons containing genes for a catabolic pathway, the initial substrates of the pathways often act as _____, while for negatively regulated operons containing genes for anabolic pathways, the end products of the pathways usually act as _____.

3. The regulatory mechanism, _____, is observed for numerous amino acid biosynthetic pathways. In this mechanism, the behavior of ribosomes impacts the continuation of transcription. Between the operator and the structural genes of the operon is the _____ region. This region encodes a leader peptide that has two or more codons for the amino acid end product of the pathway, and it has an _____, a rho-independent transcription termination site that can form a terminator hairpin. If the ribosome successfully translates the leader peptide sequence of the mRNA, then the terminator hairpin forms and stops continuation of transcription of the operon. This can only happen if the amino acid is readily available and activated by the appropriate _____ _____. If the amino acid is not readily available, the ribosome will briefly stall during translation of the leader peptide sequence. If this occurs while the leader region is still being transcribed, it prevents formation of the terminator hairpin. Therefore, transcription continues, the mRNA is translated, the enzymes of the pathway are synthesized, and the needed amino acid is synthesized. Such a regulatory process is only possible when the processes of transcription and translation are _____ _____. Therefore, it is only seen in _____ organisms and not in _____ organisms.

4. Some operons are regulated by _____ proteins, which bind the operator and prevent transcription. This type of regulation of gene expression is called _____ control. Other operons are regulated by activator proteins (e.g., _____ activator protein, which regulates the lactose operon and other operons in *E. coli*), which promote transcription. This type of regulation is called _____ operon control.

5. Many biosynthetic enzymes are _____ enzymes whose levels are reduced in the presence of end products called _____. In this type of regulatory system, the newly synthesized repressor protein is inactive and referred to as the _____. It is activated by the _____.

6. When arabinose is absent the *ara* operon is _____ by AraC; however, when _____ is present, this interaction is prevented and AraC _____ expression.

7. In response to environmental conditions, a membrane-bound _____ _____ protein acts on a _____ protein by phosphorylating it. This _____ _____ system can regulate gene expression. Longer pathways that use transfer of phosphoryl groups to control gene expression are called _____ systems.

8. Intercellular communication important in gene expression, known as _____ _____, is mediated in procaryotes by the release of _____ _____ that act as _____, in some cases through two-component regulatory systems.

9. In eukaryotes, many _____ transcription factors are required, that includes stimulation by _____ that bind _____ and repressors that bind _____.

130

MULTIPLE CHOICE

For each of the questions below select the *one best* answer.

1. Which of the following is NOT a regulatory mechanism used to control the lactose operon in *E. coli*?
    a.   induction
    b.   catabolite repression
    c.   attenuation
    d.   All of the above are used to regulate the lactose operon.

2. Which of the following is NOT a regulatory mechanism used to control the tryptophan operon in *E. coli*?
    a.   repression
    b.   catabolite repression
    c.   attenuation
    d.   All of the above are used to regulate the tryptophan operon in *E. coli.*

3. Which of the following is least likely to mediate rapid responses to changes in environmental conditions?
    a.   metabolic channeling
    b.   adjustment of enzyme activity
    c.   regulation of gene expression
    d.   All of the above are equally likely to respond to rapid environmental changes.

4. Which of the following will most likely conserve the greatest amount of energy for the cell?
    a.   metabolic channeling
    b.   adjustment of enzyme activity
    c.   regulation of gene expression
    d.   All of the above conserve nearly equal amounts of energy.

5. Which of the following is the inducer for the lactose operon of *E. coli*?
    a.   catabolite activator protein (CAP)
    b.   3′, 5′ cyclic adenosine monophosphate (cAMP)
    c.   lactose (allolactose)
    d.   cyclic AMP receptor protein (CRP)

6. Which of the following is NOT a level at which gene expression can be controlled?
    a.   transcription
    b.   translation
    c.   posttranslation
    d.   All can be points of control.

7. Which of the following is a role for riboswitches?
    a.   stalling ribosomes during translation
    b.   inducing expression of ribosomal proteins
    c.   differential folding of mRNA leader sequences
    d.   transmitting signals from environmental sensor proteins

8. Riboswitches lead to alternative folding of mRNAs during transcription and lead to the formation of:
    a.   termination loops
    b.   antitermination loops
    c.   both (a) and (b)
    d.   neither (a) nor (b)

9. When alternate sigma factors are available to RNA polymerase?
    a.   gene expression changes
    b.   binding to enhancers is blocked
    c.   operator sequences are not used
    d.   rho factors are needed for expression

10. Which of the following is NOT an example of a global regulatory system?
    a.   stimulon
    b.   regulon
    c.   modulon
    d.   alteron

TRUE/FALSE

_____ 1. If a particular energy source is unavailable, the enzymes required for its utilization are needed and energy is expended to produce them.

_____ 2. In the presence of both glucose and lactose, the lactose repressor is not bound to the operator; however, the genes of the lactose operon are still not expressed: catabolite repression affects the binding of RNA polymerase to the promoter, but does not involve the operator.

_____ 3. Although DNA replication and cell division are separate processes, they are tightly coordinated. Therefore, if a drug or a gene mutation inhibits DNA synthesis, cell division is also blocked.

_____ 4. Even at high rates of cell division (doubling time less than 60 minutes), DNA replication for the next doubling is not initiated until the previous round of cell division has been completed.

_____ 5. β-galactosidase is a repressible enzyme.

_____ 6. One example of the regulation of gene expression by small RNA molecules (sRNAs) is the binding of heterogeneous nuclear RNA molecules (hnRNAs) to target mRNA, thus blocking their translation.

_____ 7. Riboswitches act at both the level of transcription and translation.

_____ 8. The corepressor for the *trp* operon is tryptophan.

_____ 9. In most cases, archaeal regulatory proteins function like those of eucaryotes.

_____ 10. One procaryotic cell cannot affect the gene expression of another cell.

_____ 11. Sporulation in *Bacillus subtilis* is controlled in part by a phosphorelay system.

_____ 12. The arabinose operon is both positively and negatively controlled at the level of transcription.

_____ 13. Sensor kinases are important intracellular proteins found in the nucleoid region.

CRITICAL THINKING

1. Look at the following two diagrams. Compare and contrast the regulatory mechanisms depicted in (A) with those in (B). Which is more likely to regulate the enzymes of a catabolic pathway and which is more likely to regulate an anabolic pathway? Explain.

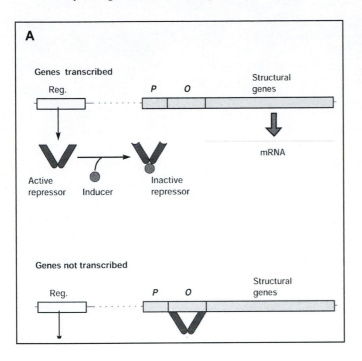

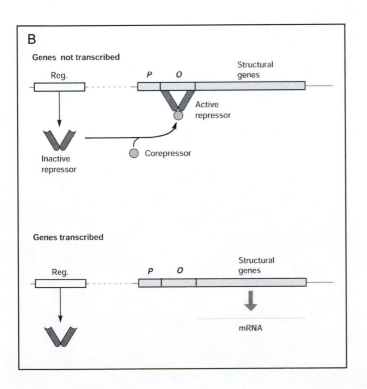

2. After inoculating a flask of minimal broth containing glucose and lactose with *E. coli* and following the growth kinetics, you obtain the following growth curve:

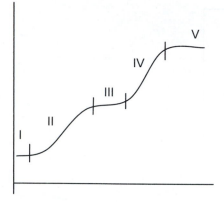

a. Describe each stage of the growth curve (I–V) in terms of growth rates, growth substrates, and general metabolic activity.

b. Diagram the lactose operon. What differences are there in the state of the operon during phase I, phase II, phase III, and phase IV of the given growth curve? Be sure to discuss the promoter and operator regions as well as the roles of the *lac* repressor and CAP in each of the first four phases.

ANSWER KEY

Terms and Definitions

1. f, 2. d, 3. i, 4. j, 5. c, 6. b, 7. q, 8. o, 9. m, 10. e, 11. h, 12. r, 13. p, 14. k, 15. a, 16. g, 17. n, 18. l

Fill in the Blank

1. glucose; lactose; diauxic 2. inducers; corepressors 3. attenuation; leader; attenuator; aminoacyl-tRNA synthetases; tightly coupled; procaryotic; eucaryotic 4. repressor; negative; catabolite; positive 5. repressible; corepressors; aporepressor; corepressor 6. repressed; arabinose; stimulates 7. sensor kinase; response-regulator; signal transduction; phosophorelay 8. quorum sensing; homoserine lactones; autoinducers 9. general; activators; enhancers; silencers

Multiple Choice

1. c, 2. b, 3. c, 4. c, 5. c, 6. d, 7. c, 8. c, 9. a, 10. d

True/False

1. F, 2. T, 3. T, 4. F, 5. F, 6. F, 7. T, 8. T, 9. F, 10. F, 11. T, 12. T, 13. F

13 Microbial Genetics: Mechanisms of Genetic Variation

CHAPTER OVERVIEW

This chapter begins with a discussion of mutation and genetic variation and includes molecular mechanisms of mutation and repair. A general discussion of bacterial recombination, plasmids, and transposable elements follows, with examination of the acquisition of genetic information by conjugation, transformation, and transduction. The way these recombination procedures are used to map the bacterial genome is explained. Finally, viral recombination and genome mapping are discussed.

CHAPTER OBJECTIVES

After reading this chapter you should be able to:
- discuss the nature and causes of mutations
- discuss the various genetic repair mechanisms and their limitations
- discuss the nature of procaryotic recombination
- distinguish horizontal gene transfer from vertical gene transfer
- compare and contrast conjugation, transformation, and transduction
- discuss how transposable elements can move genetic material between bacterial chromosomes and within a chromosome to cause changes in the genome and the phenotype of the organism
- discuss the use of conjugation, transformation, and transduction to map the bacterial genome
- discuss the methods used to map viral genes

CHAPTER OUTLINE
I. Mutations and Their Chemical Basis
    A. Mutation overview
        1. A mutation is a stable, heritable change in the genomic nucleotide sequence; this can be a single base change (point mutation), several base changes, or larger insertions, deletions, inversions, duplications, and translocations
        2. Mutations can arise in two ways:
            a. Spontaneous mutations arise occasionally in the absence of any added agent
            b. Induced mutations are the result of exposure to a mutagen (physical or chemical agent)
    B. Spontaneous mutations
        1. Arise occasionally in all cells without exposure to external agents; they are often the result of errors in replication, lesions to the DNA, insertion of DNA segments into genes
        2. Errors in replication can be due to tautomeric shifts, which cause base substitutions
            a. Transition mutation—substitution of one purine for another, or of one pyrimidine for another
            b. Transversion mutation—substitution of a purine for a pyrimidine or vice versa
        3. Lesions in the structure of DNA; the loss of a nitrogenous base creating an apurinic or apyrimidinic site can cause spontaneous mutations
        4. Insertion of DNA segments (e.g., insertion sequences and transposons) can cause spontaneous mutations that usually inactivate genes

5. Some spontaneous mutations appear to be directed or adaptive mutations, suggesting that either the process is not entirely random or that hypermutation rapidly produces many mutations through the activation of special mutator genes

C. Induced mutations
   1. Mutations can be induced by agents that damage DNA, alter its chemistry, or interfere with its functioning
   2. Base analogs are structurally similar to normal nitrogenous bases and can be incorporated into DNA during replication, but exhibit base-pairing properties different from the bases they replace
   3. Specific mispairing occurs when a mutagen is a DNA-modifying agents that changes a base's structure and thereby alters its pairing characteristics (e.g., alkylating agents)
   4. Intercalating agents, which become inserted between the stacked bases of the helix, distort the DNA and thus induce single nucleotide pair insertions or deletions
   5. Many mutagens (e.g., UV radiation, ionizing radiation, some carcinogens) can severely damage DNA so that it cannot act as a replication template; this could be lethal without repair mechanisms that restore the DNA

D. Effects of mutations
   1. Forward mutation—a conversion from the most prevalent gene form (wild type) to a mutant form
   2. Reversion mutation—a second mutation event that makes the mutant appear to be a wild type again
      a. Back mutation (true reversion)—conversion of the mutant nucleotide sequence back to the wild-type sequence
      b. Suppressor mutation—a reestablishment of the wild-type phenotype by a second mutation that overcomes the effect of the first mutation; can be in the same gene or a different gene, but does not restore the original sequence
   3. Point mutations affect only one base pair and are more common than large deletions or insertions
      a. Silent mutations are alterations of the base sequence that do not alter the amino acid sequence of the protein because of code degeneracy
      b. Missense mutations are alterations of the base sequence that result in the incorporation of a different amino acid in the protein; at the level of protein function, the effect may range from complete loss of activity to no change in activity
      c. Nonsense mutations are alterations that produce a translation termination codon; this results in premature termination of protein synthesis; location of the mutation within the protein will determine the extent of change in function
      d. Frameshift mutations are insertions or deletions of one or two base pairs that thereby alter the reading frame of the codons
      e. Conditional mutations are expressed only under certain environmental conditions
      f. Biochemical mutations result in changes in the metabolic capabilities of a cell; auxotrophs cannot grow on minimal media because they have lost a biosynthetic capability and require supplements; prototrophs are wild-type organisms that can grow on minimal media
      g. Resistance mutations result in acquired resistance to some pathogen, chemical, or antibiotic
   4. Mutations can also occur in regulatory sequences and in tRNA and rRNA genes; all can give observable phenotypes

II. Detection and Isolation of Mutants
   A. Mutant detection
      1. Visual observation of changes in colony characteristics
      2. Auxotrophic mutants can be detected by replica plating on media with and without the growth factor required; mutants are those growing with the factor but not without it

137

B. Mutant selection is achieved by finding the environmental condition under which the mutant will grow but the wild type will not (useful for isolating auxotrophic revertants, resistance mutants, and substrate utilization mutations)

C. Carcinogenicity testing
1. Many cancer-causing agents (carcinogens) are also mutagens, therefore tests for mutagenicity can be used as a screen for carcinogenic potential
2. The Ames test is a widely used mutagenicity test; it detects an increase in reversion of special strains of *Salmonella typhimurium* from histidine auxotrophy to prototrophy after exposure to a potential carcinogen

III. DNA Repair
A. Proofreading by DNA polymerases immediately repairs many replication errors
B. Excision repair
1. Corrects damage that causes distortions of DNA (e.g., thymine dimers, apurinic or apyrimidinic sites, damaged or unnatural DNA)
2. For nucleotide excision repair, the damaged area is excised, producing a single-stranded gap, and then the gap is filled in by DNA polymerase I, and DNA ligase joins the new fragment into the existing DNA strand
3. For base excision repair, DNA glycosylases remove the damaged base, and this signals AP nucleases to mark the damaged DNA, which is then excised and repaired by DNA polymerase I and ligase
C. Direct repair of thymine dimers and alkylated bases occurs through photoreactivation or the action of alkyl- or methyltransferases, respectively
D. Mismatch repair
1. The mismatch repair system corrects replication errors that result in mismatched base pairs; newly replicated DNA is detected by a lack of DNA methylation
2. The mismatch is detected by MutS and repaired through excision by MutH
E. Recombinational repair
1. Recombination with an undamaged molecule, if available, is used to restore DNA that has damage in both strands through the action of RecA protein; an undamaged molecule can be available in rapidly dividing cells where there is a copy of the chromosome that has not yet segregated into daughter cells
F. The SOS response
1. SOS repair is a type of recombination repair that depends on the RecA protein; it is used to repair excessive damage that halts replication; it is an error-prone process that results in many mutations
2. RecA derepresses the synthesis of a variety of DNA repair genes; very serious damage is treated by translesion DNA synthesis that is highly error prone

IV. Creating Genetic Variability
A. Recombination is a process by which one or more nucleic acid molecules are rearranged or combined to produce a new nucleotide sequence; mutant and wild-type alleles (alternate forms of a gene) can be exchanged
B. Recombination in eucaryotes occurs during meiosis and results from crossing-over between homologous chromosomes (chromosomes containing identical sequences of genes)
C. Horizontal gene transfer in procaryotes
1. Horizontal (or lateral) gene transfer moves genes from one mature, independent organism to another (compare this to vertical gene transfer—transmission of genes from parents to offspring)
2. Exogenote—donor DNA that enters the bacterium by one of several mechanisms
3. Endogenote—the genome of the recipient
   a. Merozygote—a recipient cell that is temporarily diploid for a portion of the genome during the gene transfer process
4. Types of horizontal exogenote transfer

            a.   Conjugation is direct transfer from donor bacterium to recipient while the two are temporarily in physical contact

            b.   Transformation is transfer of a naked DNA molecule

            c.   Transduction is transfer mediated by a bacteriophage

       5.   Intracellular fates of exogenote

            a.   Integration into the host chromosome

            b.   Independent functioning and replication of the exogenote without integration (a partial diploid clone develops)

            c.   Survival without replication (only the one cell is a partial diploid)

            d.   Degradation by host nucleases (host restriction)

            e.   Most linear DNA fragments are not stably maintained unless integrated into the endogenote

   D.   Recombination at the molecular level

       1.   General recombination usually involves a reciprocal exchange in which a pair of homologous sequences breaks and rejoins (double-stranded break model) in a crossover; nonreciprocal general recombination involves the incorporation of a single strand into the chromosome to form a stretch of heteroduplex DNA

       2.   Site-specific recombination is the nonhomologous insertion of DNA into a chromosome; often occurs during viral genome integration into the host, a process catalyzed by enzymes specific for the virus and its host

       3.   Transposition is a kind of recombination that occurs throughout the genome and does not depend on sequence homology

V.   Transposable Elements

   A.   Transposition is the movement of pieces of DNA around in the genome; transposons are segments of DNA that can move about chromosomes, "jumping genes"

   B.   Insertion sequences (IS elements) contain genes only for those enzymes required for transposition (e.g., transposase); they are bound on both ends by inverted terminal repeat sequences

   C.   Composite transposons carry other genes in addition to those needed for transposition (e.g., for antibiotic resistance, toxin production, etc.)

   D.   Transposition can occur by two mechanisms:

       1.   Simple transposition is a cut-and-paste process involving transposase-catalyzed excision of a transposon and insertion into a new target site

       2.   Replicative transposition is a mechanism during which a replicated copy of the transposon inserts at the target site on the DNA, while the original copy remains at the parental site

   E.   Effects of transposable elements

       1.   Insertional mutagenesis can cause deletion of genetic material at or near the target site, arrest of translation or transcription due to stop codons or termination sequences located on the inserted material, and activation of genes near the point of insertion due to promoters located on the inserted material

       2.   Fusion of plasmids and insertion of F plasmids into chromosomes

       3.   Generation of plasmids with resistance genes

   F.   Conjugative transposons can move between bacteria through the process of conjugation

VI.   Bacterial Plasmids

   A.   Plasmids are small, circular DNA molecules that replicate independently within host cells; episomes are plasmids that can exist with or without being integrated into the host chromosomes

   B.   Conjugative plasmids can transfer copies of themselves to new hosts during conjugation; F factor plays a key role in bacterial conjugation

VII. Bacterial Conjugation

   A.   The transfer of genetic information via direct cell-cell contact; this process is mediated by fertility factors (F plasmids)

   B. $F^+ \times F^-$ mating

       1.   In *E. coli* and other gram-negative bacteria, an F plasmid moves from the donor (F^+) to a recipient (F^-) while being replicated by the rolling-circle mechanism

        a.   The displaced strand is transferred via a sex pilus and then copied to produce double-stranded DNA; the donor retains the other parental DNA strand and its complement; thus the recipient becomes F^+ and the donor remains F^+

        b.   Chromosomal genes are not transferred

  C.   Hfr conjugation

     1.   F plasmid integration into the host chromosome results in an Hfr (high frequency of recombination) strain of bacteria

     2.   The mechanics of conjugation of Hfr strains are similar to those of F^+ strains

     3.   The initial break for rolling-circle replication is at the integrated plasmid's origin of transfer site

        a.   Part of the plasmid is transferred first

        b.   Chromosomal genes are transferred next

        c.   The rest of the plasmid is transferred last

     4.   Complete transfer of the chromosome takes approximately 100 minutes, but the conjugation bridge does not usually last that long; therefore, the entire F factor is not usually transferred, and the recipient remains F^-

  D.   F′ conjugation

     1.   When an integrated F plasmid leaves the chromosome incorrectly, it may take with it some chromosomal genes from one side of the integration site; this results in the formation of an abnormal plasmid called an F′ plasmid

     2.   The F′ cell (cell harboring an F′ plasmid) retains its genes, although some of them are in the chromosome and some are on the plasmid; in conjugation, an F′ cell behaves as an F^+ cell, mating only with F^- cells

     3.   The chromosomal genes included in the plasmid are transferred with the rest of the plasmid, but other chromosomal genes usually are not

     4.   The recipient becomes an F′ cell, and a partially diploid merozygote

  E.   Other examples of bacterial conjugation

     1.   Less is known about conjugative transfer in gram-positive bacteria

     2.   No sex pilus is formed; however, cells may directly adhere to each other using special plasmid-encoded proteins

VIII. DNA Transformation

  A.   Transformation—a naked DNA molecule from the environment is taken up by the cell and incorporated into its chromosome in some heritable form

  B.   A competent cell is one that is capable of taking up DNA and therefore acting as a recipient; only a limited number of species are naturally competent; the mechanics of the natural transformation process differ from species to species

  C.   Species that are not normally competent (such as *E. coli*) can be made competent by calcium chloride treatment and other methods that make the cells more permeable to DNA

IX. Transduction

  A.   Transduction is the transfer of bacterial genes by viruses (bacteriophages); it occurs as the result of the reproductive cycle of the virus

     1.   Lytic cycle—a viral reproductive cycle that ends in lysis of the host cell; viruses that use this cycle are called virulent bacteriophages

     2.   Lysogeny—a reproductive cycle that involves maintenance of the viral genome (prophage) within the host cell (usually integrated into the host cell's chromosome), without immediate lysis of the host; with each round of cell division, the prophage is replicated and inherited by daughter cells; bacteriophages reproducing by this mechanism are called temperate phages; certain stimuli (e.g., UV radiation) can trigger the switch from lysogeny to the lytic cycle

  B.   Generalized transduction

     1.   Transfer of *any* portion of the bacterial genome; occurs during the lytic cycle of virulent and temperate bacteriophages

     2.   The phage degrades the host chromosome into randomly sized fragments

3. During assembly, fragments of host DNA of the appropriate size can be mistakenly packaged into a phage head (generalized transducing particle)
4. When the next host is infected, the bacterial genes are injected and a merozygote is formed
   a. Preservation of the transferred genes requires their integration into the host chromosome
   b. Much of the transferred DNA does not integrate into the host chromosome, but is often able to survive and be expressed; the host is called an abortive transductant

C. Specialized transduction
   1. Transfer of only *specific* portions of the bacterial genome; carried out only by temperate phages that have integrated their DNA into the host chromosome at a specific site in the chromosome
      a. The integrated prophage is sometimes excised incorrectly and contains portions of the bacterial DNA that was adjacent to the phage's integration site on the chromosome
      b. The excised phage genome is defective because some of its own genes have been replaced by bacterial genes; therefore, the bacteriophage cannot reproduce
      c. When the next host is infected, the donor bacterial genes are injected, leading to the formation of a merozygote
   2. Low-frequency transduction lysates—lysates containing mostly normal phages and just a few specialized transducing phages
   3. High-frequency transduction lysates—lysates containing a relatively large number of specialized transducing phages; created by coinfecting a host cell with a helper phage (normal phage) and a transducing phage; the helper phage allows the transducing phage to replicate, thus increasing the number of transducing phages in the lysate

X. Mapping the Genome
   A. Hfr mapping involves the use of an interrupted mating experiment to locate the relative position of genes
      1. The conjugative bridge is broken and the Hfr × F⁻ mating is stopped at various times after initiation
      2. While the bridge is intact, chromosome transfer occurs at a constant rate
      3. The order and timing of gene transfer directly reflects the order of genes on the chromosome
      4. Interrupted mating is good for mapping genes that are 3 minutes or more apart; however, the instability of the conjugation bridge makes it nearly impossible to map genes that are very distant; several Hfr strains with different integration sites are used to generate overlapping maps, which can then be pieced together to form the entire genome map
   B. Transformation mapping—the frequency with which two genes simultaneously transform a recipient cell (cotransformation) indicates the distance between the genes; the higher the frequency of cotransformation, the closer the two genes are; overlapping maps can be pieced together to complete the genome map
   C. Generalized transduction maps—as with transformation mapping, the frequency of cotransduction indicates the distance of two genes from each other
   D. Specialized transduction maps provide distances from integration sites, which themselves must be mapped by conjugation mapping techniques

XI. Recombination and Genome Mapping in Viruses
   A. Recombination maps are generated from crossover frequency data obtained when cells are infected with two or more phage particles simultaneously
   B. Heteroduplex mapping—wild-type and mutant chromosomes are denatured and allowed to reanneal together; homologous regions pair normally, but mutant regions form bubbles that can be seen in electron micrographs; generates a physical map of the viral genome
   C. Restriction endonuclease mapping—locates deletions and other mutations by examining the electrophoretic mobility (size) of the fragments generated
   D. Sequence mapping—small phage genomes can be directly sequenced to map genes

TERMS AND DEFINITIONS

Place the letter of each term in the space next to the definition or description that best matches it.

_____ 1. Alterations in the base sequence of the genomic nucleic acid

_____ 2. Mutations that are only expressed under certain environmental conditions

_____ 3. Physical or chemical agents that can cause mutation

_____ 4. Mutations that result in purine-purine or pyrimidine-pyrimidine substitutions

_____ 5. Mutations that result in purine-pyrimidine or pyrimidine-purine substitutions

_____ 6. A mutational reversion assay that is used to determine if a compound is carcinogenic

_____ 7. Mutations that result in a change in the reading frame

_____ 8. The most prevalent gene form in a population

_____ 9. A mutation from the most prevalent gene form in the population

_____10. A mutation that restores the wild-type phenotype

_____11. A second mutation that overcomes the effect of the first mutation, but does not restore the wild-type sequence of nucleotides

_____12. A mutation that involves only one base pair

_____13. A mutation that does not alter the amino acid sequence of the resulting protein

_____14. A mutation that changes the amino acid sequence of the resulting protein by substitution of one amino acid for another

_____15. A mutation that causes premature termination of the synthesis of the protein product

_____16. A population of cells that are genetically identical

a. Ames test
b. back (reversion) mutation
c. clone
d. conditional mutations
e. conjugation
f. crossing-over
g. endogenote
h. exogenote
i. forward mutation
j. frameshift mutations
k. horizontal gene transfer
l. lysogeny
m. merozygote
n. missense mutation
o. mutagens
p. mutations
q. nonsense mutation
r. point mutation
s. prophage
t. recombination
u. replicon
v. sex pilus
w. silent mutation
x. suppressor mutation

y.   transduction
z.   transformation
aa. transition mutations
bb. transposable element
cc. transversion mutations
dd. wild type

_____17.   The process that occurs when genetic material from
           two organisms is combined, forming a genotype that
           differs from that of either parent
_____18.   The piece of donor DNA in a recombination event
_____19.   The recipient DNA in a recombination event
_____20.   A cell that is temporarily diploid for a portion of the
           genome during genetic transfer processes
_____21.   Process that leads to chromosome exchange during
           meiosis
_____22.   A piece of DNA that can move between chromosomes
           or within a single chromosome
_____23.   Transfer of genetic information via direct cell-cell
           contact
_____24.   Transfer of genetic information by uptake of a naked
           DNA molecule from the environment
_____25.   Transfer of genetic information via viruses
_____26.   The relationship between a phage and its host in which
           the phage genome exists in the host and is replicated
           without destroying the host cell
_____27.   The latent form of the virus genome that remains
           within the host without destroying it
_____28.   Transfer of genes between independent, mature
           organisms
_____29.   DNA molecule or sequence that has an origin of
           replication replicated as a single unit
_____30.   Bacterial structure that joins a donor and recipient
           together and may serve as the channel for DNA
           transfer during conjugation

FILL IN THE BLANK

1. Spontaneous mutations arise without exposure to external agents. They are often the result of replication errors but can also arise when DNA is damaged. For instance, it is possible for a nucleotide to lose its nitrogenous base producing either an _____ site or an _____ site.

2. Induced mutations arise upon exposure to external agents called _____ that directly damage DNA, alter its chemistry, or interfere with repair mechanisms. For instance, _____ _____ (e.g., 5-bromouracil) are similar to normal nitrogenous bases and can be incorporated into a polynucleotide chain during replication. Some chemicals (e.g., acridine orange) insert themselves between the stacked bases of a DNA double helix. These are called _____ _____. Alkylating agents (e.g., nitrosoguanidine) alter the structure of nitrogenous bases and change their base-pairing characteristics, resulting in _____ _____.

3. One repair mechanism, _____ _____, is able to repair damaged DNA for which there is no remaining template. The _____ protein is important in this type of repair. One example of this type of repair system is _____ repair, which is induced when DNA damage is so great that DNA synthesis is stopped. Although this repair mechanism may allow the bacteria to survive, it is error prone and produces_____.

4. Mismatched pairs that are not detected during replication by the _____ activity of DNA polymerase are usually subsequently corrected by _____ repair system. This involves excising nucleotides from one strand and replacing them; thus, this postreplication repair mechanism is also a type of _____ repair. In order for this repair system to work, it must be able to distinguish old DNA strands from new. This distinction is possible because old strands have methyl groups on the bases as the result of a process called _____ _____.

5. Mutations can alter phenotype in several different ways. _____ mutations change the cellular or colonial characteristics. _____ mutations, when expressed, result in the death of the organism. In diploid organisms, these are usually only recovered if they are recessive; in haploid organisms the mutation must be _____ if it is to be recovered.

6. Mutations that inactivate a metabolic pathway are called _____ mutations. A microorganism with this kind of mutation is often unable to grow on minimal medium, and for growth it requires an adequate supply of the pathway's end product. Such mutants are called _____, while microbial strains that can grow on minimal medium are called _____.

7. _____ mutations are those that cause a shift in the _____ frame of a gene. When the mutation occurs early in a gene, virtually the entire _____ region is altered resulting in the synthesis of a truncated or nonfunctional protein. A second _____ mutation shortly downstream from the first may restore the _____ frame and thereby minimize the phenotypic effect. This second mutation is a good example of a _____ mutation.

8. Thymine dimers can be split apart into separate thymines with the help of visible light in a photochemical reaction catalyzed by the enzyme photolyase. This is called _____, and since it does not remove and replace nucleotides, it is relatively free of _____.

9. Mutations that appear to have been chosen by the organism so that it is better adapted to its environment are called _____ or _____ mutations. It is hypothesized that they may be the result of _____ followed by selection of favorable mutants.

10. The most common form of recombination is _____ recombination, which usually involves a reciprocal exchange between a pair of homologous DNA sequences. Integration of viral genomes into bacterial chromosomes can occur by another type of recombination known as _____ recombination, in which the viral genetic material is not homologous with the host DNA. In _____ recombination, recombination accompanies replication of genetic material and does not depend on sequence homology.

11. Bacterial recombination normally takes place when a piece of donor DNA, the _____, enters the cell and becomes a stable part of the recipient's genome, the _____. During replacement of host genetic material, the recipient becomes a diploid for a portion of the genome and is referred to as a _____.

12. There are two types of transposable elements. The simplest are _____ _____, which contain only the genes needed for transposition bounded at both ends by inverted repeats. One of these genes codes for

the enzyme _____. The second type of transposable element is a _____ _____, which contains genes other than those required for transposition. Some transposons bear transfer genes and can move between bacteria through the process of conjugation. These are called _____ _____.

13. Proteins produced by bacteria that destroy other bacteria are called _____.

14. The F (fertility) factor carries genes for synthesis of a ____ _____ and for plasmid transfer. During mating of donor and recipient strains, a process called _____, the F factor replicates by the rolling-circle mechanism and one copy moves to the recipient.

15. Transfer of genetic information by uptake of a naked DNA molecule from the environment is called _____. In order to take up a naked DNA molecule, a cell must be _____, which may only occur at certain stages in the life cycle of the organism.

16. The transfer of bacterial genes by viruses is called_____. When any part of the bacterial chromosome is transferred, it is called unrestricted, or _____ _____, and the phages that transfer the DNA are called _____ _____ phages. Some temperate phages can incorrectly excise from the host chromosome when switching from lysogeny to the lytic cycle, and these may carry genes that were adjacent to the integration site. Since the genes that may be carried are restricted to those located near specific integration sites, this is called restricted, or _____ ,_____.

17. Conjugation involving _____ strains is frequently used to map the relative locations of bacterial genes. The technique involves disruption of the conjugation bridge in what is called an _____ _____ experiment.

18. During bacterial transformation, _____ _____ is formed. This is a short stretch of DNA for which one strand is from the donor and the other is from the recipient.

19. Some phage-infected bacteria only produce phages under certain environmental conditions. These bacteria are said to be _____ or _____. The phages infecting these bacteria are called _____ phages, and the latent form of the phage genome that remains in the host without destroying it is called the _____.

20. The lysate produced following induction of lysogenized bacteria is called a ____ _____ lysate, because it contains only a few transducing phages. Each transducing phage is defective and can only integrate into a new host genome if a normal phage, called the _____ phage is in the same cell. The lysate produced following induction of bacteria lysogenized by both a defective phage and a normal phage is called a _____ _____lysate, because it contains a roughly equal number of normal phages and transducing phages.

MULTIPLE CHOICE

1. Which of the following repair mechanisms corrects damage that causes distortions in the DNA double helix (e.g., thymine dimers, apurinic sites, apyrimidinic sites) by removing and replacing a short stretch of nucleotides in the damaged strand?
    a.   photoreactivation
    b.   mismatch repair
    c.   excision repair
    d.   recombination repair

2. Resistance mutations can confer resistance to which of the following?
    a.   pathogens
    b.   chemicals
    c.   antibiotics
    d.   All of the above are correct.

3. A particular mutation results in the substitution of cytosine for thymine in one strand of the DNA. Upon subsequent DNA replication, one of the daughter cells receives a GC pair in this position instead of an AT pair. What is this type of mutation called?
    a.   transversion
    b.   transition
    c.   frameshift
    d.   insertion

4. Which of the following can cause transition and transversion mutations?
    a.   incorporation of a base analog that exhibits different base-pairing properties from those of the base it replaces
    b.   chemical modification of an existing base in the DNA so that during the next round of replication it will base pair differently from the unmodified base
    c.   Both of the above are correct.
    d.   None of the above is correct.

5. Back mutations that restore the wild-type phenotype can occur by which of the following mechanisms?
    a.   true reversion back to the wild-type nucleotide sequence
    b.   mutation that results in a different nucleotide sequence from that of the wild type, but that restores the amino acid sequence in the protein to the wild-type sequence
    c.   a second mutation that overcomes the effect of the first mutation; the first

mutation is not changed, but the function of the protein is restored
    d.   All of the above can restore the wild-type phenotype.

6. What are transposable elements that contain genes other than those required for transposition called?
    a.   insertion sequences
    b.   plasmids
    c.   composite transposons
    d.   None of the above is correct.

7. In which way do transposable elements differ from temperate bacteriophages or from plasmids?
    a.   Transposable elements lack a viral life cycle.
    b.   Transposable elements are unable to reproduce autonomously.
    c.   Transposable elements are unable to exist apart from the chromosome.
    d.   All of the above are ways that transposable elements differ from bacteriophages or plasmids.

8. What impact can transposons have on the host cell?
    a.   They can cause mutations.
    b.   They can block transcription or translation.
    c.   They can turn genes on or off.
    d.   All of the above are possible impacts of transposable elements.

9. Which of the following is NOT true about Hfr × F⁻ matings?
    a. The recipient can become F⁺ (or Hfr) if the mating lasts long enough for the entire bacterial chromosome to be transferred.
    b. The recipient usually remains F⁻ because the connection usually breaks before the entire bacterial chromosome can be transferred.
    c. The recipient may become F′ if more than half of the plasmid is transferred.
    d. All of the above are true about Hfr × F⁻ matings.

10. When an F factor leaves an Hfr chromosome, it occasionally picks up some bacterial genes. What is the resulting plasmid called?
    a. F⁺ plasmid
    b. F⁻ plasmid
    c. Hfr plasmid
    d. F′ plasmid

11. Which of the following has NOT been used to map chromosomal locations of bacterial genes?
    a. Hfr × F⁻ conjugation
    b. F′ × F⁻ conjugation
    c. F⁺ × F⁻ conjugation
    d. All of the above have been used to map bacterial genes.

12. Which of the following represents the best description of host restriction?
    a. the inability to take up an exogenote during transformation
    b. the inability to integrate an exogenote into the host chromosome
    c. the degradation of an exogenote by host nucleases
    d. the inability to express the genes located on an exogenote

13. Which of the following is NOT a possible fate for an exogenote?
    a. integration into the host chromosome
    b. expression of the genes and replication of the exogenote without integration into the host chromosome
    c. survival of the exogenote without integration or replication
    d. All of the above are possible fates for an exogenote.

TRUE/FALSE

_____ 1. Missense mutations may play an important role in providing new variability to drive evolution because they are often not lethal and, therefore, they remain in the gene pool.

_____ 2. Nonsense mutations that cause premature termination of translation always severely affect the phenotypic expression of the gene by resulting in the production of a nonfunctional gene product.

_____ 3. Point mutations are more common than large deletions or insertions.

_____ 4. Bacteria that are partial diploids, containing nonintegrated, transduced DNA, are called abortive transductants.

_____ 5. A plasmid that can exist independent of the host chromosome but that cannot be integrated into the host chromosome is called an episome.

_____ 6. The F factor is a conjugative plasmid that is particularly efficient at initiating conjugation with appropriate recipient cells.

_____ 7. Nonconjugative plasmids can move between bacteria during conjugation if conjugation is initiated by another plasmid that is conjugative.

_____ 8. Multiple-drug-resistant plasmids are usually produced when a single plasmid accumulates several transposons, each carrying one or more antibiotic resistance genes.

_____ 9. Transposable elements have only been found in procaryotes and do not appear to play a major role in eucaryotic genetics.

_____ 10. In an $F^+ \times F^-$ mating, the recipient becomes F^+ after mating has been completed.

_____ 11. In an $F^+ \times F^-$ mating, the donor becomes F^- after mating has been completed.

_____ 12. In an $F^+ \times F^-$ mating, chromosomal genes are frequently transferred.

_____ 13. All R factor plasmids are nonconjugative.

_____ 14. Host restriction refers to the ability of some organisms to degrade exogenotes that enter the cell.

_____ 15. During transposition, the original transposon is replicated and remains at the parental site in the chromosome, while the copy moves to a new site.

_____ 16. Even if two viruses simultaneously enter a host cell, no recombination can occur between the two viral genomes.

_____ 17. Viruses can be mapped by recombination and heteroduplex mapping techniques.

_____ 18. Bacterial recombination is a two-way process in which DNA is exchanged between the two cells involved.

CRITICAL THINKING

1. Hfr × F⁻ matings have been used to map the bacterial genome. Demonstrate with diagrams how this is done. Why is more than one Hfr strain needed in order to generate a complete map of the bacterial chromosome?

2. A strain of bacteria is protrophic. How would you isolate from this strain one that requires the amino acid leucine (i.e., it is a leucine auxotroph)?

3. You have a bacterial strain that is a tryptophan auxotroph and sensitive to the antibiotic streptomycin. You expose this strain to a mutagen. How would you isolate mutants that no longer require tryptophan (i.e., strains that have reverted to prototrophy)? How would you isolate mutants that are resistant to streptomycin? How would you isolate mutants that no longer require leucine *and* that are resistant to streptomycin?

ANSWER KEY

Terms and Definitions

1. p, 2. d, 3. o, 4. aa, 5. cc, 6. a, 7. j, 8. dd, 9. i, 10. b, 11. x, 12. r, 13. w, 14. n, 15. q, 16. c, 17. t, 18. h, 19. g, 20. m, 21. f, 22. bb, 23. e, 24. z, 25. y, 26. l, 27. s, 28. k, 29. u, 30. v

Fill in the Blank

1. apurinic; apyrimidinic 2. mutagens; base analogs; intercalating agents; specific mispairing 3. recombination repair; recA; SOS; mutations 4. proofreading; mismatch; excision; DNA methylation 5. Morphological; Lethal; conditional 6. biochemical; auxotrophs; prototrophs 7. Frameshift; reading; coding; frameshift; reading; suppressor 8. photoreactivation; error 9. directed; adaptive; hypermutation 10. general; site-specific; replicative 11. exogenote; endogesnote; merozygote 12. insertion sequences; transposase; composite transposon; conjugative transposons 13. bacteriocins 14. sex pilus; conjugation 15. transformation; competent 16. transduction; generalized transduction; generalized transducing; specialized transduction 17. Hfr; interrupted mating 18. heteroduplex DNA 19. lysogens; lysogenic; temperate; prophage 20. low-frequency transduction; helper; high-frequency transduction

Multiple Choice

1. c, 2. d, 3. b, 4. c, 5. d, 6. c, 7. d, 8. d, 9. c, 10. d, 11. c, 12. c, 13. d

True/False

1. T, 2. F, 3. T, 4. T, 5. F, 6. T, 7. T, 8. T, 9. F, 10. T, 11. F, 12. F, 13. F, 14. T, 15. T, 16. F, 17. T, 18. F

14   Recombinant DNA Technology

CHAPTER OVERVIEW

This chapter focuses on practical applications of the microbial genetic principles discussed in previous chapters. Although we have been altering the genetic makeup of organisms for centuries and nature has been doing it even longer, only recently have we been able to manipulate DNA directly using genetic engineering or recombinant DNA technology. The potential benefits of these techniques are great and affect such diverse areas as medicine, agriculture, and industry. However, the use of these techniques is not without risks, and these risks must be considered in any discussion of this technology.

CHAPTER OBJECTIVES

After reading this chapter you should be able to:

• discuss the use of recombinant DNA technology to genetically engineer various organisms
• describe PCR, real-time PCR, and RT-PCR and discuss their usefulness to genetic engineering
• discuss the key role played by restriction endonucleases and DNA ligase in genetic engineering
• discuss how plasmids, phages, cosmids, and artificial chromosomes are used as vectors for insertion and expression of foreign genes in an organism
• discuss the use of both procaryotes and eucaryotes as target organisms for foreign gene insertion
• discuss the contributions already made by the use of recombinant DNA technology
• discuss the risks and the ethical problems associated with the use of recombinant DNA technology

CHAPTER OUTLINE

I.   Introduction
    A.   Genetic engineering is the deliberate modification of an organism's genetic information by directly changing its nucleic acid
    B.   Recombinant DNA technology is the collection of methods used to accomplish genetic engineering
    C.   The generation of a large number of genetically identical DNA molecules is called cloning
    D.   Biotechnology is defined in this text as those processes in which living organisms are manipulated, particularly at the molecular genetic level, to form useful products
II.   Historical Perspectives
    A.   Arber and Smith (late 1960s) discovered restriction endonucleases, which cleave DNA at specific sequences; Boyer (1969) first isolated the restriction endonuclease *Eco*RI
    B.   Baltimore and Temin (1970) independently discovered reverse transcriptase; this enzyme can be used to construct a DNA copy, called complementary DNA (cDNA), of any RNA molecule
    C.   Jackson, Symons, and Berg (1972) generated the first recombinant DNA molecules by using DNA ligase to join DNA fragments together; Cohen and Boyer (1973) produced the first recombinant plasmid (vector), which was introduced into and replicated within a bacterial host

D. Southern (1975) developed a blotting procedure for detecting specific DNA fragments, using radioactive DNA hybridization probes; this is useful in isolating particular genes of interest; nonradioactive, enzyme-linked, or chemiluminescent probes can now replace the earlier radioactive probes; they are faster and safer

E. By the late 1970s, procedures for rapidly sequencing DNA molecules, synthesizing oligonucleotides, and expression of eucaryotic genes in bacteria had been developed

III. Synthetic DNA
   A. Oligonucleotides are short pieces of DNA or RNA
   B. DNA oligonucleotides can be made by adding one nucleotide at a time to a growing chain
   C. Oligonucleotides can be used in site-directed mutagenesis; in this process, an oligonucleotide containing the desired sequence change is used as a primer for DNA polymerase, which then replicates the remainder of the target gene and produces a new gene copy with the desired mutation; this can then be introduced into a new host

IV. The Polymerase Chain Reaction (PCR)
   A. PCR is used to synthesize large quantities of a specific DNA fragment without cloning it
   B. Synthetic DNA molecules with sequences identical to those flanking the target sequence are used as primers for DNA synthesis; replication is carried out in successive cycles using a heat-stable DNA polymerase
   C. Since its initial discovery, PCR has been automated and improved; real-time PCR can be used to quantitate the amount of target genes in the sample by monitoring the kinetics of amplification using fluorescent signals; mRNA can be amplified and quantified by creating cDNA prior to PCR using reverse transcriptase (RT-PCR)
   D. PCR has proven valuable in molecular biology, medicine (e.g., PCR-based diagnostic tests), and in biotechnology (e.g., use of DNA fingerprinting in forensic science; production of insulin)

V. Gel Electrophoresis
   A. Agarose or polyacrylamide gels are used to separate DNA fragments based on size
   B. DNA fragments are pulled through the gel by an electric current; DNA fragments of similar size form bands within the gel

VI. Cloning vectors and creating recombinant DNA
   A. Recombinant DNA technologies require propagation of specific DNA fragments by cloning into DNA vectors that will replicate in a host organism; the four major types of vectors are: plasmids, phages, cosmids, and artificial chromosomes
   B. Plasmids
      1. Replicate autonomously and are easy to purify; introduced by conjugation or transformation
      2. The origin of replication (*ori*) allows the plasmid to replicate in host cells and determines how many copies of the plasmid that a cell will contain; some plasmids called shuttle vectors have two origins of replication specific for different hosts
      3. Plasmids used for biotechnology typically have a selectable marker such as an antibiotic-resistance gene so that only cells containing the plasmid can grow under certain conditions (e.g., presence of the antibiotic)
      4. The multicloning site or polylinker is a region of the plasmid that has several unique restriction sites; this allows the circular plasmid to opened up to insert DNA fragments for cloning
   C. Phage vectors are phage genomes engineered to include restriction sites useful for cloning; once DNA is inserted into the vector the phage can be used to infect host cells
   D. Cosmids were created to clone larger fragments of DNA, and contain selectable markers, polylinkers, and *cos* sites from λ phage that allow for viral packaging; once the phage is introduced into a host, it replicates as a plasmid
   E. Artificial chromosomes were created to clone extraordinarily large pieces of DNA; bacterial artificial chromosomes (BACs) and yeast artificial chromosomes (YACs) have been engineered to include the sequences needed to act like natural chromosomes when inserted into host organisms

VII. Construction of Genomic Libraries
   A. Genomic libraries are valuable when cloning a gene that has an unknown sequence; all of the DNA sequences of an entire genome should be represented in the library
   B. Making a genomic library:

1. The DNA of an organism is fragmented by endonuclease cleavage and all resulting fragments are cloned into a vector
2. Plasmid or cosmid vectors are used to insert the clones into host cells; each cell receives only one vector and hence only one cloned fragment of the genome; the population of host cells taken together includes every fragment of the genome, with only one per cell
3. The clone containing the desired fragment can be identified by using a nucleic acid hybridization probe; when no sequence information is available, the desired clone is identified by expression in the host, often reversing auxotrophy or another deficiency in a process called phenotypic rescue
4. Phage vectors are used to infect a lawn of host bacteria on an agar plate, forming viral plaques (each with a different genomic DNA insert) that can be screened using a nucleic acid hybridization probe
5. Once identified, the vector is extracted, and the desired fragment is purified

C. In eucaryotes, it is best to create a cDNA library (no introns) rather than a genomic library

VIII. Inserting Recombinant DNA into Host Cells
A. Transformation and electroporation are popular means to insert recombinant DNA into host microbes; the hosts typically have been engineered to lack RecA and restriction enzymes
B. Electroporation is a procedure in which target cells are mixed with DNA and are then exposed briefly to high voltage; this works with bacteria, mammalian cells, and plant cell protoplasts
C. Genes of interest can be inserted directly into animal cells by microinjection; if the genes are stably incorporated into fertilized eggs, the resulting organism is called a transgenic animal
D. The gene gun is used to shoot DNA-coated microprojectiles into plant and animal cells
E. Special plasmids or viruses can also be used to insert desired genes into eucaryotic cells

IX. Expressing Foreign Genes in Host Cells
A. To express a foreign (heterologous) gene in a host cell, the gene must:
1. Have a promoter that is recognized by the host RNA polymerase
2. Have leader sequences that allow for ribosome binding
B. Expression vectors are designed to provide the above features; in addition they have useful restriction endonuclease sites and regulatory sequences that can be used to control expression of the foreign gene

X. Applications of Genetic Engineering
A. Medical applications
1. Production of medically important proteins, including somatostatin, human growth hormone, human insulin, interferon, interleukin-2, blood-clotting factor VIII, monoclonal antibodies (produced in transgenic plants and mice), and synthetic vaccines
2. Other uses already developed or being investigated include diagnostic probes for certain infectious diseases and genetic disorders
3. Gene therapy aims to replace defective alleles with functioning copies through infection with engineered viruses; these could be used to correct genetic disorders or cure difficult diseases
4. Biotechnology also can be used to alter cells harvested from patients or for stem cell therapies
B. Agricultural applications
1. Introduction of new desirable traits (e.g., increased growth rate) into farm animals and crop plants
2. Currently, many crops grown by farmers in the U.S. are genetically modified (e.g., corn, soybeans, cotton, canola, potato, squash, and tomato), often to resist the action of pesticides used to control weeds or to enhance resistance to pests

XI. Social Impact of Recombinant DNA Technology
A. Benefits are inherent in applications, but many risks and philosophical questions of the technology must be considered
B. Substantial controversy surrounds the use of stem cell therapies, particularly embryonic stem cells
C. Ecological concerns include possible ecosystem disruption caused by environmental release of recombinant organisms (genetic pollution); genetically engineered plants may spread their recombinant genes to create "super-weeds"

D. Recombinant DNA technologies also can be used to generate weapons for biological warfare or terrorism

TERMS AND DEFINITIONS

Place the letter of each term in the space next to the definition or description that best matches it.

_____ 1. Enzymes that recognize and cleave DNA at specific base pair sequences

_____ 2. A DNA copy of an mRNA that is produced by reverse transcriptase

_____ 3. A carrier of foreign DNA into the cloning host

_____ 4. A piece of detectably labeled nucleic acid that hybridizes with complementary DNA fragments and is used to locate them

_____ 5. The phenomenon of movement of charged molecules in an electrical field; it is used to separate nucleic acid fragments (and/or proteins)

_____ 6. A vector that has sequences necessary for packaging into bacteriophage lambda capsids

_____ 7. A vector that has all of the necessary transcription and translation start and stop signals, and that has nearby useful restriction endonuclease sites to enable the insertion of foreign DNA fragments in proper orientation

_____ 8. The process in which a high-voltage electric current induces target cells to take up DNA

_____ 9. The process in which recombinant phage DNA is taken up directly by the target cell without using complete phage particles

_____ 10. A vector with all of the features necessary for chromosomal replication in yeast; it can carry very large pieces of foreign DNA into a host organism

_____ 11. A short piece of DNA, often synthesized for use as a probe or primer

_____ 12. A technique that can be used to amplify and quantify mRNA

_____ 13. Target genes in a sample can be quantified by monitoring the kinetics of the reaction

_____ 14. A type of vector that can be used to clone large fragments of DNA

a. bacterial artificial chromosome
b. complementary DNA (cDNA)
c. cosmid
d. electrophoresis
e. electroporation
f. expression vector
g. oligonucleotide
h. probe
i. real-time PCR
j. restriction enzymes (endonucleases)
k. RT-PCR
l. transfection
m. vector
n. yeast artificial chromosomes

FILL IN THE BLANK

1. Enzymes called _____ _____ cleave DNA at specific sequences.
2. Successful isolation of recombinant clones is dependent on the availability of suitable _____, which can be obtained in a variety of ways. Frequently they are constructed from_____ clones, which are produced from isolated mRNA molecules by the action of the enzyme reverse transcriptase. Once a _____ has been created, it is labeled, usually with a radioactive label such as ^{32}P. Radioactive labels are easily detected by _____, in which energy released by the radioactive isotope causes formation of dark-silver grains on a sheet of photographic film.

3. A common technique called _____ can be used separate DNA fragments based on _____ by screening the fragments through an _____ _____. The DNA fragments will form _____ that can be used to determine the _____ of the fragments.

4. A collection of techniques called _____ _____ _____ can be used to deliberately modify an organism by directly changing its genome (a process called _____ _____.) The development of these techniques resulted from the discovery of several enzymes, including _____ enzymes, DNA ligase, and reverse transcriptase.

5. The problem of recombinant gene expression in host cells is overcome with the help of special cloning vectors called _____ vectors that contain the necessary _____ and _____ start signals.

6. Genetic engineering has contributed and will continue to contribute to the fields of _____, _____, and _____, in addition to its contributions to basic research.

7. When making a genomic library, DNA is fragmented using _____ _____ and then the fragments are _____ into a _____. The library can be screened by using a hybridization _____ or by _____ in a host.

8. A procedure in which target cells are mixed with DNA and briefly exposed to high voltage in order to introduce the DNA into the target cell is called _____. The DNA can be inserted directly into animal cells by _____.

9. In many instances, a gene is cloned by finding it in a set of cloned fragments representing the entire genome of an organism. Such a set of DNA fragments is called a _____ _____.

10. Short pieces of either DNA or RNA are called _____. They are useful as _____ for identifying genes of interest, in creating mutants with known defects by _____ _____, and as primers for amplifying DNA by the _____ _____ _____ .

MULTIPLE CHOICE

For each of the questions below select the *one best* answer.

1. Which of the following enzymes is used to produce complementary DNA (cDNA)?
   a. restriction endonucleases
   b. DNA polymerase
   c. reverse transcriptase
   d. DNA ligase

2. Which of the following is NOT normally used as a cloning vector?
   a. transposon
   b. plasmid
   c. cosmid
   d. bacteriophage

3. Which of the following is NOT effective for inserting foreign DNA into a plasmid?
   a. Cut plasmid and foreign DNA with the same enzyme to produce the same sticky ends, which can then be used to hold the fragment of foreign DNA in place for insertion.
   b. Cut plasmid and foreign DNA with an enzyme to produce blunt ends, and then add complementary tails by using terminal transferase to produce sticky ends.
   c. Cut plasmid and foreign DNA with an enzyme that produces blunt ends, and then use T4 DNA ligase to do a blunt-end ligation.
   d. All of the above are effective for inserting foreign DNA into a plasmid.

4. Which of the following is currently produced by recombinant DNA technology for use in humans?
   a. somatostatin
   b. human interferon
   c. human insulin
   d. All of the above are being produced by recombinant DNA technology for use in humans.

5. Which of the following sequences of steps cannot be used to clone a desired DNA fragment?
   a. Cleave the DNA; isolate the fragment; clone the isolated fragment.
   b. Cleave the DNA; clone all the resulting fragments; isolate a clone containing the desired fragment.
   c. Synthesize the desired fragment; clone the synthesized fragment.
   d. All of the above can be used to clone a desired DNA fragment.

6. Which of the following processes can be used to create mutants with known sequence alterations?
   a. polymerase chain reaction
   b. site-directed mutagenesis
   c. genomic library mutagenesis
   d. none of the above

7. Which of the following cloning vectors carries the least amount of foreign DNA?
   a. artificial chromosome (e.g., YAC or BAC)
   b. bacteriophage
   c. cosmid
   d. plasmid

8. Which of the following cloning vectors can carry the most foreign DNA?
   a. artificial chromosome (e.g., YAC or BAC)
   b. bacteriophage
   c. cosmid
   d. plasmid

9. Which of the following is a way to detect DNA fragments of interest on a Southern blot?
   a. enzyme-linked probes
   b. chemiluminescent probes
   c. radioactive probes
   d. any of the above

10. Which of the following is not a characteristic of plasmids used for biotechnology?
    a. replicate with host chromosome
    b. small circular DNA
    c. contain a multiclonal site
    d. have a selectable marker

TRUE/FALSE

_____ 1. The Southern blotting technique was named after the person who developed the procedure, E. M. Southern.

_____ 2. In electrophoresis, DNA fragments separate according to size, with the smallest fragments migrating the farthest.

_____ 3. Regardless of the specific technique used in recombinant DNA technology, one of the keys to successful cloning is choosing the right vector.

_____ 4. Cosmids are so named because they can be used to express foreign genes in a variety of different cloning hosts.

_____ 5. It is not necessary to remove introns from eucaryotic genes before cloning them in a procaryotic organism, because when the eucaryotic RNA transcript is produced in the procaryote, the introns are removed by the same posttranscriptional processing inherent in the eucaryotic cell of origin.

_____ 6. The gene gun is so named because it shoots DNA-coated microprojectiles directly into the target cells.

_____ 7. Although the term "biotechnology" can be used in many ways, this text uses the term to refer to the manipulation of organisms, particularly at the molecular genetic level, to form useful products.

_____ 8. The polymerase chain reaction is used to detect small DNA fragments in a Southern blot.

_____ 9. Typical PCR reactions are useful in quantifying genes.

_____ 10. When genomic libraries are made with phage vectors, viral plaques on bacterial lawns are probed.

CRITICAL THINKING

1. Expression vectors have been extensively modified to allow for efficient expression of recombinant genes. List as many of these modifications as possible and explain the advantages of each.

2. Crop plants have been genetically engineered to resist common pesticides. However, there has been some resistance to the use of these organisms in the field. What are the major objections? How would you argue against these objections? If you agree with the objections, propose modifications that could be made to overcome them.

3. Gene therapy aims to replace defective alleles with functioning genes to repair genetic diseases. What would be a suitable method for introducing these genes into humans? What are some of the problems that may be encountered with gene therapy? Are there any social consequences to this kind of treatment in the long term? Are there ethical considerations in using gene therapy technology inappropriately?

ANSWER KEY

Terms and Definitions

1. j, 2. b, 3. m, 4. h, 5. d, 6. c, 7. f, 8. e, 9. l, 10. n, 11. g, 12. k, 13. i, 14. a

Fill in the Blank

1. restriction endonucleases 2. probes; cDNA; probe; autoradiography 3. electrophoresis; size; agarose gel; bands; sizes 4. recombinant DNA technology; genetic engineering; restriction 5. expression; transcription; translation 6. medicine; industry; agriculture 7. restriction endonucleases; cloned; vector; probe; expression 8. electroporation; microinjection 9. genomic library 10. oligonucleotides; probes; site-directed mutagenesis; polymerase chain reaction

Multiple Choice

1. c, 2. a, 3. d, 4. d, 5. d, 6. b, 7. d, 8. a, 9. d, 10. a

True/False

1. T, 2. T, 3. T, 4. F, 5. F, 6. T, 7. T, 8. F, 9. F, 10. T

15   Microbial Genomics

CHAPTER OVERVIEW

This chapter introduces genomics, a revolutionary new discipline in the biological sciences. Techniques important to the study of genomes are discussed. Bioinformatics, functional genomics, and comparative genomics are detailed. Proteomics theory and techniques are discussed. The chapter then gives numerous examples of the types of patterns already being discerned in the analysis of the microbial genomes thus far sequenced. Finally, metagenomic analysis of environmental communities is introduced.

CHAPTER OBJECTIVES

After reading this chapter you should be able to:

- define genomics and bioinformatics
- compare and contrast structural genomics, functional genomics, and comparative genomics
- describe methods of sequencing DNA and the whole-genome shotgun method for sequencing a genome
- describe the types of analyses done for functional genomics and proteomics
- discuss some of the insights gained thus far by the analysis of microbial genomes
- discuss metagenomics

CHAPTER OUTLINE

I.   Introduction
      A.   Genomics is the study of the molecular organization of genomes, their information content, and the gene products they encode
           1.   Structural genomics—study of the physical nature of genomes (i.e., determine and analyze the DNA sequence of the genome)
           2.   Functional genomics—study of the way the genome functions
           3.   Comparative genomics—compares genomes from different organisms; helps discern patterns in function and regulation and provides information about microbial evolution
      B.   Genomics will enable scientists to get a holistic view of microbial genetics, gene expression patterns, microbial communities, and evolutionary relationships
II.   Determining DNA Sequences
      A.   Sanger method
           1.   Uses dideoxynucleoside triphosphates (ddNTPs) in DNA synthesis; these lack a 3′-hydroxyl and terminate DNA synthesis
           2.   Single strands of DNA are mixed with a primer, DNA polymerase I, four deoxynucleoside triphosphates (one is labeled), and a small amount of one of the ddNTPs; DNA synthesis begins with primer but terminates each time a ddNTP is added to the chain
           3.   Four reactions are run, each with a different ddNTP; these reactions generate DNA fragments of different length because the site at which the ddNTP is inserted is random
           4.   Newly synthesized DNA fragments are separated electrophoretically on a polyacrylamide gel or with capillary electrophoresis; the gel can autoradiographed if radioactive ddNTPs were used or monitored with a laser if fluorescent ddNTPs were used; the sequence is then read from the autoradiogram or chromatographic trace

B.   Automated systems now exist, which make rapid sequencing possible

III. Whole-Genome Shotgun Sequencing
- A.   Sequencing a genome by the whole-genome shotgun approach is a multi-step process
  1. Library construction—chromosomes are broken into gene-sized fragments, inserted into plasmids, and transformed into special *E. coli* strains
  2. Random sequencing—the cloned fragments are sequenced
  3. Fragment alignment and gap closure—DNA fragments are clustered and assembled into longer stretches of sequence by comparing nucleotide sequence overlaps between fragments; this produces larger contiguous sequences called contigs; the contigs are aligned in the proper order to form the completed genome sequence; gaps in the sequence are filled
  4. Editing—sequence is proofread to resolve any ambiguities in the sequence

IV. Bioinformatics
- A.   The field concerned with the management and analysis of biological data using computers
- B.   Genome annotation is done once the sequence is obtained; annotation involves identifying open reading frames (ORFs), determining potential amino acid sequences and comparison to known proteins; such comparison allows tentative assignment of gene function as well as identification of transposable elements, operons, and repeat sequences, and the detection of various metabolic pathways

V. Functional Genomics
- A.   Functional genomics is focused on how genes and genomes operate; two or more genes in the genome of a single organism that arise through duplication of a common ancestral gene are called paralogs, and between genomes are called orthologs
- B.   Translated amino acid sequences can be analyzed for motifs (short patterns) that represent functional units within the protein
- C.   Genomic analyses also can provide information about the phylogenetic relationships among microbes
- D.   Evaluation of RNA-level gene expression: microarray analysis
  1. DNA microarrays (gene chips)—solid supports (e.g., glass) that have DNA attached in highly organized arrays of spots; in commercial chips, the array may consist of many expressed sequence tags (an expressed gene product made from cDNA) covering every ORF of an organism
  2. The mRNA or cDNA to be analyzed (target mixture) is isolated, labeled with fluorescent reporter groups, and incubated with the DNA chip; fluorescence at an address on the chip indicates that the DNA probe on the chip is bound to a mRNA or cDNA in the target mixture; analysis of the hybridization pattern shows which genes are being transcribed
  3. Using this procedure, the characteristic expression of whole sets of genes during differentiation or in response to environmental changes can be observed; patterns of gene expression can be detected and functions can be tentatively assigned based on expression

VI. Comparative Genomics
- A.   Comparisons of genomes leads to new insights in microbial biology
- B.   Comparative genomics can help determine the origin and prevalence of phenotypic traits, detect lateral gene transfers, and develop new vaccines

VII. Proteomics
- A.   Study of genome function at the level of translation
  1. Proteome—entire collection of proteins that an organism produces; proteomics is the study of the proteome
  2. Much interest is in functional proteomics (determining the function of proteins, how they interact with each other, and how they are regulated);
     - a.   Two-dimensional electrophoresis is used to resolve thousands of proteins in a mixture; proteins are first separated based on charge qualities and then by size
     - b.   Mass spectrometry is used to tentatively identify the proteins isolated by two-dimensional electrophoresis; N-terminal amino acid sequencing can be used to determine ORFs when the genome sequence is available

3. Structural proteomics attempts to directly determine the three-dimensional structures of many proteins and then uses that information to predict the structures of other proteins and protein complexes based on their amino acid sequence (protein modeling)

VIII. Insights from Microbial Genomes

A. Identification of genes with unknown functions is important given that as many as half of a genome's ORFs may not match proteins of known functions; genes can be "knocked out" (removed by recombination) and the resulting phenotype examined to determine the function of the deleted gene

B. Genomic analysis of pathogenic microbes provides information about the evolution of virulence, host-pathogen interactions, potential treatments, and vaccines (rational drug design)

1. Findings of particular interest

a. *Staphylococcus aureus* (causative agent of a wide variety of diseases ranging from food poisoning to abscesses; it is becoming increasingly resistant to antibiotics)—several of the antibiotic-resistance genes are located on plasmids and transposons

b. *Chlamydiae* (bacteria that have a unique life cycle, are often referred to as energy parasites, and lack peptidoglycan)—have a genome that is similar to many other bacteria and even contains some genes for ATP synthesis and peptidoglycan synthesis; however, these bacteria lack a gene long held to be required for septum formation during cell division

c. *Treponema pallidum* (the causative agent of syphilis, which has not been cultured outside the human body)—has a genome that shows it is metabolically crippled, both anabolically and catabolically; it also has a family of genes that encode surface proteins, suggesting that this bacterium may be able to change its surface proteins and thereby avoid attack by the host immune system

d. *Mycobacterium tuberculosis* (the causative agent of tuberculosis)—has a very large genome with more than 250 genes devoted to lipid metabolism; it also has a large number of regulatory genes, suggesting that the infection process is very complex; other genes may enable the bacterium to change its antigens and thus elude the host immune system

e. *Mycobacterium leprae* (the causative agent of leprosy)—has a much smaller genome than *M. tuberculosis* and about half of its genome is devoid of functional genes

f. *Rickettsia prowazekii* (a bacterium thought to be related to the ancient bacterium that gave rise to eucaryotic mitochondria)—has sequences consistent with this hypothesis

C. Genomic analysis of extremophiles

1. Genomics is providing a way to determine how extremophiles thrive in harsh conditions

2. *Dienococcus radiodurans* is extremely tolerant of radiation damage to its genome; not only are DNA repair systems upregulated after irradiation, but a host of its genes, suggesting coordination of a complex network of processes

IX. Environmental Genomics

A. Environmental genomics, or metagenomics, is being used to study microbial diversity in natural systems; fewer than 1% of the microbes in the environment can grow in the laboratory, so genetic techniques are used to directly detect and enumerate microbial populations

B. The genomes of entire microbial communities can be sequenced and assembled, giving a picture of their species composition and functionality; new species (phylotypes) are detected, unique genes catalogued, and new functions ascribed to taxa

TERMS AND DEFINITIONS

_____ 1. The study of the molecular organization of genomes, their information content, and the gene products they encode.

_____ 2. The study of the physical nature of genomes

_____ 3. The study of the way genomes function

_____ 4. The comparison of genomes from different organisms to discern patterns of gene function and regulation and microbial evolution

_____ 5. Identification and localization of genes in a genome, and the determination of their function by comparison to gene sequences in databases

_____ 6. A reading frame sequence that is not interrupted by a stop codon; if larger than 100 codons, it is thought to encode a protein

_____ 7. The field concerned with the management and analysis of biological data using computers

_____ 8. The entire collection of proteins that an organism produces

_____ 9. The study of the array of proteins an organism can produce

_____ 10. The study of the function of different proteins, how they interact with each other, and how they are regulated

_____ 11. The process of determining the structure of various proteins and then using that information to predict the structure of other proteins and protein complexes based on their amino acid sequence

a. annotation
b. bioinformatics
c. comparative genomics
d. functional genomics
e. functional proteomics
f. genomics
g. open reading frame (ORF)
h. proteome
i. proteomics
j. structural genomics
k. structural proteomics

FILL IN THE BLANK

1. The most widely used sequencing technique was developed by Frederick Sanger. It uses dideoxynucleotides and is called the _____ technique. Since the development of this technique, it has been automated and is now used to sequence entire genomes, usually by a method developed by Craig Venter, Hamilton Smith and others called _____ _____ _____.

2. Analysis of vast amounts of genome data requires sophisticated computers and computer software; these analytical procedures are part of the field of _____.

3. The study of the way a genome functions is called _____ _____. It begins with _____ of the genome, which identifies genes and tentatively assigns functions to them. One important aspect of understanding the function of a genome, is to determine under what conditions each gene in the genome is expressed. One of the best ways to evaluate gene expression is through the use of _____ _____, which are highly organized arrays of DNA on a solid support (e.g., glass or silicon). Commercially made arrays often use short sequences (~25 base pairs in length) that are unique to a gene, rather than the entire gene sequence. These short sequences are called _____ _____ _____, and they are derived from cDNA molecules.

4. The proteome is often analyzed by _____ _____, followed in many cases by mass spectrometry. Together, these techniques often provide more information about genes expression than the use of _____ _____.

5. The use of genomics to study microbial diversity in natural systems is called _____. The genomes of entire communities can be _____ and then _____ to give a picture of the functionality of the entire community.

DISCOVERIES FROM GENOME ANALYSIS

From the list of microorganisms below select the organism that best matches the description of the discovery.

_____ 1. Its genome is very similar to that of mitochondria, suggesting that aerobic respiration in eucaryotes arose from one of its ancestors.

_____ 2. Its genome contains more than 250 genes for lipid metabolism; the identification of genes encoding surface and secretory proteins may help vaccine development.

_____ 3. Its genome is extremely tolerant of radiation damage.

_____ 4. Known for a variety of antibiotic-resistance genes on plasmids or transposons

_____ 5. The loss of metabolic genes from its genome may explain why it hasn't been cultivated outside a host.

_____ 6. Its genome contains genes for ATP synthesis even though it obtains most of its ATP from its host. Its genome also contains plantlike genes, which suggest it might have originally infected plantlike hosts.

a. *Chlamydia trachomatis*
b. *Mycobacterium tuberculosis*
c. *Rickettsia prowazekii*
d. *Treponema pallidum*
e. *Dienococcus radiodurans*
f. *Staphylococcus aureus*

MULTIPLE CHOICE

For each of the questions below select the *one best* answer.

1. Which of the following is NOT a step in whole-genome shotgun sequencing?
   a. library construction
   b. sequencing of randomly produced fragments
   c. fragment alignment and closure
   d. editing
   e. All of the above are steps in whole-genome shotgun sequencing.

2. Which of the following is a general pattern of genome organization discerned by comparisons of genomes?
   a. There is very little variation in genome organization in bacteria and archaea.
   b. There has been considerable horizontal gene transfer, especially of housekeeping or operational genes.
   c. Most parasitic organisms have more genes than do free-living organisms.
   d. All of the above patterns have been observed.

3. A complex mixture of proteins can be separated using two-dimensional electrophoresis. What is the basis for the separation?
   a. charge differences (isoelectric focusing)
   b. size differences
   c. both (a) and (b)
   d. neither (a) nor (b)

4. Which type of genomic analysis provides information about microbial evolution?
   a. structural genomics
   b. functional genomics
   c. comparative genomics
   d. none of the above

5. Translated amino acid sequences can be analyzed for motifs. What do these represent?
   a. functional units
   b. transcriptional controls
   c. paralogs
   d. orthologs

6. Microarray analysis is NOT appropriate for which of the following?
   a. monitoring individual gene expression
   b. tentatively assigning gene functions
   c. observing patterns of gene expression
   d. determining phylogenetic relationships

7. What percentage of environmental microbes grow in the laboratory?
   a. 1%
   b. 20%
   c. 60%
   d. nearly 100%

TRUE/FALSE

_____ 1. The genome of *M. genitalium* is one of the smallest of any free-living organism.

_____ 2. There has been a great deal of horizontal gene transfer between genomes in both *Bacteria* and *Archaea*.

_____ 3. One of the ultimate goals of genomic analysis is to model a cell on a computer and make predictions about how it would respond to environmental changes.

_____ 4. It is unlikely that genomic analysis will provide any information useful for understanding pathogenicity or for developing treatments for infectious disease.

_____ 5. The new field of pharmacogenomics should produce many new drugs to treat disease.

_____ 6. Open reading frames are known to be functional genes.

CRITICAL THINKING

1. In order for computers to identify open reading frames (ORFs) and other features of a genome, they must be programmed to do so. What features of a nucleotide sequence would be important for identifying ORFs? Explain your choices. Would the features be the same for both eukaryotic and procaryotic organisms? Explain.

ANSWER KEY

Terms and Definitions

1. f, 2. j, 3. d, 4. c, 5. a, 6. g, 7. b, 8. h, 9. I, 10. e, 11. k

Fill in the Blank

1. Sanger; whole-genome shotgun sequencing 2. bioinformatics 3. functional genomics; annotation; DNA microarrays (chips); expressed sequence tags 4. two-dimensional electrophoresis; DNA microarrays (chips) 5. metagenomics; sequenced; assembled

Discoveries from Genome Analysis

1. c, 2. b, 3. e, 4. f, 5. d, 6. a

Multiple Choice

1. e, 2. b, 3. c, 4. c, 5. a, 6. d, 7. a

True/False

1. T, 2. T, 3. T, 4. F, 5. T, 6. F

16 The Viruses: Introduction and General Characteristics

CHAPTER OVERVIEW

Viruses are small, acellular entities that usually possess only a single type of nucleic acid and must use the metabolic machinery of a living host in order to reproduce. Viruses have been and continue to be of tremendous importance because many human diseases have a viral etiology. The study of viruses has contributed greatly to our knowledge of molecular biology, and the blossoming field of genetic engineering is based on discoveries in the field of virology. This chapter focuses on the general properties of viruses, the development of the science of virology, and the methodology used to study viruses.

CHAPTER OBJECTIVES

After reading this chapter you should be able to:

- define viruses and discuss the implications of the concepts embodied in the definition
- discuss the composition and types of viral capsids
- discuss the variety found in viral genomes
- discuss the various requirements for culturing viruses
- discuss the methodology employed for virus purification and enumeration
- describe the way in which viruses are classified

CHAPTER OUTLINE

I.   Early Development of Virology
   A.   Many epidemics of viral diseases occurred before anyone understood the nature of their causative agents
   B.   Progress in preventing viral diseases preceded the discovery of viruses: for instance, in the early 18th century, intentional inoculation of children with material from smallpox lesions was practiced in some countries in order to prevent the disease; Edward Jenner (1798) published case reports of successful attempts to prevent disease (smallpox) by vaccination
   C.   The word *virus,* which is Latin for poison, was used to describe diseases of unknown origin; filtering devices, which trapped bacteria but not viruses, were used by several scientists (Ivanowski, Beijerinck, Loeffler, Frosch, and Reed) to study a number of infectious agents; their recognition of an entity that was filterable (i.e., passed through a filter) led to the modern use of the term virus
   D.   The role of viruses in causing malignancies was established by Ellerman and Bang (1908), who showed that leukemia in chickens was caused by a virus, and Peyton Rous (1911), who showed that muscle tumors in chickens were caused by a virus
   E.   The existence of bacterial viruses was established by the work of Frederick Twort (1915), who first isolated bacterial viruses, and Felix d'Herelle (1917), who devised a method for enumerating them,

demonstrated that they could reproduce only in live bacteria, and called them bacteriophages (or phages)

    F.   W.M. Stanley (1935) helped demonstrate the chemical nature of viruses when he crystallized the tobacco mosaic virus and showed that it was mostly composed of protein; subsequently, F. C. Bawden and N. W. Pirie (1935) separated tobacco mosaic virus particles into protein and nucleic acid components

II.   General Properties of Viruses

    A.   They are infectious agents that have a simple, acellular (not composed of cells) organization, consisting of one or more molecules of DNA or RNA enclosed in a coat of protein, and sometimes in more complex layers

    B.   Virions (virus particle) contain either DNA or RNA, but not both

    C.   They are obligate intracellular parasites that can only reproduce within host cells; outside of host cells, virions are inert

III.   The Structure of Viruses

    A.   Virion size—ranges from 10 nm to 400 nm

    B.   General structural properties

        1.   Nucleocapsid—the nucleic acid plus the surrounding capsid (protein coat that surrounds the genome); for some viruses this may be the whole virion; other viruses may possess additional structures

        2.   Three types of capsid symmetry: icosahedral, helical, complex

        3.   Viral capsids are constructed from many copies of one or a few types of proteins (protomers), which are assembled together with the viral genome

        4.   Enveloped viruses have an outer membranous layer (envelope) surrounding the nucleocapsid; naked viruses lack an envelope

    C.   Helical capsids—hollow tube with a protein wall shaped as a helix or spiral; may be either rigid or flexible

    D.   Icosahedral capsids—regular polyhedron with 20 equilateral triangular faces and 12 vertices; appears spherical; constructed of capsomers (ring or knob-shaped units), each usually made of five (pentamers or pentons) or six protomers (hexamers or hexons); viral capsids self-assemble

    E.   Viruses with capsids of complex symmetry

        1.   Poxviruses are large (200 to 400 nm) with an ovoid exterior shape

        2.   Some bacteriophages have complex, elaborate shapes composed of heads (icosahedral symmetry) coupled to tails (helical symmetry); the structure of the tail regions is particularly variable; such viruses are said to have binal symmetry

    F.   Viral envelopes and enzymes

        1.   Envelopes are membrane structures surrounding some (but not all) viruses

            a.   Lipids and carbohydrates are usually derived from the host membranes

            b.   Proteins are virus-specific

            c.   Many have protruding glycoprotein spikes (peplomers)

        2.   Although viruses lack true metabolism, they may carry one or more enzymes, many of which are involved in viral nucleic acid replication; in some cases the enzyme is capsid-associated, but in most cases the enzyme is located within the capsid

    G.   Viral genomes

        1.   Viral genome may be either RNA or DNA, single- or double-stranded

        2.   DNA viruses

            a.   Most use double-stranded DNA (dsDNA) as genome; the DNA can be linear or circular

            b.   Many have one or more unusual bases (e.g., hydroxymethylcytosine instead of cytosine)

        3.   RNA viruses—most have single-stranded RNA (ssRNA) as their genome

            a.   Plus-strand viruses have a genomic RNA with the same sequence as the viral mRNA; the genomic RNA molecules may have other features (5′ cap, poly-A tail, etc.) common to mRNA and may direct the synthesis of proteins immediately after entering the cell

            b.   Negative-strand viruses have a genomic RNA complementary to the viral mRNA

        c.   Many have segmented genomes (genomes consisting of more than one molecule); each being unique and frequently encoding a single protein

IV.   Virus Reproduction
    A.   Common steps in viral life cycles
        1.   Attachment to the host cell
        2.   Entry of the nucelocapsid or viral genome into the host cell
        3.   Expression of viral genes in host cell; virus takes control of cell's machinery and produces viral proteins and genomes
        4.   New virions self-assemble into mature virions
        5.   Virions released by budding or cell lysis

V.   The Cultivation of Viruses
    A.   Cultivation requires a suitable host
    B.   Hosts for animal viruses
        1.   Suitable host animals
        2.   Embryonated eggs
        3.   Tissue (cell) cultures—monolayers of animal cells
            a.   Cell destruction can be localized if infected cells are covered with a layer of agar; the areas of localized cell destruction are called plaques
            b.   Viral growth does not always result in cell lysis to form a plaque; microscopic (or macroscopic) degenerative effects can sometimes be seen; these are referred to as cytopathic effects
    C.   Bacterial and archaeal viruses are usually cultivated in broth or agar cultures of suitable, young, actively growing host cells; broth cultures are usually clear, while plaques (clearings) form in bacterial lawns on agar plates
    D.   Plant viruses can be cultivated in plant tissue cultures, cultures of separated plant cells, plant protoplast cultures, or whole plants; in whole plants they may cause localized necrotic lesions or generalized symptoms of infection

VI. Virus Purification and Assays
    A.   Virus purification
        1.   Centrifugation of cell homogenates containing virus particles
            a.   Differential centrifugation separates according to size
            b.   Gradient centrifugation separates according to buoyant density or to sedimentation rate (size and density), and is more sensitive to small differences among various viruses
        2.   Differential precipitation with ammonium sulfate or polyethylene glycol separates viruses from other components of the mixture
        3.   Denaturation and precipitation of contaminants with heat, pH, or even organic solvents can sometimes be used
        4.   Enzymatic degradation of cellular proteins and/or nucleic acids can sometimes be used because viruses tend to be more resistant to these types of treatment
    B.   Virus assays
        1.   Particle counts
            a.   Direct counts can be made with an electron microscope
            b.   Indirect counts can be made using methods such as hemagglutination (virus particles can cause red blood cells to clump together or agglutinate)
        2.   Measures of infectivity
            a.   Plaque assays involve plating dilutions of virus particles on a lawn of host cells; clear zones result from viral damage to the cells; results are expressed as plaque-forming units (PFUs); similar measurements can be made by counting the number of pocks or necrotic lesions
            b.   Infectious dose assays are an end-point method for determining the smallest amount of virus needed to cause a measurable effect, usually on 50% of the exposed target units; results are expressed as infectious dose (ID_{50}) or lethal dose (LD_{50})

VII. Principles of Virus Taxonomy
   A. In 1971, the International Committee for Taxonomy of Viruses developed a uniform classification system, which places the greatest weight on these properties:
      1. Nucleic acid type
      2. Nucleic acid strandedness (double or single stranded)
      3. The sense of ssRNA genomes
      4. The presence or absence of an envelope
      5. The host
   B. In addition, other characteristics (capsid symmetry, diameter of capsid or nucleocapsid, number of capsomeres in icosahedral viruses, immunological properties, gene number and genomic map, intracellular location of virus replication, presence or absence of a DNA intermediate in the replication of ssRNA viruses, type of virus release, and disease caused by the virus) are considered in the complementary Baltimore classification scheme

TERMS AND DEFINITIONS

Place the letter of each term in the space next to the definition or description that best matches it.

_____ 1. The complete virus particle as it exists extracellularly
_____ 2. The localized areas of cellular destruction and lysis that appear as clear zones on confluent lawns of host cell growth
_____ 3. Microscopic or macroscopic degenerative changes or abnormalities in infected host cells or tissues
_____ 4. Centrifugation of a suspension at various speeds in order to separate particles of different sizes
_____ 5. Centrifugation of a suspension through a medium whose density varies from top to bottom; particles will separate on the basis of size and/or density
_____ 6. The clumping of red blood cells in the presence of a virus suspension
_____ 7. The viral nucleic acid plus the surrounding protein coat
_____ 8. The protein coat surrounding the viral nucleic acid
_____ 9. Viral genetic material that is divided into separate parts, which may be packaged together or separately; all parts are needed to establish a productive infection
_____ 10. Glycoproteins on the envelope of a virus that project outward from the surface of the virion
_____ 11. A membranous structure that surrounds the nucleocapsid of some virions
_____ 12. The localized areas of destruction occurring on plants infected by a virus

a. capsid
b. cytopathic effects
c. differential centrifugation
d. envelope
e. gradient centrifugation
f. hemagglutination
g. necrotic lesions
h. nucleocapsid
i. plaques
j. segmented genome
k. spikes or peplomers
l. virion

KEY SCIENTISTS AND THE HISTORY OF VIROLOGY

Match the following scientists with their discoveries.

____ 1. Began vaccinating humans to prevent smallpox

____ 2. Developed a porcelain filter that could remove bacteria from samples and enabled scientists to distinguish viruses from bacteria

____ 3. Demonstrated that the causative agent of tobacco mosaic disease would pass through a filter designed to retain bacteria

____ 4. Demonstrated that the filterable causative agent of tobacco mosaic disease could reproduce in a susceptible host

____ 5. Demonstrated that the causative agent of hoof-and-mouth disease in cattle was filterable and reproductively active

____ 6. Demonstrated that the causative agent of yellow fever in humans was filterable and could be transmitted by mosquitoes

____ 7. Demonstrated that leukemia in chickens was caused by a filterable agent

____ 8. Demonstrated that muscle tumors in chickens were caused by a filterable agent

____ 9. Demonstrated that bacteria could be infected by viruses

____ 10. Firmly established the existence of bacterial viruses and developed a plaque assay for enumerating them

____ 11. Crystallized tobacco mosaic virus and found that it was largely or completely composed of protein

____ 12. Separated tobacco mosaic virus particles into protein and nucleic acid

a. Bawden and Pirie
b. Beijerinck
c. Chamberland
d. d'Herelle
e. Ellerman and Bang
f. Ivanowski
g. Jenner
h. Loeffler and Frosch
i. Reed
j. Rous
k. Stanley
l. Twort

FILL IN THE BLANK

1. A complete virus particle, or _____, consists of one or more molecules of _____ or _____ enclosed in a coat of _____.

2. Viruses can be cultured by inoculating a layer of cells and then covering the cells with a thin layer of agar to limit their spread, so that only adjacent cells are infected by newly produced virions. As a result, localized areas of cellular destruction and lysis called _____ are formed. This phenomenon is the basis of a method for enumerating viruses called a ____ _____. In this procedure, each _____ is assumed to have arisen from the reproduction of one virion. Therefore, the number of _____ counted at a given dilution can be used to calculate the number of infectious virions, or_____ _____.

3. Virions range in size from about 10 nm to 300–400 nm. The smallest viruses are little larger than _____, while the largest viruses are about the same size as small _____ and can be seen in the light microscope.

4. Viruses that are regular polyhedrons with 20 faces are said to have _____ symmetry, while those shaped like hollow cylinders have _____ symmetry. Some _____viruses exhibit the former type of symmetry in the head region and the latter type in the tail. These viruses are said to have _____ symmetry.

5. Many viruses have an outer membrane layer called an _____ surrounding the nucleocapsid. These viruses often have glycoprotein spikes or _____ protruding from their outer surface, which may be involved in virus attachment to host cells.

6. Viral capsids are constructed from many copies of one or a few types of protein subunits called _____.

7. Icosahedral viruses are constructed from ring- or knob-shaped units called _____ that can either be _____, which are located at the vertices, or _____, which are located at the edges or the triangular faces.

8. When biological effects are not readily quantified, an end-point dilution is often useful. The normal end point used is the dose necessary to affect 50% of the cultures. If the end point measured is host damage, that dilution is called the _____ _____; if the end point is the death of the culture or organism, it is called the _____ _____.

MULTIPLE CHOICE

For each of the questions below select the *one best* answer.

1. Which of the following is NOT true about the word *virus?*
   a. It is of Latin origin and means *poison.*
   b. It is currently used to describe small, acellular entities that cause disease.
   c. It was used to describe disease-causing agents long before those agents were characterized.
   d. All of the above are true about the word *virus.*

2. In which of the following ways do viruses differ from living microbial cells?
   a. Viruses are acellular.
   b. Viruses have RNA or DNA, but usually not both within the same virion.
   c. Viruses are obligate intracellular parasites that cannot reproduce independently of host cells.
   d. All of the above are ways in which viruses differ from living microbial cells.

3. What is the type of centrifugation where a suspension is centrifuged at various speeds in order to separate particles of various sizes?
   a. isopycnic centrifugation
   b. density centrifugation
   c. differential centrifugation
   d. variable centrifugation

4. In gradient density centrifugation, which of the following is true?
   a. Particles will continue to settle toward the bottom of the tube if the centrifugation is continued.
   b. Particles will come to rest when the density of the surrounding medium is equal to the density of the particle, even if the centrifugation is continued longer.
   c. The smallest particle will sediment the fastest.
   d. Large particles will sediment farther than small particles.

5. What viruses have a single-stranded RNA genome with the same base sequence as the viral mRNA?
   a. plus-strand RNA viruses
   b. minus-strand RNA viruses
   c. complementary-strand RNA viruses
   d. None of the above is correct.

6. Which of the following is a viral genome that exists as several separate, nonidentical molecules packaged together or separately?
   a. diploid genome
   b. segmented genome
   c. polyploid genome
   d. fractionated genome

TRUE/FALSE

_____ 1. In the extracellular phase, viruses possess few if any enzymes, and cannot reproduce independently of living cells.

_____ 2. Like bacteria and eucaryotic microorganisms, viruses can be cultured on artificial media.

_____ 3. Bacteriophages (phages) are viruses that have the typical procaryotic appearance of bacteria.

_____ 4. In viruses with helical symmetry, the nucleic acid is wound in a spiral and lies in the hollow interior formed by the protein subunits.

_____ 5. All capsids with helical symmetry tend to be rigid.

_____ 6. In viral envelopes, the lipids usually come from the host cell, while the proteins are coded for by viral genes.

_____ 7. The presence or absence of an envelope is not a useful characteristic in classifying viruses because any given virus may at one time have an envelope and at another time may not have an envelope.

_____ 8. In viral classification the greatest weight is given to the intracellular location of viral replication.

_____ 9. For some viruses with segmented RNA genomes, the segments may be packaged into separate virion structures.

_____ 10. Viruses can be counted directly with a transmission electron microscope.

_____ 11. Most DNA viruses have single-stranded DNA as their genome, and most RNA viruses have double-stranded RNA as their genome.

CRITICAL THINKING

1. Viruses are thought by some scientists to be among the most primitive of living organisms. Others suggest that they are not living organisms. Take a stand on this debate and defend your position. What would be the most likely arguments from those taking the opposing viewpoint? How would you counter their arguments?

2. Consider the sequence shown below. It represents a segment of the mRNA produced by a virus. Suppose the virus was a double-stranded DNA virus; what would the corresponding genomic sequence be? Suppose the virus was a plus-strand RNA virus; what would the corresponding genomic sequence be? Finally, suppose the virus was a negative-strand RNA virus; what would the corresponding genomic sequence be?

5'CUGGAUCA3'

ANSWER KEY

Terms and Definitions

1. l, 2. i, 3. b, 4. c, 5. e, 6. f, 7. h, 8. a, 9. j, 10. k, 11. d, 12. g

Key Scientists and the History of Virology

1. g, 2. c, 3. f, 4. b, 5. h, 6. i, 7. e, 8. j, 9. l, 10. d, 11. k, 12. a

Fill in the Blank

1. virion; DNA; RNA; protein 2. plaques; plaque assay; plaque; plaques; plaque-forming units (PFUs)
3. ribosomes; bacteria 4. icosahedral; helical; complex; binal 5. envelope; peplomers 6. protomers
7. capsomers; pentamers; hexamers 8. infectious dose (ID_{50}); lethal dose (LD_{50})

Multiple Choice

1. d, 2. d, 3. c, 4. b, 5. a, 6. b

True/False

1. T, 2. F, 3. F, 4. T, 5. F, 6. T, 7. F, 8. F, 9. T, 10. T, 11. F

17   The Viruses: Viruses of Bacteria and *Archaea*

CHAPTER OVERVIEW

This chapter focuses on the characteristics of the bacterial viruses (bacteriophages). It begins with the classification of bacterial and archaeal viruses, and then details the infectious cycle of those DNA viruses that cause destruction (lysis) of host cells. RNA phages are discussed, and the chapter continues with information about phages that can set up a stable residence within the host cell. These phages are called temperate phages, and the process is referred to as lysogeny. Finally, bacteriophage genomes are discussed.

CHAPTER OBJECTIVES

After reading this chapter you should be able to:

- describe the four phases of the viral life cycle
- discuss the differences between DNA phages and RNA phages in terms of their life cycles and their interactions with their hosts
- discuss the establishment and maintenance of lysogeny by temperate phages
- discuss the nature of the bacteriophage genome

CHAPTER OUTLINE

I.   Classification of Bacterial and Archaeal Viruses
    A.   More than 2,000 viral species have been catalogued; only about 40 of these are archaeal viruses and these mainly fall into new taxa based on morphology
    B.   Capsid structure is a critical for identifying viruses; however, lateral gene transfer among viruses confounds classification schemes
    C.   Most bacteriophages possess dsDNA, although other types are known
II. Virulent Double-Stranded DNA Phages
    A.   Virulent phages only reproduce via the lytic cycle, which culminates with the host cell bursting and releasing virions
    B.   The one-step growth experiment
        1.   Reproduction is synchronized so that events during replication can be observed
            a.   Bacteria are infected and then diluted so that the released phages will not immediately find new cells to infect
            b.   The released phages are then enumerated
        2.   Several distinct phases are observed in the viral replication cycle
            a.   Latent period—no release of virions detected; represents the shortest time required for virus reproduction and release; the early part of this period is called the eclipse period, and during this period no infective virions can be found even inside infected cells
            b.   Rise period (burst)—rapid lysis of host cells and release of infective phages; burst size is the number of infective virions released per infected cell
            c.   Plateau period—no further release of infective virions
    C.   Adsorption and penetration
        1.   Viruses attach to specific receptor sites (proteins, lipopolysaccharides, teichoic acids, etc.) on the host cell
        2.   Many viruses inject DNA into the host cell, leaving an empty capsid outside

D. Synthesis of phage nucleic acids and proteins
    1. mRNA molecules transcribed early in the infection (early mRNA) are synthesized using host RNA polymerase; early proteins, made at the direction of early mRNA molecules, direct the synthesis of protein factors and enzymes required to take over the host cell
    2. Transcription of viral genes then follows an orderly sequence due to the modification of the host RNA polymerase and changes in sigma factors
    3. Later in the infection viral DNA is replicated
        a. Synthesis of viral DNA sometimes requires the initial synthesis of alternate bases; these are sometimes used to protect the phage DNA from host enzymes (restriction endonucleases) that would otherwise degrade the viral DNA and thereby protect the host
        b. For some bacteriophages, concatemers (long chains) of the DNA genome are formed; these are later cleaved during assembly
E. Assembly of phage particles
    1. Late mRNA molecules (those made after viral nucleic acid replication) direct the synthesis of capsid proteins and other proteins involved in assembly (e.g., scaffolding proteins) and release of the virus
    2. Assembly proceeds sequentially by subassembly lines, which assemble different structural units (e.g., baseplate, tail tube); these are then put together to make the complete virion
    3. DNA packaging is accomplished with a protein complex called the packasome, which includes the terminase complex that fills in gaps at the end of concatemers
F. Release of phage particles—many phages lyse their host by damaging the cell wall or the cytoplasmic membrane with T4 lysozyme and holin enzyme

III. Single-Stranded DNA Phages
A. φX174 (+strand DNA virus—virus genome that has the same sequence as the viral mRNA)
    1. The single-stranded genome is converted to a double-stranded replicative form (RF) by the host DNA polymerase
    2. The RF directs synthesis of more RF, RNA, and +strand DNA genome
    3. Phages are released by lysis of host cell
B. Filamentous phages (e.g., fd)
    1. Phage DNA enters via the host cell's sex pilus and an RF is synthesized
    2. The RF directs mRNA synthesis and DNA replication via the rolling-circle method
    3. Phages are released without lysing the host cell; instead they are released by a secretory process

IV. Reproduction of RNA Phages
A. Single-stranded RNA phages
    1. RNA replicase is used to replicate the RNA genome
        a. The RNA genome is usually plus stranded and can act as mRNA to direct the synthesis of the replicase during an initial step after penetration
        b. The +strand RNA is then converted to dsRNA, the replicative form
        c. Replicative form is then used as a template for production of multiple copies of the +strand RNA, both genomic and mRNA molecules
    2. Capsid proteins are made, and +strand RNA is packaged into new virions
    3. One or more lysis proteins then function to release the phage
B. The dsRNA phage φ6 is an enveloped virus that infects *Pseudomonas phaseolicola*
    1. It attaches to the pilus as it retracts and uses viral enzymes to pass through the cell wall and cell membrane
    2. Viral RNA-dependent RNA polymerase acts as a transcriptase, making viral mRNA, and as a replicase, making new +strand RNA genome segments that serve as templates for generating the complimentary –strand
    3. A nonstructural viral protein P12 works to mature the virions and add the envelope

V. Temperate Bacteriophages and Lysogeny
A. Temperate phages are capable of lysogeny, a nonlytic relationship with their hosts (virulent phages lyse their hosts)

1. In lysogeny, the viral genome (called a prophage) remains in the host (usually integrated into the host chromosome) but does not kill (lyse) the host cell; the cells are said to be lysogenic (or are called lysogens)
2. It may switch to the lytic cycle at some later time; this process is called induction
B. Most bacteriophages are temperate; it is thought that being able to lysogenize bacteria is advantageous; supporting this is the observation that certain conditions favor the establishment of lysogeny
C. Lysogenic conversion is a change that is induced in the host phenotype by the presence of a prophage, and that is not directly related to the completion of the viral life cycle; examples include:
1. Modification of lipopolysaccharide structure in infected *Salmonella*
2. Production of diphtheria toxin only by lysogenized strains of *Corynebacterium diphtheriae*
D. Establishment of lysogeny (bacteriophage lambda)
1. Two sets of viral promoters are available to host RNA polymerase
2. A repressor protein (lambda repressor) may be made from genes adjacent to one of these promoters
3. If the lambda repressor binds to its target operator before the other promoter is used, that promoter is blocked and lysogeny is established
4. If genes associated with that second promoter (*Cro* protein) are expressed before the lambda repressor can bind to the operator, the lytic cycle is established
5. Induction (the termination of lysogeny and entry into the lytic cycle) will occur if the level of lambda repressor protein decreases; this is usually in response to environmental damage to host DNA
6. For lambda and most temperate phages, if lysogeny is established, the viral genome integrates into the host chromosome; however, some temperate phages can establish lysogeny without integration
VI. Bacteriophage Genomes
A. Bacteriophage genomes are mosaics with blocks of shared genes; all blocks likely have a shared evolutionary history
B. Blocks of genes likely spread through nonhomologous recombination that also allows for exchange of genes between viruses and their hosts, in some cases (shiga-like toxins in *E. coli*) contributing virulence factors

TERMS AND DEFINITIONS

Place the letter of each term in the space next to the definition or description that best matches it.

_____ 1. The term that describes the time immediately after infection during which no infective virions are released

_____ 2. The term that describes the time immediately after infection during which no infective viruses can be found even inside the infected cell

_____ 3. The term that describes the time during which there is a rapid lysis of host cells and release of infective virions

_____ 4. The number of infective virions produced per infected cell

_____ 5. The place on the surface of a host cell where a phage can attach

_____ 6. The mRNA that is made before viral nucleic acid has been replicated

_____ 7. The mRNA that is made after viral nucleic acid has been replicated

_____ 8. An enzyme that cleaves DNA at specific points, thereby destroying it

_____ 9. An RNA-dependent RNA polymerase

_____ 10. Viruses that can only establish a lytic infection

_____ 11. An infection that kills the cell by causing it to burst open and release virus particles

_____ 12. Viruses that can either establish a lytic infection or lysogeny

_____ 13. An infection that does not kill the host and that maintains the viral genome in a dormant state; this dormant genome is replicated when the host genome is replicated

_____ 14. A change in the phenotype of a lysogenized cell that is not directly related to the completion of the virus life cycle

_____ 15. A protein encoded within the lambda genome that establishes and maintains the lysogenic state

_____ 16. A dsDNA or dsRNA that is produced by a phage with a single-stranded genome; it acts as a template for the synthesis of mRNA and genomic nucleic acid

_____ 17. The switching of an infected bacterial lysogen to the active production of viral progeny

_____ 18. An infected bacterium carrying a dormant prophage in the lysogenic state

a. burst size
b. early mRNA
c. eclipse period
d. induction
e. lambda repressor
f. late mRNA
g. latent period
h. lysogen
i. lysogenic conversion
j. lysogeny
k. lytic infection
l. receptor site
m. replicative form
n. restriction enzyme
o. rise period
p. RNA replicase
q. temperate phages
r. virulent bacteriophages

FILL IN THE BLANK

1. The phage life cycle can be studied in a _____ _____experiment. In the _____ period, which immediately follows virus addition, there is no release of virions. Early in this period, no infective virions are found even inside the infected cells; this is called the _____ period. Subsequently, there is rapid lysis of host cells and release of infective virions. This is called the _____ period or _____. Finally, a plateau is reached and no more virions are released.

2. Virus-specific mRNA that is synthesized before viral nucleic acid is replicated is called _____ mRNA, while that produced after viral replication is called _____ mRNA.

3. The DNA of T-even phages contains _____ instead of cytosine. This helps protect it from host _____ enzymes that would otherwise destroy the viral DNA, thereby preventing infection.

4. T4 DNA replication produces _____, long strands of several genome copies linked together.

5. Phages that are capable only of the lytic infectious cycle are called _____ phages, while those that are capable of both lytic infection and lysogeny, a dormant state, are called _____ phages. The bacteria carrying these dormant phages are called _____.

6. The latent form of the genome that exists when a phage establishes lysogeny is called a _____. This can be integrated into the host chromosome or may exist independently. The viral enzyme that catalyzes integration is called _____.

7. A temperate phage may cause a change in the phenotype of the host that is not directly related to the completion of its life cycle. This process is called _____ _____

8. Under some conditions, *E. coli* cells infected with lambda phage will change from the lysogenic state and enter the lytic cycle. A decrease in the level of _____ _____ triggers the switch, which is referred to as _____.

9. Phages having single-stranded genomes begin their replicative process by synthesizing a double-stranded molecule called a _____ _____. The virally encoded enzyme _____ _____ catalyzes this synthesis in ssRNA viruses.

10. Two of the first proteins to appear after infection with lambda are the _____ _____, which blocks transcription of proteins required for the lytic cycle, and _____ _____, which blocks establishment of lysogeny. The fate of the virus in the host depends on which protein first reaches high levels. Lysogeny is established if _____ _____ wins the race.

11. The self-assembly of complex bacteriophages such as T4 is aided by _____ _____, which function in the assembly of the procapsid.

12. Bacteriophage genomes are _____ with blocks of genes that likely spread through _____ _____.

MULTIPLE CHOICE

For each of the questions below select the *one best* answer.

1. Which of the following is least important in classifying bacteriophages?
   a. phage morphology
   b. host range
   c. type of nucleic acid (DNA or RNA)
   d. strandedness of nucleic acid (single or double stranded)

2. Which of the following does NOT serve as phage receptor sites?
   a. lipopolysaccharides
   b. teichoic acids
   c. proteins
   d. All of the above serve as phage receptor sites.

3. In T-even phages, which of the following makes the initial contact with the appropriate receptor site?
   a. tail fiber
   b. base plate
   c. collar
   d. tail tube

4. What is the function of virus-specific enzymes produced before viral nucleic acid replication?
   a. degradation of host DNA
   b. modification of host RNA polymerase to recognize viral promoters
   c. production of any unusual bases required by the virus for DNA replication
   d. All of the above are functions performed by virus-specific enzymes.

5. The sequence of genes in each T4 virus within a population starts with a different gene at the 5′ end. If each of these linear pieces is coiled into a circle, the gene sequences are identical. What is the term used to describe T4 DNA?
   a. linear circle
   b. linearly permuted
   c. circularly permuted
   d. linearly circular

6. Which of the following is NOT a function of the replicative form (dsDNA) of the ssDNA phage φX174?
   a. synthesis of more RF copies
   b. synthesis of minus-strand DNA
   c. synthesis of plus-strand DNA
   d. synthesis of mRNA

7. Which of the following is NOT a translation product of late mRNA?
   a. phage structural proteins
   b. phage assembly proteins that are not incorporated in the capsid
   c. phage proteins needed to replicate the phage nucleic acid
   d. phage release proteins

8. What is the number of infective virions released from a single infected cell called?
   a. infective dose
   b. burst size
   c. multiplicity of infection
   d. burst plateau

9. Which of the following is responsible for the establishment and maintenance of lysogeny in cells infected with bacteriophage lambda?
   a. lactose repressor
   b. lambda repressor
   c. lambda lysogeny protein
   d. lysogeny maintenance protein

10. What is unique about bacteriophage φ6, a phage that infects *Pseudomonas phaseolicola*?
    a. It has a dsRNA genome.
    b. It has a membranous envelope.
    c. Both of the above are correct.
    d. None of the above is correct.

11. Which enzyme is responsible for the escape of the lambda genome from the host chromosome when induction occurs?
    a. integrase
    b. RNA replicase
    c. excisionase
    d. lambda repressor

TRUE/FALSE

_____ 1. In the one-step growth experiment, greatly diluting the culture after initial infection prevents the spread of progeny viruses to other cells.

_____ 2. When the single-stranded DNA phage φX174 infects a cell, transcription must take place before replication can occur.

_____ 3. The single-stranded genome of RNA phages serves both as a template for its own replication and as mRNA.

_____ 4. Generally, a low multiplicity of infection (MOI) favors lysogeny, while a high MOI favors lytic infection.

_____ 5. Filamentous fd phages are released by secretion through the host plasma membrane, leaving the host relatively undamaged and able to continue to release more phage particles.

_____ 6. The tail tube of a complex bacteriophage may interact with the plasma membrane to form a pore through which the DNA passes.

_____ 7. Noncapsid proteins that aid in the assembly of virion structures are referred to as scaffolding proteins.

_____ 8. The four major types of phage morphology are: tailless icosahedral; phages with contractile tails; phages with noncontractile tails; and filamentous phages.

_____ 9. Bacteriophage genomes diverged as a whole from a common ancestor.

PHAGE REPLICATION

Fill in the requested information in the table below.

Stage or Event	Description of Stage or Event		
	dsDNA phages (e.g., T4)	ssDNA phages (e.g., φX174)	ssRNA phages (e.g., MS2)
Adsorption to host cell and penetration			
mRNA synthesis			
Genome synthesis			
Protein synthesis			
Assembly			
Release			

CRITICAL THINKING

1. Consider the results of a one-step growth experiment presented in the following graph. Label the brackets on the figure and explain what each represents. Under ideal conditions, how much time is associated with each period, and how many phage particles are released?

(a) _____
_____
_____

(b) _____
_____
_____

(c) _____
_____
_____
_____

(d) _____
_____
_____
_____

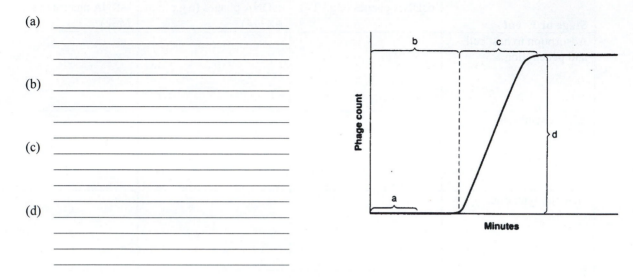

2. When the T4 genome is packaged into its head from long multigenome concatemers, the packaged DNA is about 2% longer than the complete genome. Using diagrams and an arbitrary gene set (A–M), show how the linear T4 DNA is circularly permuted.

ANSWER KEY

Terms and Definitions

1. g, 2. c, 3. o, 4. a, 5. l, 6. b, 7. f, 8. n, 9. p, 10. r, 11. k, 12. q, 13. j, 14. i, 15. e, 16. m, 17. d, 18. h

Fill in the Blank

1. one-step growth; latent; eclipse; rise; burst 2. early; late 3. hydroxymethylcytosine; restriction
4. concatemers 5. virulent; temperate; lysogens 6. prophage; integrase 7. lysogenic conversion
8. lambda repressors; induction 9. replicative form; RNA replicase 10. lambda repressor; cro protein;
lambda repressor 11. scaffolding proteins 12. mosaics; nonhomologous recombination

Multiple Choice

1. b, 2. d, 3. a, 4. d, 5. c, 6. b, 7. c, 8. b, 9. b, 10. c, 11. c

True/False

1. T, 2. F, 3. T, 4. F, 5. T, 6. T, 7. T, 8. T, 9. F

18 The Viruses: Eucaryotic Viruses and Other Acellular Infectious Agents

CHAPTER OVERVIEW

This chapter focuses on the characteristics of viruses that infect eucaryotes. Mammalian viruses are emphasized because they are causative agents of many human diseases. Other viruses, such as plant viruses and insect viruses, are also discussed. The chapter concludes with a discussion of infectious agents that are even simpler than viruses: the viroids, virusoids, and prions.

CHAPTER OBJECTIVES

After reading this chapter you should be able to:

- compare and contrast viruses that infect eucaryotes with those that infect procaryotes
- describe the various ways that viruses of eucaryotes can harm their host organisms
- discuss the establishment of persistent virus infections
- discuss the mechanisms by which viruses may contribute to the development of certain cancers
- discuss the importance of plant viruses and the technical difficulties that have hindered rapid progress in understanding them
- discuss fungal, algal, and protozoan viruses
- discuss the potential use of insect viruses for pest control
- discuss the nature and significance of viroids, virusoids, and prions

CHAPTER OUTLINE

I. Taxonomy of Eucaryotic Viruses
   A. Morphology is the most important characteristic for classification
   B. The physical and chemical nature of the virion, especially the nucleic acids, also are important for classification
II. Reproduction of Vertebrate Viruses
   A. Adsorption of virions
      1. Virions attach to specific receptors, usually cell surface proteins or glycoproteins that are required by the cell for normal cell functioning; receptor distribution often varies over the surface of the host cells and lipid rafts seem to be involved in both virion entry and assembly
      2. Viral surface proteins and/or enzymes may mediate virus attachment to the cellular receptors
   B. Penetration and uncoating
      1. Shortly after adsorption, viruses penetrate and enter the host cell; virus uncoating (removal of the capsid and release of viral nucleic acid) occurs during or shortly after penetration
      2. There appear to be two common modes of entry
         a. Fusion of the viral envelope with the host cytoplasmic membrane results in deposition of the nucleocapsid core within the cell
         b. Endocytosis; lysosomal enzymes and low endosomal pH often trigger the uncoating process
      3. Once in the cytoplasm the nucleic acid may function while still attached to capsid components or may only function after completion of uncoating
   C. Genome replication and transcription in DNA viruses

1. Expression of early viral genes (usually catalyzed by host enzymes) is devoted to taking over the host cell; this may involve halting synthesis of host DNA, RNA, and protein or, in some cases, these processes may be stimulated
2. Later, viral DNA replication occurs, usually in the nucleus
3. Some examples
    a. Parvoviruses (ssDNA)—have a very small genome with overlapping genes; use host enzymes for all biosynthetic process; genome ends fold over to make double-stranded regions to prime DNA polymerase
    b. Herpesviruses (dsDNA)—host RNA polymerase is used to transcribe early genes; DNA replication is catalyzed by viral DNA polymerase
    c. Poxviruses (dsDNA)—viral RNA polymerase synthesizes early mRNA; one of the early gene products is viral DNA polymerase, which replicates the viral genome
    d. Hepadnaviruses (circular dsDNA)—use reverse transcriptase to replicate the DNA genome via an RNA intermediate
D. Genome replication, transcription, and protein synthesis in RNA viruses
    1. Transcription in RNA viruses (except retroviruses)
        a. Positive-strand RNA viruses use their genome as mRNA
        b. Negative-strand RNA viruses use viral RNA-dependent RNA polymerase (transcriptase) to synthesize mRNA, using the genome as the template
        c. dsRNA viruses use viral RNA-dependent RNA polymerase to synthesize mRNA
    2. Replication in RNA viruses (except retroviruses)
        a. ssRNA viruses use viral replicase (an RNA-dependent RNA polymerase) to convert ssRNA into dsRNA (replicative form); the replicative form serves as template for genome synthesis
        b. dsRNA viruses—in some cases, the viral mRNA molecules associate with special proteins to form a large complex; replicase then uses these mRNA molecules as templates for synthesis of the dsRNA genome
    3. Retroviruses make a dsDNA copy (called proviral DNA) from their ssRNA genome using the enzyme reverse transcriptase (RNA-dependent DNA polymerase)
        a. The proviral DNA is integrated into the host chromosome
        b. The integrated proviral DNA can then direct the synthesis of mRNA
    4. Translation of viral several genes from a single mRNA can involve cleavage of a genome-sized polypeptide, synthesis of subgenomic mRNAs, or ribosomal frameshifting
E. Assembly of virus capsids
    1. Capsid proteins are synthesized by host cell ribosomes under the direction of viral late genes
    2. Empty procapsids are produced and the nucleic acid is inserted by an unknown mechanism
    3. Enveloped virus nucleocapsids are assembled similarly (except for poxvirus nucleocapsids, which are assembled by a complex process that begins with enclosure of some of the cytoplasmic matrix by construction of a membrane, followed by movement of viral DNA into the center of the immature virus)
F. Virion release
    1. Naked viruses are usually released when the host cell lyses
    2. Enveloped viruses are usually released by the following mechanism
        a. Virus-encoded proteins are incorporated into the host cell's plasma membrane (some viruses use the host's nuclear membrane, endoplasmic reticulum, Golgi apparatus, or other membranes)
        b. The nucleocapsid buds outward, forming the envelope during release; actin cytoskeleton microfilaments can aid virion release (e.g., poxviruses)
III. Cytocidal Infections and Cell Damage
A. Viruses often damage their host cells, in some cases causing cell death; if death occurs the infection is cytocidal
B. Seven mechanisms for causing cell damage have been identified
    1. Inhibition of host DNA, RNA, and protein synthesis
    2. Lysosome damage, leading to release of hydrolytic enzymes into the cell

3. Plasma membrane alteration, leading to host immune system attack on the cell or leading to cell fusion
4. Toxicity from high concentrations of viral protein
5. Formation of inclusion bodies that may cause direct physical disruption of cell structure
6. Chromosomal disruptions
7. Malignant transformation to a tumor cell

IV. Persistent, Latent, and Slow Virus Infections
   A. Persistent infections—long-lasting infections
      1. Chronic virus infection—the virus is usually detectable, but clinical symptoms are mild or absent for long periods
      2. Latent virus infections—the virus stops reproducing and remains dormant for a period before becoming active again; during latency no symptoms, antibodies, or viruses are detectable
   B. Causes of persistence and latency are probably multiple
      1. Viral genome integrates into host chromosome
      2. Virus becomes less antigenic
      3. Virus mutates to less virulent and slower-reproducing form
      4. Deletion mutation produces defective interfering (DI) particles, which cannot reproduce but slow normal virus reproduction and thereby reduce host damage and establish a chronic infection
   C. Slow virus infections are those that cause progressive, degenerative diseases with symptoms that increase slowly over a period of years

V. Viruses and Cancer
   A. Cancer is a disease where there is abnormal cell growth (neoplasia) and spread of the abnormal cells throughout the body (metastasis); tumors are growths or lumps of tissue that can be benign (nonspreading) or malignant (cancerous)
   B. Carcinogenesis is a complex, multistep process that involves a triggering event and the activity of oncogenes
   C. The viral etiology of human cancers is difficult to establish because Koch's postulates can only be satisfied for diseases by experimenting on humans; however, viruses have been implicated in a number of human cancers
      1. Epstein-Barr virus (EBV)—a herpesvirus that may cause:
         a. Burkitt's lymphoma; found mostly in central and western Africa
         b. Nasopharyngeal carcinoma; found in Southeast Asia
         c. Infectious mononucleosis; found in the rest of the world
         d. Evidence suggests that host infection with malaria is necessary for EBV to cause Burkitt's lymphoma; this is supported by the low incidence of Burkitt's lymphoma in the U.S. where malaria is rare
      2. Hepatitis B virus may be associated with one form of liver cancer
      3. Human papillomavirus has been linked to cervical cancer
      4. Human T-cell lymphotropic viruses (the retroviruses HTLV-1 and HTLV-2) are associated with adult T-cell leukemia and hairy-cell leukemia, respectively
   D. Viruses may cause cancer by a variety of mechanisms
      1. Viruses may carry one or more cancer-causing genes (oncogenes)
      2. Viruses may produce a regulatory protein that activates cell division
      3. Viruses may insert a promoter or enhancer next to a cellular proto-oncogene (a cellular gene that regulates cell growth and reproduction), causing abnormal expression of this gene and thereby deregulating cell growth
      4. Viruses may inhibit the activity of tumor suppressors (e.g., p53) that regulate cell cycling

VI. Plant Viruses
   A. Have not been well studied, primarily because they are difficult to cultivate and purify
   B. Virion morphology—does not differ significantly from that of animal viruses or bacteriophages; most are RNA viruses
   C. Plant virus reproduction (using tobacco mosaic virus as an example)

1. Penetration through the plant cuticle is often helped by the action of sucking insects such as aphids
2. The virus uses either a cellular or a virus-specific RNA-dependent RNA polymerase
3. The virus produces proteins, which then spontaneously assemble
4. Viral spread is through the plant vascular system or to adjacent cells through plasmodesmata
5. The virus causes many cytological changes, such as the formation of inclusion bodies and the degeneration of chloroplasts

VII. Viruses of Fungi and Protists
   A. Most viruses of higher fungi (mycoviruses) are dsRNA viruses that cause latent infections
   B. Viruses of lower fungi are dsRNA or dsDNA viruses that cause lysis of infected cells
   C. Algal viruses have dsDNA genomes and polyhedral capsids
   D. Protozoan viruses can be dsDNA or dsRNA viruses; one group (Mimivirus) has a very large genome

VIII. Insect Viruses
   A. Members of many virus families are known to infect insects
   B. Infection is often accompanied by formation of granular or polyhedral inclusion bodies
   C. Some may persist as latent infections
   D. Current interest in most insect viruses focuses on their use for biological pest control; they have several advantages over chemical toxins:
      1. They are invertebrate-specific and, therefore, should be safe
      2. They have a long shelf-life and high environmental stability
      3. They are well suited for commercial production because they reach high concentrations in infected insects

IX. Viroids and Virusoids
   A. Viroids are circular ssRNA molecules that lack a capsid and cause a variety of plant diseases; the ssRNA molecules do not act as mRNAs (do not encode genes) and appear to be replicated by host RNA polymerases by a rolling-circle mechanism
   B. Pathogenicity of viroids may result from RNA silencing, where viroid RNA forms dsRNA by pairing with host mRNAs; these dsRNA are destroyed by host cell defenses and, hence, certain host genes are not translated, leading to disease
   C. Virusoids are circular ssRNA molecules (satellite RNAs) that encode proteins and need helper viruses for infectivity; the hepatitis D virusoid requires the help of the hepatitis B virus

X. Prions
   A. Prions are proteinaceous infectious particles (PrP) that are *not* associated with a nucleic acid
   B. Genes that encode PrP have been identified in normal animal tissue
      1. It is hypothesized that abnormal PrP causes prion diseases by inducing a change from the normal conformation of the cellular PrP to the abnormal form
      2. This new abnormal PrP then causes other normal cellular PrP molecules to change to the abnormal form
   C. Prions cause progressive, degenerative central nervous system disorders, including scrapie (sheep and goats), bovine spongiform encephalopathy (mad cow disease), Kuru (found only in the Fore, an eastern New Guinea tribe that practices ritual cannibalism), Creutzfeldt-Jakob disease (CJD), fatal familial insomnia, and Gerstmann-Strassler-Scheinker Syndrome (human diseases)

TERMS AND DEFINITIONS

Place the letter of each term in the space next to the definition or description that best matches it.

_____ 1. An enzyme that makes a DNA copy of an RNA molecule

_____ 2. Capsid prior to incorporation of virus genome

_____ 3   An infection that results in cell death

_____ 4. Clusters of subunits or virions within the host nucleus or cytoplasm

_____ 5. Infectious diseases that develop very slowly

_____ 6. Abnormal cell growth and reproduction

_____ 7. Genes whose expression or abnormal expression leads to the development of cancer

_____ 8   Causative agents of plant disease that consist of small, circular RNA molecules but lack protein capsids

_____ 9. Causative agents of progressive degenerative diseases in humans and other animals that consist of small proteinaceous particles but apparently lack nucleic acid genomes

_____ 10. Virus particles that cannot replicate and that slow normal virus reproduction thereby reducing host damage and establishing a chronic infection

_____ 11. Genes expressed early in a viral infection; they are often involved in taking over host cell functions

_____ 12. Circular ssRNA molecules that code proteins but need helper viruses for infectivity

a. cytocidal infection
b. defective interfering (DI) particles
c. early genes
d. inclusion bodies
e. neoplasia
f. oncogenes
g. prions
h. procapsid
i. reverse transcriptase
j. slow virus diseases
k. viroids
l. virusoids

FILL IN THE BLANK

1. One unique group of positive-strand RNA viruses synthesizes dsDNA, called _____ DNA, using the enzyme _____ _____ and their RNA genome as the template. The enzyme used has a component called _____ _____,which degrades the RNA template after synthesis of the first DNA strand. Because the flow of genetic information is "backwards" in these viruses (i.e., RNA to DNA rather than DNA to RNA as in cellular organisms), they are called _____.

2. Most ssRNA viruses use a viral _____ to synthesize dsRNA, which directs the formation of new genomes. This dsRNA molecule is referred to as the _____ _____.

3. The numerous diseases collectively referred to as _____ are characterized by abnormal cell growth called _____, which results in the formation of a lump of tissue called a _____. Cells in malignant _____ are able to spread throughout the body by a process known as _____.

4. Because their genome is complementary to their mRNA, _____ -strand RNA viruses must employ a virion-associated _____ to synthesize mRNA.

5. Some infections have a rapid onset and last for a relatively short time; they are called _____ infections. Others are long-asting infections called _____ infections. There are several types of these infections. In _____ virus infections, the virus is detectable and clinical symptoms are mild or absent for long periods (e.g., hepatitis B and HIV infections). In _____ virus infections, the virus stops reproducing and remains dormant before becoming active again (e.g., herpes simplex virus infections). A small group of viruses cause extremely slow-developing infections. These infections are called _____ virus diseases (e.g., subacute sclerosing panencephalitis).

6. Small ssRNA molecules that have no capsid and cause disease in plants are called _____, while small proteinaceous particles that have no apparent nucleic acid and cause progressive degenerative diseases in humans and other animals are called _____.

7. Many insect viruses produce _____ _____ (clustering of subunits or virions in the nucleus or cytoplasm of the host cell), which often aid in viral transmission.

8. Most _____ virions are released upon host cell lysis. In contrast, during _____ virus reproduction, release and envelope formation occur simultaneously, without lysing the host cell.

CHARACTERISTICS OF ANIMAL VIRUSES

Complete the table by filling in the missing information for each virus family:

Virus Family	Envelope (yes or no)	Nucleic Acid Type (DNA or RNA; single or double stranded; positive strand or negative strand)	Representative Virus Genus
Adenoviridae			
Arenaviridae			
Baculoviridae			
Bunyaviridae			
Calciviridae			
Coronavirdae			
Herpesvidae			
Iridoviridae			
Orthomyxoviridae			
Papovaviridae			
Parmyxoviridae			
Parvoviridae			
Picornaviridae			
Poxviridae			
Reoviridae			
Retroviridae			
Rhabdoviridae			
Togaviridae			

MULTIPLE CHOICE

For each of the questions below select the *one best* answer.

1. What type of molecule is the most common animal virus receptor?
   a. lipoprotein
   b. glycoprotein
   c. phosphoprotein
   d. teichoic acid

2. Where are poliovirus receptors found?
   a. in cells of all tissues
   b. in spinal cord anterior horn cells only
   c. in nasopharynx, gut, and spinal cord anterior horn cells
   d. in gut cells only

3. Which of the following is NOT used by viruses to kill host cells?
   a. inhibition of host DNA, RNA, and protein synthesis
   b. lysosomal damage
   c. alteration of host cell membranes
   d. formation of inclusion bodies
   e. All of the above are used by viruses to kill host cells.

4. Which of the following is NOT commonly used by animal viruses to enter host cells?
   a. capsid remains outside the cell, as the nucleic acid is injected through the plasma membrane
   b. fusion of envelope with plasma membrane
   c. endocytosis
   d. All of the above are ways that animal viruses enter host cells.

5. Enveloped viruses, except poxviruses, acquire their envelopes in which of the following ways?
   a. during assembly but before release
   b. during release
   c. after release
   d. during penetration

6. Which of the following diseases is/are thought to be produced by prions rather than viruses?
   a. kuru
   b. scrapie
   c. Creutzfeldt-Jakob disease
   d. All of the above are thought to be produced by prions.

7. In some viruses, deletion mutations result in the production of altered viruses that cannot reproduce but can slow normal virion reproduction. What are these called?
   a. deletion particles
   b. replication incompetent viruses
   c. defective interfering particles
   d. None of the above is correct.

8. Some cancer cells revert to a more primitive or less differentiated state. What is this called?
   a. anaplasia
   b. neoplasia
   c. metastasis
   d. carcinogenesis

9. Epstein-Barr virus (EBV) is one of the best-studied human cancer viruses. What does it cause?
   a. Burkitt's lymphoma
   b. nasopharyngeal carcinoma
   c. infectious mononucleosis
   d. All of the above are caused by EBV.

10. Which of the following is NOT a way in which viruses cause cancer?
    a. introduction of an oncogene carried by the virus
    b. introduction of a viral promoter or enhancer next to a previously unexpressed gene that controls cell division
    c. production of a regulatory protein that activates cellular genes controlling cell division
    d. All of the above are ways in which viruses cause cancer.

11. Which of the following is the most important agent of plant virus transmission?
    a. soil nematodes feeding on plant roots
    b. parasitic fungi
    c. pollen
    d. insects that feed on plant leaves

12. For viruses that enter the cell by endocytosis, what usually triggers uncoating?
    a. lysosomal enzymes
    b. low endosomal pH
    c. Both (a) and (b) are correct.
    d. Neither (a) nor (b) is correct.

13. Which of the following makes some insect viruses well suited for commercial production as biological pest control agents?
    a. They have a long shelf-life and are environmentally stable.
    b. They are specific for invertebrates.
    c. They reach extremely high concentrations in larval tissue.
    d. All of the above make insect viruses suitable biological pest control agents.

14. How do enveloped viruses acquire their envelopes?
    a. by budding through the plasma membrane
    b. by budding through internal cellular membranes
    c. For some viruses (a) is correct; for others viruses (b) is correct.
    d. Neither (a) nor (b) is correct.

TRUE/FALSE

_____ 1. The high degree of specificity of animal viruses for their hosts makes host range the most important criterion for classifying animal viruses.

_____ 2. All viruses, except parvoviruses, use some virus-specific enzymes for replication. Therefore, some viral proteins must enter the cell with the viral nucleic acid.

_____ 3. Unlike bacteriophages, most animal viruses do not cause degradation of host DNA.

_____ 4. All of the DNA viruses use host RNA polymerase to transcribe at least early genes.

_____ 5. The parvovirus genome is so small that it must use overlapping genes (i.e., a single sequence of bases read in different reading frames) to encode the three proteins that are its only gene products.

_____ 6. Poxviruses have a complex infectious cycle in which partial uncoating occurs followed by early gene expression. One of these early gene products is an enzyme that completes the uncoating process prior to DNA replication.

_____ 7. RNA viruses are more uniform in their replication strategies than DNA viruses.

_____ 8. Unlike other enveloped viruses that use altered cellular membranes as their envelope source, herpesviruses use the nuclear membrane in envelope formation.

_____ 9. Some plant viruses that require insect vectors for transmission can actually be cultivated in insect cells.

_____ 10. All of the known plant viruses are RNA viruses.

_____ 11. Tobacco mosaic virus mRNA arises by complex processing of genomic RNA, even though the genome is plus stranded and could conceivably function as mRNA directly.

_____ 12. EBV causes fewer cancers in the U.S. compared to Africa; this is because malaria may be a required precondition for EBV-induced cancer, and there is little malaria in the U.S.

_____ 13. Prion-mediated pathogenesis may involve conformational change in the prion protein (PrP) to an abnormal form.

_____ 14. Poxviruses use actin microfilaments to escape from host cells without damaging the cell.

_____ 15. The tobacco mosaic virus is the only virus incapable of self-assembly.

_____ 16. All known mycoviruses of higher fungi have dsRNA genomes, whereas the viruses of lower fungi have either dsRNA or dsDNA genomes.

_____ 17. Baculoviruses and other viruses are finding uses as biological control agents for insect pests.

CRITICAL THINKING

1. Compare and contrast the replication and transcription strategies of plus-strand and minus-strand viruses. Be sure to include retroviruses in your discussion.

2. Transposable elements and retroviruses have certain structural features in common. Furthermore, certain effects on their hosts are also similar. Discuss these similarities and show how the study of transposable elements has contributed to our understanding of retrovirus-mediated carcinogenesis.

ANSWER KEY

Terms and Definitions

1. i, 2. h, 3., a 4. d, 5. j, 6. e, 7. f, 8. k, 9. g, 10. b 11. c 12. l

Fill in the Blank

1. proviral; reverse transcriptase; ribonuclease H; retroviruses 2. replicase; replicative form 3. cancer; neoplasia; tumor; tumors; metastasis 4. negative; transcriptase 5. acute; persistent; chronic; latent; slow 6. viroids; prions 7. inclusion bodies 8. naked; enveloped

Multiple Choice

1. b, 2. c, 3. e, 4. a, 5. b, 6. d, 7. d, 8. a, 9. d, 10. d, 11. d, 12. c, 13. d, 14. c

True/False

1. F, 2. F, 3. T, 4. F, 5. T, 6. T, 7. F, 8. T, 9. T, 10. F, 11. T, 12. T, 13. T, 14. T, 15. F, 16. T 17. T

19 Microbial Evolution, Taxonomy, and Diversity

CHAPTER OVERVIEW

Microorganisms are tremendously diverse in size, shape, physiology, and lifestyle. This chapter introduces the general principles of microbial taxonomy and presents an overview of the current classification scheme accepted by most microbiologists. Subsequent chapters will examine the various groups of microorganisms in greater detail.

CHAPTER OBJECTIVES

After reading this chapter you should be able to:

- discuss microbial evolution, the RNA world, and endosymbiotic theory
- discuss the three domains of living organisms (*Bacteria, Archaea*, and *Eucarya*)
- discuss the rationale behind the science of taxonomy
- discuss the meaning of the word *species* and the basis for grouping organisms into species
- discuss the ways of classifying organisms
- discuss the various characteristics used in taxonomy and explain why nucleic acid sequences are probably the best indicators of microbial phylogeny and relatedness
- discuss the classification scheme used in *Bergey's Manual of Systematic Bacteriology*
- discuss the dynamic nature of bacterial taxonomy and the new types of data that are contributing to the changes being made

CHAPTER OUTLINE

I. Microbial Evolution
   A. Earth is about 4.6 billion years old; what appear to be the fossilized remains of 3.5 billion-year-old procaryotic cells have been found in stromatolites and sedimentary rocks
   B. It is speculated in the RNA world theory that the first self-replicating molecules were RNAs that also had catalytic activity (ribozymes)
      1. Stromatolites are layered or stratified rocks that are formed by incorporation of mineral sediments into microbial mats
      2. The earliest procaryotes were probably anaerobic and lithotrophs
      3. Cyanobacteria probably developed oxygenic photosynthesis 3.0 billion years ago forming layered rocks called stromatolites
   C. The work of Carl Woese and his collaborators suggests that organisms fall into one of three domains based on nucleotide sequence similarities among small subunit ribosomal RNAs (SSU rRNAs)
      1. *Eucarya*—contains all eucaryotic organisms; have a complex membrane-delimited organelle structure
      2. *Bacteria*—comprise the majority of procaryotes; cell walls contain muramic acid; membrane lipids contain ester-linked, straight-chain fatty acids

3. *Archaea*—procaryotes that lack muramic acid, have lipids with ether-linked, branched, aliphatic chains, lack thymidine in the T arm of tRNA molecules, have distinctive RNA polymerases, and have ribosomes with a different composition and shape than those observed in *Bacteria*

D. Universal phylogenetic trees are constructed by aligning SSU rRNA gene sequences from a variety of organisms and determining their similarity using statistical analyses; the resulting tree demonstrates evolutionary relatedness; the universal tree has a single root, a commune ancestor to all life

E. The first event in the development of eucaryotes was the formation of the nucleus (possibly by fusion of ancient bacteria and archaea)

F. The endosymbiotic origin of mitochondria and chloroplasts
  1. Endosymbiotic theory states that mitochondria and chloroplasts arose from procaryotes, α-proteobacteria, and cyanobacteria, respectively, that lived within eucaryotic cells, eventually losing parts of their genome to the host cell and thereby losing the ability to survive independently; organelles have both circular chromosomes related to bacterial genomes and independent transcription and translation systems
  2. The hydrogen hypothesis states that the progenitor of mitochondria was an α-proteobacterium that produced hydrogen through fermentation that was required by the host; the hydrogenosome may be another relic of this early symbiont
  3. The serial endosymbiotic theory states that eucaryotes developed in a series of discrete endosymbiotic steps

G. Evolutionary processes
  1. Microevolution is small changes through genetic variation that lead to speciation or extinction (macroevolution); the rate of evolution is not steady, but likely is punctuated by periods of relatively rapid speciation (punctuated equilibrium)

II. Introduction to Microbial Classification and Taxonomy
A. Taxonomy is the science of biological classification
  1. Classification is the arrangement of organisms into groups (taxa)
  2. Nomenclature refers to the assignment of names to taxonomic groups
  3. Identification refers to the determination of the taxon to which a particular isolate belongs
  4. Systematics is the scientific study of organisms with the ultimate object of characterizing and arranging them in an orderly manner

B. Natural classification schemes arrange organisms into groups based on shared biological characteristics; for higher organisms, morphology is the central defining characteristic; for procaryotes, a polyphasic approach using phenotype, genotype, and phylogenetics is needed

C. Phenetic classification—groups organisms based on mutual similarity of their phenotypic characteristics

D. Phylogenetic classification—groups organisms based on evolutionary history as determined by the fossil record for higher organisms or SSU rRNA gene sequences for procaryotes

E. Genotypic classification—groups organisms based on the genetic similarity of whole genomes

F. Numerical taxonomy
  1. Information about the properties of an organism is converted to a form suitable for numerical analysis and is compared by means of a computer
  2. The presence or absence of at least 50 (preferably several hundred) characters should be compared (morphological, biochemical, and physiological characters)
  3. An association coefficient is determined between characters possessed by two organisms
    a. Simple matching coefficient—proportion of characters that match whether present or absent
    b. Jaccard coefficient—ignores characters that both organisms lack
  4. These values are arranged to form a similarity matrix; organisms with great similarity are grouped together into phenons; the significance of the phenons is not always obvious, but phenons with an 80% similarity often are equivalent to bacterial species
  5. A treelike diagram called a dendrogram is used to display the results of numerical taxonomic analysis

III. Taxonomic Ranks
   A.   The taxonomic ranks (in ascending order) are: species, genus, family, order, class, phylum, and domain; however, microbiologists often use informal names that are descriptive (e.g., methanogens, purple bacteria, lactic acid bacteria)
   B.   The basic taxonomic group is the species
      1.   Procaryotic species are not defined on the basis of sexual reproductive compatibility (as for higher organisms) but rather are based on phenotypic and genotypic differences
         a.   Currently, a procaryotic species is defined as a collection of strains that share many stable properties and differ significantly from other groups of strains
         b.   This definition is not completely satisfactory and other definitions have been proposed
      2.   A strain is a population of organisms that is distinguishable from at least some other populations in a taxonomic category; a strain is thought to have descended from a single organism or pure culture isolate
         a.   Biovars—strains that differ biochemically or physiologically
         b.   Morphovars—strains that differ morphologically
         c.   Serovars—strains that differ in antigenic properties
         d.   The type strain is usually the first studied (or most fully characterized) strain of a species; it does not have to be the most representative member
      3.   A genus is a well-defined group of one or more species that is clearly separate from other genera
   C.   In the binomial system of nomenclature devised by Carl von Linné (Carolus Linnaeus), the genus name is capitalized while the specific epithet is not; both terms are italicized (e.g., *Escherichia coli*); after first usage in a manuscript the first name will often be abbreviated to the first letter (e.g., *E. coli*)
IV. Major Characteristics Used in Taxonomy and Phylogeny
   A.   Classical characteristics
      1.   Morphological characteristics—easy to analyze, genetically stable, and usually do not vary greatly with environmental changes; often are good indications of phylogenetic relatedness
      2.   Physiological and metabolic characteristics—directly related to enzymes and transport proteins (gene products) and therefore provide an indirect comparison of microbial genomes
      3.   Ecological characteristics—include life-cycle patterns, symbiotic relationships, ability to cause disease, habitat preferences, and growth requirements
      4.   Genetic analysis—includes the study of chromosomal gene exchange through transformation and conjugation; these processes only rarely cross genera; one must take care to avoid errors that result from plasmid-borne traits
   B.   Molecular characteristics
      1.   Nucleic acid base composition
         a.   G + C content can be determined from the melting temperature (T_m); T_m, is the temperature at which the two strands of a DNA molecule separate from one another as the temperature is slowly increased
         b.   G + C content is taxonomically useful because variation within a genus is usually less than 10% but variation between genera is quite large, ranging from 25 to 80%
      2.   Nucleic acid hybridization
         a.   The temperature of incubation controls the degree of sequence homology needed to form a stable hybrid; the percentage of stable hybrids formed in a mixture of DNA from two organisms reflects the degree of sequence homology
         b.   This technique is useful for studying the relatedness of closely related organisms
      3.   Nucleic acid sequencing
         a.   rRNA gene sequences are most ideal for comparisons because they contain both evolutionarily stable and evolutionarily variable sequences; PCR is used to amplify rRNA genes for sequencing and comparison to large sequence databases
         b.   Oligonucleotide signature sequences are short, conserved nucleotide sequences that are specific for a phylogenetically defined group of organisms

   c. Multilocus sequence typing (MLST) compares the sequences of several housekeeping genes, giving finer phylogenetic resolution and avoiding confusion from lateral gene transfers

   d. Genomic fingerprinting uses PCR primers directed at repetitive sequences in procaryotic genomes to generate a distinct series of DNA amplicons that enables identification to the species level

  4. Amino acid sequencing is useful because it directly reflects the genetic information of the organism

   a. Determination of the amino acid sequence of the protein is the most direct approach

   b. Indirect approaches include comparison of electrophoretic mobility, determination of immunological cross-reactivity, and comparison of enzymatic properties

V. Assessing Microbial Phylogeny

 A. Molecular chronometers estimate evolutionary relatedness in time based on the assumption of a constant rate of change, which is not a correct assumption; however, the rate of change may be constant within certain genes

 B. Phylogenetic trees

  1. Made of branches that connect nodes, which represent taxonomic units such as species or genes; rooted trees provide a node that serves as the common ancestor for the organisms being analyzed

  2. Developed by comparing molecular sequences; differences in sequences are expressed as evolutionary distance; organisms are then clustered to determine relatedness; alternatively, relatedness can be estimated by parsimony analysis assuming that evolutionary change occurs along the shortest pathway with the fewest changes to get from ancestor to the organism in question

VI. The Major Divisions of Life

 A. Several different phylogenetic trees have been proposed relating the major domains; some trees do not support a three-domain pattern

 B. One of the most important difficulties in constructing a tree is widespread, frequent horizontal gene transfer; a more correct tree may resemble a web or network with many lateral branches linking various trunks

 C. Molecular versus organismal trees—molecular trees are based on gene sequences, while organismal trees reflect the organization requirements of botanists and zoologists at the expense of phylogenetics

 D. Kingdoms—the three-domain system favored by most biologists was predated by a kingdom system that was based on cell type, cell organization, and nutritional type

  1. Whittaker's five-kingdom system was the first to gain wide acceptance

   a. *Animalia*—multicellular, nonwalled eucaryotes with ingestive nutrition

   b. *Plantae*—multicellular, walled eucaryotes with photoautotrophic nutrition

   c. *Fungi*—multicellular and unicellular, walled eucaryotes with absorptive nutrition

   d. *Protista*—unicellular eucaryotes with various nutritional mechanisms

   e. *Monera* (*Procaryotae*)—all procaryotic organisms

  2. Most biologists do not accept Whittaker's system because it does not distinguish bacteria from archaea; the kingdom *Protista* is not taxonomically useful; and the boundaries between *Protista, Plantae,* and *Fungi* are ill-defined

  3. A number of alternatives have been suggested, including a six-kingdom system and a two-empire, eight-kingdom system

 E. Higher-level classification of the *Eucarya*—a hybrid system of classification mixes morphological, biochemical, and molecular phylogenetic analyses of eucaryotic organisms to define six phylogenetically coherent clusters

VII. *Bergey's Manual of Systematic Bacteriology*—contains a description of all procaryotic species

 A. The First Edition of *Bergey's Manual of Systematic Bacteriology*—primarily phenetic

 B. The Second Edition of *Bergey's Manual of Systematic Bacteriology*—largely phylogenetic rather than phenetic; consists of five volumes

VIII. A Survey of Procaryotic Phylogeny and Diversity

A. Volume 1 (of 2nd edition of *Bergey's Manual*): The *Archaea* and the Deeply Branching and Phototrophic *Bacteria*
   1. *Archaea*—divided into two phyla
      a. *Crenarchaeota*—a diverse phylum that contains thermophilic and hyperthermophilic organisms as well as some organisms that grow in oceans at low temperatures as picoplankton
      b. *Euryarchaeota*—contains primarily methanogenic and halophilic procaryotes and also thermophilic, sulfur-reducing procaryotes
   2. *Bacteria*
      a. *Aquificae*—a phylum containing autotrophic bacteria that use hydrogen as an energy source; most are thermophilic
      b. *Thermatogae*—a phylum containing anaerobic, thermophilic fermentative, gram-negative bacteria; have unusual fatty acids
      c. "Deinococcus-Thermus"—this phylum includes bacteria with extraordinary resistance to radiation and thermophilic organisms
      d. *Chloroflexi*—this phylum consists of bacteria often called green nonsulfur bacteria; some carry out anoxygenic photosynthesis, while others are respiratory, gliding bacteria; they have unusual peptidoglycans and lack lipopolysaccharides in their outer membranes
      e. *Cyanobacteria*—a phylum consisting of oxygenic photosynthetic bacteria
      f. *Chlorobi*—this phylum contains anoxygenic photosynthetic bacteria known as the green sulfur bacteria
B. Volume 2: The Proteobacteria—devoted to a single phylum called *Proteobacteria*, which consists of a diverse array of gram-negative bacteria divided into five classes
C. Volume 3: The Low G+C Gram-Positive Bacteria—devoted to a single phylum called *Firmicutes*; all have a G+C content <50%; with the exception of the mycoplasmas, which lack a cell wall, all are gram positive; most are heterotrophs; includes genera that produce endospores
D. Volume 4: The High G + C Gram-Positive Bacteria—describes the phylum *Actinobacteria*; G + C content >50–55%; includes filamentous bacteria (actinomycetes) and bacteria with unusual cell walls (mycobacteria)
E. Volume 5: The Planctomycetes, Spirochaetes, Fibrobacteres, Bacteroidetes, and Fusobacteria—an assortment of deeply branching phylogenetic groups that are not necessarily related to one another although all are gram negative
   1. *Planctomycetes*—this phylum contains bacteria with unusual features, including cell walls that lack peptidoglycan and cells with a membrane-enclosed nucleoid; they divide by budding and produce appendages called stalks
   2. *Chlamydiae*—this phylum contains obligate-intracellular pathogens having a unique life cycle; they lack peptidoglycan
   3. *Spirochaetes*—a phylum composed of helically shaped bacteria with unique morphology and motility
   4. *Bacteroidetes*—this phylum contains a number of ecologically significant bacteria

TERMS AND DEFINITIONS

Place the letter of each term in the space next to the definition or description that best matches it.

_____ 1. The science of biological classification
_____ 2. The scientific study of organisms to ultimately characterize them and arrange them in an orderly manner
_____ 3. A population of organisms that descends from a single organism
_____ 4. A classification system based on evolutionary relationships
_____ 5. A classification system based on mutually similar attributes
_____ 6. A classification system based on the general similarity of organisms, in which computers are used to calculate association coefficients
_____ 7. The temperature at which the two strands of a double-stranded DNA molecule will separate from each other
_____ 8. The phenomenon in which two strands of nucleic acid associate with each other because they share some degree of sequence homology
_____ 9. Layered or stratified rocks that are formed by incorporation of minerals into microbial mats
_____ 10. Organisms with great similarity that are grouped together by numerical taxonomy methods
_____ 11. A general term for any group into which organisms are placed
_____ 12. The evolutionary development of a species
_____ 13. Sequences of nucleic acids and proteins that change with time
_____ 14. A taxonomy based on a wide range of phenotypic and genotypic information

a. melting temperature (*Tm*)
b. molecular chronometers
c. nucleic acid hybridization
d. numerical taxonomy
e. phenetic classification
f. phenon
g. phylogenetic (phyletic) classification
h. phylogeny
i. polyphasic taxonomy
j. strain
k. stromatolites
l. systematics
m. taxon
n. taxonomy

FILL IN THE BLANK

1. Procaryotic strains that are characterized by biochemical or physiological differences are called _____; those that differ morphologically are called _____; and those that differ antigenically are called _____.

2. The most desirable classification system is one in which organisms are arranged into groups whose members share many characteristics and in which the biological nature of organisms is reflected as much as possible. These kinds of systems are called _____ classification systems. The two most common of these are _____ systems, based on evolutionary relatedness, and _____ systems, based on mutual similarity.

3. In _____ _____ a large number of characteristics are determined. After character analysis, association coefficients are calculated and arranged to form a _____ matrix, which permits comparison of each organism with every other organism in the matrix. Organisms with great mutual similarity are grouped together in _____ and separated from dissimilar organisms. The results are often summarized in a treelike diagram called a _____.

4. The _____ _____ coefficient (S_{SM}), the most commonly used coefficient in bacteriology, is the proportion of characteristics that match regardless of whether the attribute is present or absent. The _____ coefficient, on the other hand, is calculated by ignoring any characteristics that are absent in both organisms.

197

5. The science of biological classification is called _____. It is divided into three parts: _____, the arrangement of organisms into groups (_____); _____, the assignment of names to organisms; and _____, the process of determining that a particular organism belongs to a recognized group. Microorganisms and other organisms are named according to the _____ _____.

6. The 16S rRNA of most major phylogenetic groups has one or more characteristic nucleotide sequences called _____ _____ sequences, which distinguish each group from other groups, even closely related ones.

7. The classification scheme favored by microbiologists divides all organisms into three _____. The procaryotes are divided into _____ and _____, and all eucaryotic organisms are placed in _____.

8. Phylogenetic relationships are illustrated in branched diagrams called _____ _____, which can either be rooted or unrooted. These diagrams are based on calculations of the amount of difference between gene sequences. The difference is often expressed as _____ _____, a quantitative indication of the number of positions that differ between two aligned macromolecules.

9. For procaryotes, one strain, the _____ strain, is used to define the characteristics of a species. Only those strains very similar to this strain are included in the species.

10. Many believe in the _____ theory that suggests that mitochondria and chloroplasts arose from engulfed procaryotes. These can no longer survive independently and are _____ that have retained at least some of their _____. In the _____ hypothesis the progenitor of mitochondria was a _____ that produced _____ that was required by the host.

MULTIPLE CHOICE

For each of the questions below select the *best* answer or answers.

1. Which of the following has NOT been used in systematics?
    a. physiology
    b. epidemiology
    c. ecology
    d. morphology
    e. All of the above have been used in systematics.

2. What is a population of organisms that descends from a single organism?
    a. a species
    b. a genus
    c. a strain
    d. a taxon

3. Which of the following gene-exchange mechanisms is not useful in classification studies?
    a. transformation
    b. transduction
    c. conjugation
    d. All of the above have been useful in classification studies.

4. In the scheme proposed by Woese, protists would be found in which of the following domains?
    a. *Eucarya*
    b. *Bacteria*
    c. *Archaea*
    d. None of the above is correct.

5. Which of the following is NOT true about the G + C content of organisms?
    a. Organisms with similar G + C content have similar base sequences.
    b. Organisms with very different G + C content have dissimilar base sequences.
    c. Only if two organisms are alike phenotypically does their similar G + C content suggest relatedness.
    d. All of the above are true about the G + C content of organisms.

6. Which of the following is NOT considered a classical characteristic for taxonomic purposes?
    a. ecological characteristics
    b. G + C content
    c. genetic analysis
    d. morphological characteristics
    e. All of the above are classical characteristics for taxonomic purposes.

7. What is the name of the genetic technique that compares the sequences of several housekeeping genes?
    a. numerical taxonomy
    b. multilocus sequence typing
    c. molecular chronometry
    d. indirect electrophoresis

8. Which of the following is not a type of genome fingerprinting?
   a. REP
   b. BOX
   c. ERIC
   d. All of the above are used for genome fingerprinting.

9. Which of the following are the most important genes for generating phylogenetic trees?
   a. SSU rRNA genes
   b. aminoacyl tRNA synthetase genes
   c. G + C-rich genes
   d. ribozyme genes

TRUE/FALSE

_____ 1. The definition of a species as "a group of interbreeding or potentially interbreeding natural populations that are reproductively isolated from other groups" is satisfactory for higher organisms but not for microorganisms.

_____ 2. In all cases where numerical taxonomy has been applied, the results have agreed with existing classification schemes.

_____ 3. Although procaryotes do not reproduce sexually, the study of chromosomal gene exchange is sometimes useful in their classification.

_____ 4. If DNA molecules are very different in sequence, they will not form stable hybrids.

_____ 5. DNA-DNA hybridization is useful for comparing closely related organisms; however, DNA-RNA hybridization using tRNA or rRNA can be used to compare more distantly related organisms because tRNA and rRNA genes have not evolved as rapidly as most other microbial genes.

_____ 6. Phenons determined by numerical taxonomy are generally equivalent and correspond to species as determined by traditional taxonomic methods.

_____ 7. The earliest procaryotes were probably anaerobic.

_____ 8. The endosymbiotic hypothesis proposes that mitochondria and chloroplasts developed from free-living procaryotes that invaded a precursor to the eucaryotes and established a stable relationship.

_____ 9. The five-kingdom classification scheme that is currently popular clearly fits best with all of the known phylogenetic relationships and is therefore unlikely to be seriously challenged in the foreseeable future.

_____ 10. Variation in G + C content among members of a particular genus usually is less than 10%.

_____ 11. Like the first edition, the second edition of _Bergey's Manual of Systematic Bacteriology_ will use a phenetic classification scheme.

_____ 12. A procaryotic species is most often defined as a collection of strains that have many stable properties in common and differ significantly from other groups of strains.

_____ 13. Proteins are direct reflections of mRNA sequences and therefore are useful for comparing genomes of different organisms.

_____ 14. The G + C content of DNA directly reflects base sequence.

_____ 15. The 16S and 5S rRNA molecules are the only macromolecules suitable for establishing phylogenetic relationships.

_____ 16. Although most microbiologists favor the three-domain system, there are alternatives such as the five-, six-, and eight-kingdom systems.

_____ 17. Oligonucleotide signature sequences short variable sequences used for fine phylogenetic resolution.

CRITICAL THINKING

1. Phylogenetic and phenetic schemes for classifying bacteria do not always agree with each other. Why not? Under what circumstances would it be more advantageous to use a phylogenetic scheme? In what situations would a phenetic scheme be better? How can this disagreement be resolved?

2. The gene sequences for SSU rRNAs are used for phylogentic analysis and treemaking. Of all the possible genes that could have become the standard, why was this gene chosen? Why would these genes be useful in determining phylogenies? How would lateral gene transfer affect the tracing of lineages through evolutionary time?

ANSWER KEY

Terms and Definitions

1. n, 2. l, 3. j, 4. g, 5. e, 6. d, 7. a, 8.c, 9. k, 10. f, 11. m. 12. h. 13. b. 14. i

Fill in the Blank

1. biovars; morphovars; serovars 2. natural; phylogenetic; phenetic 3. numerical taxonomy; similarity; phenons; dendrogram 4. simple matching; Jaccard 5. taxonomy; classification; taxa; nomenclature; identification; binomial system; 6. oligonucleotide signature 7. domains; *Bacteria; Archaea, Eucarya* 8. phylogenetic trees; evolutionary distance 9. type 10. endosymbiotic; organelles; genome; hydrogen; α-proteobacterium; hydrogen

Multiple Choice

1. e, 2. c, 3. b, 4. a, 5. a, 6. b, 7. b, 8. d, 9. a

True/False

1. T, 2. F, 3. T, 4. T, 5. T, 6. F, 7. T, 8. T, 9. F, 10. T, 11. F, 12. T, 13. T, 14. F, 15. F, 16. T, 17. F

20 The *Archaea*

CHAPTER OVERVIEW

This chapter summarizes the properties of a diverse group of organisms known as the *Archaea*. These organisms are very different from bacteria and from eucaryotes, but also share many qualities. The chapter describes some of the major characteristics associated with each of the major groups of archaea.

CHAPTER OBJECTIVES

After reading this chapter you should be able to:

- discuss the morphological and physiological diversity of the archaea
- describe the membrane lipid composition and cell walls of the archaea
- discuss the general genetic, molecular, and metabolic characteristics of the archaea
- discuss the habitats that are typical for the archaea
- discuss the classification scheme for the archaea used in *Bergey's Manual*
- discuss the unique cofactors used by methanogenic and sulfate-reducing archaea
- describe the structural, chemical, and metabolic adaptations that allow the archaea to grow in extreme environments

CHAPTER OUTLINE

I. Introduction to the Archaea
    A. The archaea are quite diverse in morphology, and physiology
        1. They may be spherical, rod-shaped, spiral, lobed, plate-shaped, irregularly shaped, or pleomorphic
        2. They may exist as single cells, aggregates, or filaments
        3. They may multiply by binary fission, budding, fragmentation, or other mechanisms
        4. They may be aerobic, facultatively anaerobic, or strictly anaerobic
        5. Nutritionally, they range from chemolithoautotrophs to organotrophs
    B. Ecology
        1. Some are mesophiles, while others are hyperthermophiles that can grow above 100°C
        2. They are often found in extreme aquatic and terrestrial habitats; recently, archaea have been found in cold environments and may constitute up to 34% of the procaryotic biomass in Antarctic surface waters; a few are symbionts in animal digestive systems
    C. Archaeal cell walls and membranes
        1. Archaeal cell walls are quite diverse and do not contain bacterial peptidoglycan (no muramic acid or D-amino acids); may contain pseudomurein, chondroitin sulfate-like polysaccharides, or protein walls
        2. The membrane lipids of archaea differ from those of other organisms in having ether linkages (rather than esters), long tetraethers, and pentacyclic rings
    D. Genetics and molecular biology
        1. The archaeal chromosomes that have been studied consist of a single, closed DNA circle as seen in bacteria, except that some are considerably smaller; *Archaea* have few plasmids; genomic analysis suggests they are as distinctive genotypically as they are in other respects
        2. Archaeal DNA replication combines features of both bacteria and eucaryotes; it is bidirectional like bacteria, but uses eucaryotic-like initiation proteins; some archaeal genomes have histone proteins like those in eucaryotes

3. Transcription in archaea uses an RNA polymerase very much like that of eucaryotes and promoter recognition is dependent on eucaryotic-like transcription factor proteins
4. Archaeal mRNA is like that of bacteria (i.e., it may be polygenic, and there is no evidence of intron-containing precursors); archaeal promoters are similar to those of bacteria
5. Archaeal tRNAs have modified bases not found in the other domains; the initiator tRNA carries methionine like eucaryotes; ribosomal structure appears to be unique but has antibiotic sensitivities similar to eucaryotes
6. Archaea use a Sec-dependent pathway for protein secretion that relies on binding to signal peptides like bacteria, but some of the protein machinery resembles that of eucaryotes

E. Metabolism
1. Carbohydrate metabolism is best understood
   a. Archaea do not use the Embden-Meyerhof pathway for glucose catabolism; however, they frequently use a reversal of that pathway for gluconeogenesis
   b. Some (halophiles and extreme thermophiles) have a complete TCA cycle while others (methanogens) do not
2. Archaeal biosynthetic pathways appear to be similar to those of other organisms
3. Autotrophy is widespread; the reductive TCA cycle and the reductive acetyl-CoA cycle are used for carbon fixation

F. Archaeal taxonomy
1. *Bergey's Manual* divides the archaea into two phyla: *Euryarchaeota* and *Crenarchaeota*
2. *Euryarchaeota* are metabolically and ecologically diverse and include methanogens, halophiles, and thermophiles
3. *Crenarchaeota* are considered a more ancient lineage—that is, they are almost entirely thermophiles and hyperthermophiles

II. Phylum *Crenarchaeota*—consists of one class divided into three orders (*Thermoproteales, Sulfolobales, and Desulfurococcales*)
A. Many are extremely thermophilic, acidophilic, and sulfur-dependent, although there is molecular evidence that *Crenarchaeota* may be widespread in mesophilic environments
1. Sulfur may be used as an electron acceptor in anaerobic respiration, or as an electron source by lithotrophs
2. Almost all are strict anaerobes
3. They grow in geothermally heated water or soils (solfatara) that contain elemental sulfur (e.g., sulfur-rich hot springs and waters surrounding submarine volcanic activity); some (e.g., *Pyrodictum* spp.) can grow quite well above the boiling point of water (optimum 105°C)
4. Some are organotrophic; others are lithotrophic
5. There are 25 genera; two of the better-studied genera are *Sulfolobus* and *Thermoproteus*

B. Genus *Sulfolobus*
1. Stain gram negative; are aerobic, irregularly lobed, spherical bacteria
2. Thermoacidophiles
3. Cell walls lack peptidoglycan but contain lipoproteins and carbohydrates
4. Oxidize sulfur to sulfuric acid; oxygen is the normal electron acceptor, but ferric iron can also be used
5. Sugars and amino acids may serve as carbon and energy sources

C. Genus *Thermoproteus*
1. Long, thin, bent or branched rods
2. Cell wall is composed of glycoprotein
3. Strict anaerobes
4. They have temperature optima from 70–97°C and pH optima from 2.5 to 6.5
5. They grow in hot springs and other hot aquatic habitats that contain elemental sulfur
6. They carry out anaerobic respiration using organic molecules as electron donors and elemental sulfur as the electron acceptor; they can also grow lithotrophically using H_2 and S^0 as electron donors and CO or CO_2 as the sole carbon source

III. Phylum *Euryarchaeota*
A. The Methanogens— consists of 26 genera

1. They are strict anaerobes that obtain energy by converting CO_2, H_2, formate, methanol, acetate, and other compounds to either methane or to methane and CO_2; there are at least five orders, which differ greatly in shape, 16S rRNA sequence, cell wall chemistry and structure, membrane lipids, and other features
2. Methanogens belonging to the order *Methanopyrales* have been suggested to be among the earliest organisms to evolve on Earth
3. Methanogenesis is an unusual metabolic process and methanogens contain several unusual cofactors
4. They thrive in anaerobic environments rich in organic matter, such as animal rumens and intestinal tracts, freshwater and marine sediments, swamps, marshes, hot springs, anaerobic sludge digesters, and within anaerobic protozoa
5. They are of great potential importance because methane is a clean-burning fuel and an excellent energy source
6. They may be an ecological problem, however, because methane is a greenhouse gas that could contribute to global warming and also because methanogens can oxidize iron, which contributes significantly to the corrosion of iron pipes

B. The Halobacteria—consists of 17 genera
1. Aerobic chemoheterotrophs with respiratory metabolism; are nonmotile or motile; found in many different cell shapes including cubes and pyramids
2. Require at least 1.5 M NaCl and have growth optima near 3–4 M NaCl (if the NaCl concentration drops below 1.5 M the cell walls disintegrate; because of this they are found in high-salinity habitats and can cause spoilage of salted foods)
3. *Halobacterium salinarum* uses four different light-utilizing rhodopsin molecules
   a. Bacteriorhodopsin uses light energy to drive outward proton transport for ATP synthesis; thus they carry out a type of photosynthesis that does not involve chlorophyll
   b. Halorhodopsin uses light energy to transport chloride ions into the cell to maintain a 4–5 M intracellular KCl concentration
   c. Two sensory rhodopsins act as photoreceptors that control flagellar activity to position the bacterium in the water column at a location of high light intensity, but one in which the UV light is not sufficiently intense to be lethal

C. The Thermoplasms—consists of three genera
1. Thermoacidophiles that lack cell walls
2. Genus *Thermoplasma*
   a. Frequently found in coal mine refuse, in which chemolithotrophic bacteria oxidize iron pyrite to sulfuric acid and thereby produce a hot acidic environment
   b. Optimum temperature for growth is 55–59°C and the optimal PH is 1 to 2
   c. Cell membrane is strengthened by large quantities of diglycerol tetraethers, lipopolysaccharides, and glycoproteins
   d. Histonelike proteins stabilize the DNA; the DNA-protein complex forms particles resembling eucaryotic nucleosomes
   e. At 59°C *Thermoplasma* takes the form of an irregular filament; the cells may be flagellated and motile
3. Genus *Picrophilus*
   a. Isolated from hot solfateric fields
   b. Has an S-layer outside the plasma membrane
   c. Irregularly shaped cocci with large cytoplasmic cavities that are not membrane bounded
   d. Aerobic and grows between 47°C and 65°C with an optimum of 60°C
   e. It grows only below pH 3.5, has an optimum of pH 0.7 and will grow at or near pH 0

D. Extremely thermophilic S^0 metabolizers—class *Thermococci*, consists of three genera
1. Strictly anaerobic, reduce sulfur to sulfide
2. Are motile by means of flagella
3. Have optimum growth temperatures around 88–100°C

E. Sulfate-reducing archaea—Class *Archaeoglobi*, consisting of one major genus, *Archaeoglobus,* and two others

1. Gram-negative, irregular coccoid cells with walls of glycoprotein subunits
2. Use a variety of electron donors (hydrogen, lactate, glucose) and reduce sulfite, sulfate, or thiosulfate to sulfide
3. Are extremely thermophilic (optimum around 83°C); they are usually found near marine hydrothermal vents
4. Contain two methanogen coenzymes

TERMS AND DEFINITIONS

Place the letter of each term in the space next to the definition or description that best matches it.

_____ 1. A peptidoglycan-like polymer in the cell walls of some methanogenic archaea

_____ 2. A strict anaerobe that obtains energy by converting CO_2, H_2, formate, methanol, acetate, and other compounds to methane

_____ 3. An organism that requires at least 1.5 M NaCl in order to grow

_____ 4. A modified cell membrane that carries out photosynthesis in the absence of chlorophyll

_____ 5. A protein that mediates photosynthesis without chlorophyll

_____ 6. An organism with a temperature growth optimum around 80°C and a pH growth optimum from 1.0 to 6.5

_____ 7. Photoreceptor that controls flagellar activity to position *Halobacterium* in the water column

_____ 8. Thermoacidophilic bacteria that lack cell walls and grow in refuse piles of coal mines

a. bacteriorhodopsin
b. extreme halophile
c. methanogen
d. pseudomurein
e. purple membrane
f. sensory rhodopsin
g. thermoacidophile
h. thermoplasma

ARCHAEAL TAXONOMY

Complete the following table.

Phylum	Representative Genera	Important Characteristics (structural, physiological, ecological)
Crenarchaeota		
Euryarchaeota		
Methanogens (Classes *Methanobacteria, Methanococci* and *Methanopyri*)		
Halobacteria (Class *Halobacteria*)		
Thermoplasms (Class *Thermoplasmata*)		
Extremely thermophilic S^0 metabolizers (Class *Thermococci*)		
Sulfate reducers (Class *Archaeoglobi*)		

FILL IN THE BLANK

1. Living organisms can be divided into three domains; *Eucarya, Bacteria,* and _____. The latter have many distinguishing characteristics. For instance, they lack peptidoglycan in their cell walls; instead, some have a peptidoglycan-like component called _____.
2. The _____ and extreme thermophiles have a complete TCA cycle, but _____ do not.
3. Many members of the phylum *Crenarchaeota* (e.g., genus *Solfolobus*) grow in geothermally heated waters having a low pH. Such organisms are called _____.

206

4. Under certain environmental conditions, some strains of *Halobacterium* synthesize a modified cell membrane called the _____ membrane, which contains the protein _____. This protein drives outward proton transport, creating a proton gradient that can be used to synthesize ATP.

5. The metabolically and ecologically diverse _____ includes methanogens, _____ and _____, while the more ancient lineage, _____, are almost entirely _____ and _____.

TRUE/FALSE

_____ 1. Archaea are all strict anaerobes.
_____ 2. *Thermoproteus* is found in hot aquatic habitats that contain elemental sulfur.
_____ 3. Halorhodopsin uses light energy to pump protons for ATP synthesis.
_____ 4. The chromosomal DNA of some archaea is smaller than the chromosomal DNA of bacteria.
_____ 5. *Archaeoglobus* uses elemental sulfur as an electron acceptor.
_____ 6. Recently, archaea have been found in cold environments such as Antarctic surface waters.
_____ 7. The genome of *Methanococcus jannaschii* has been completely sequenced and has been found to be genotypically indistinct from the bacteria.
_____ 8. *Picrophilus* has a pH optimum below 1 and can even grow at or near pH 0.
_____ 9. Autotrophic archaea use the reductive TCA cycle and the acetyl-CoA cycle to fix carbon.

MULTIPLE CHOICE

For each of the questions below select the *one best* answer.

1. Which of the following is used by archaea as a mechanism for reproduction?
   a. binary fission
   b. budding
   c. fragmentation
   d. All of the above are correct.
3. Which of the following is true about archaeal mRNA?
   a. Polygenic mRNA has been found.
   b. mRNA splicing has not been found.
   c. Both (a) and (b) are correct.
   d. Neither (a) nor (b) is correct.
4. Which of the following is true about archaeal ribosomes?
   a. They are 70S, like bacterial ribosomes.
   b. Their shape differs from both bacterial and eucaryotic ribosomes.
   c. They have antibiotic sensitivities similar to those of eucaryotic ribosomes.
   d. All of the above are true.
5. Which of the following groups does not have a functional TCA cycle?
   a. methanogens
   b. extreme halophiles
   c. extreme thermophiles
   d. All of the above have a functional TCA cycle.
6. Which is the largest group of archaea?
   a. the methanogens
   b. the halobacteria

   c. the thermoplasms
   d. the thermococci
7. Which of the following is NOT true about methanogens?
   a. They are potentially of great importance because methane is an even-burning fuel and an excellent energy source.
   b. They may contribute to the greenhouse effect and global warming.
   c. They may cause corrosion in iron pipes.
   d. All of the above are true about methanogens.

8. What happens to the cell wall of *Halobacterium* if the NaCl concentration drops to about 1.5 M?
   a. The cell wall becomes more rigid.
   b. The cell wall loses permeability.
   c. The cell wall disintegrates.
   d. All of the above are correct.
10. Which of the following is true about *Sulfolobus?*
   a. Oxygen is the normal electron acceptor.
   b. Ferric ion can be used as an electron acceptor.
   c. Both (a) and (b) are correct.
   d. Neither (a) nor (b) is correct.

11. Which of the following is true about the genus *Thermoplasma?*
   a. They are spherical at temperatures below 59°C.
   b. They are filamentous at temperatures at or above 59°C.
   c. Both (a) and (b) are correct.
   d. Neither (a) nor (b) is correct.
12. Which of the following do not have cell walls?
   a. halobacteria
   b. thermoplasms
   c. thermococci
   d. None of the above have cell walls.

CRITICAL THINKING

1. In the future, certain members of the archaea may provide a pollution-free source of energy. Explain.

ANSWER KEY

Terms and Definitions

1. d, 2. c, 3. b, 4. e, 5. a, 6. g, 7. f, 8. h

Fill in the Blank

1. *Archaea*; pseudomurein 2. halophiles; methanogens 3. thermoacidophiles 4. purple; bacteriorhodopsin 5. *Euryarchaeota*; halophiles; thermophiles; *Crenarchaeota*; thermophiles; hyperthermophiles

True/False

1. F, 2. T, 3. F, 4. T, 5. F, 6. T, 7. F, 8. T, 9.T

Multiple Choice

1. d, 2. b, 3. c, 4. d, 5. a, 6. a, 7. d, 8. c, 9. d, 10. c, 11. c, 12. b

21 Bacteria: The Deinococci and Nonproteobacteria Gram Negatives

CHAPTER OVERVIEW

This chapter is devoted to some of the more interesting and important bacterial phyla from volumes 1 and 5 of *Bergey's Manual of Systematic Bacteriology*. The distinguishing characteristics, morphology, reproduction, physiology, and ecology of each phylum are included. The taxonomy of each phylum is summarized and representative species are discussed.

CHAPTER OBJECTIVES

After reading this chapter you should be able to:

- discuss the deeply branching bacterial phyla *Aquificae* and *Thermotogae*
- discuss the deinococci, focusing on their extraordinary resistance to desiccation and radiation
- compare and contrast the phyla of photosynthetic bacteria
- discuss the unique structural features of *Planctomycetes*
- discuss the unique lifestyle of the Chlamydiae
- discuss the unique structural features and motility of the *Spirochaetes*
- discuss the important metabolic and ecological characteristics of the *Bacteroidetes*
- discuss gliding motility

CHAPTER OUTLINE

I. *Aquificae* and *Thermotogae*
    A. Phylum *Aquificae*—thought to represent the deepest (oldest) branch of bacteria; two of its best studied genera are *Aquifex* and *Hydrogenobacter*
        1. Hyperthermophilic with optimum of 85°C and maximum of 95°C
        2. Chemolithoautotrophic—generate energy by oxidizing electron donors such as hydrogen, thiosulfate, and sulfur with oxygen as the electron acceptor
    B. Phylum *Thermotogae*—second-deepest branch of the bacteria; best studied are members of the genus *Thermotoga*
        1. Hyperthermophiles with an optimum of 80°C and a maximum of 90°C
        2. Gram-negative rods with an outer sheath-like envelope (like a toga) that can balloon out from the ends of the cell
        3. Grow in active geothermal areas (e.g., marine hydrothermal vents and terrestrial solfataric springs)
        4. Chemoheterotrophs with a functional glycolytic pathway; can grow anaerobically on carbohydrates and protein digests
II. Deinococcus-Thermus
    A. Consists of one class, two orders, and three genera; genus *Deinococcus* is the best studied
    B. Genus *Deinococcus*
        1. Spherical or rod-shaped; often associated in pairs or tetrads
        2. Aerobic, mesophilic, catalase positive, and usually able to produce acid from only a few sugars
        3. They stain gram positive but have a layered cell wall and an outer membrane like gram-negative bacteria; have L-ornithine in their peptidoglycan and lack teichoic acid

4. Have a plasma membrane with large amounts of palmitoleic acid rather than phosphatidylglycerol phospholipids
5. Can be isolated from ground meat, feces, air, freshwater, and other sources but their natural habitat is not known
6. Extraordinarily resistant to desiccation and radiation; their genome structure and unusual ability to repair chromosomal damage (even fragmentation) probably accounts for this ability
7. Genome consists of two circular chromosomes, a mega plasmid, and a small plasmid; the chromosome exists as a compact, ring-like structure and this may help maintain continuity even after radiation-induced breaks occur; accumulation of Mn(II) may protect against radiation-induced toxic oxygen species

III. Photosynthetic Bacteria
   A. Both oxygenic and anoxygenic photosynthesis are observed in photosynthetic bacteria
      1. Cyanobacteria carry out oxygenic photosynthesis, using water as an electron source for the generation of NADH and NADPH
      2. Green and purple bacteria carry out anoxygenic photosynthesis, using reduced molecules other than water as electron sources for the generation of NADH and NADPH
         a. Purple sulfur bacteria use reduced sulfur compounds as electron sources and accumulate sulfur granules *within* their cells
         b. Green sulfur bacteria use reduced sulfur compounds as electron sources and deposit sulfur granules *outside* their cells
         c. Purple nonsulfur bacteria use organic molecules as electron sources
   B. There is a correlation between the type of photosynthetic pigments, oxygen relationships, and ecological distribution
      1. Purple and green bacteria are anaerobes and use bacteriochlorophyll
         a. Grow better in deeper, anaerobic zones of aquatic habitats
         b. Their bacteriochlorophylls absorb shorter wavelengths of light, which penetrate to these deeper zones
      2. Cyanobacteria have chlorophyll *a*, which absorbs longer wavelengths of light; these bacteria are found primarily at the surface of bodies of water
   C. *Bergey's Manual* divides the photosynthetic bacteria into six groups distributed into five bacterial phyla; there appears to have been considerable horizontal transfer of photosynthetic genes between the five phyla:
      1. Phylum *Chlorobi*—green sulfur bacteria
      2. Phylum *Chloroflexi*—green nonsulfur bacteria
      3. Phylum *Cyanobacteria*
      4. Phylum *Proteobacteria*—purple sulfur bacteria (γ-proteobacteria) and purple nonsulfur bacteria (α-proteobacteria and β-proteobacteria)
      5. Phylum *Firmicutes*—heliobacteria
   D. Phylum *Chlorobi*—green sulfur bacteria
      1. Consists of one class, one order, and one family; representative genera include *Chlorobium, Prosthecochloris,* and *Pelodictyon*
      2. Obligately anaerobic photolithoautotrophs that use hydrogen sulfide, elemental sulfur, and hydrogen as electron sources; elemental sulfur produced by sulfide oxidation is deposited outside the cell
      3. Photosynthetic pigments are located in ellipsoidal vesicles called chlorosomes, which are attached to the plasma membrane but are not continuous with it; the chlorosome membrane is not a normal lipid bilayer; chlorosomes have accessory bacteriochlorophylls but the reaction center bacteriochlorophyll is located in the plasma membrane
      4. Flourish in anaerobic, sulfide-rich zones of lakes; although they lack flagella and are nonmotile, some species have gas vesicles to adjust their depth in water for adequate light and hydrogen sulfide; species without gas vesicles are found in sulfide-rich mud at the bottom of lakes and ponds.
      5. Morphologically diverse (rods, cocci, or vibrios; grow singly, in chains, or in clusters); are grass-green or chocolate-brown in color

E. Phylum *Chloroflexi*—green nonsulfur bacteria
1. A deep ancient branch in the bacterial tree
2. Genus *Chloroflexus*—major representative of the *photosynthetic* green nonsulfur bacteria
   a. Filamentous, gliding bacteria
   b. Thermophilic; often isolated from neutral to alkaline hot springs where they grow in the form of orange-reddish mats
   c. Ultrastructure and photosynthetic pigments are like green bacteria, but their metabolism is similar to that of the purple nonsulfur bacteria
   d. Can carry out anoxygenic photosynthesis with organic compounds as carbon sources or can grow aerobically as a chemoheterotroph
3. Genus *Herpetosiphon*—represents nonphotosynthetic members of phylum *Chloroflexi*; contains gliding, rod-shaped filamentous bacteria; aerobic chemoorganotrophs with respiratory metabolism; isolated from fresh water and soil
F. Phylum *Cyanobacteria*
1. This largest and most diverse group of photosynthetic bacteria consists of true procaryotes; however, their photosynthetic systems resemble those of eucaryotes; in addition they are metabolically flexible
   a. Have chlorophyll *a* and photosystem II, and carry out oxygenic photosynthesis
   b. Photosynthetic pigments are in thylakoid membranes lined with particles called phycobilisomes (contain phycobilin pigments), which transfer energy to photosystem II
   c. Fix carbon dioxide by the Calvin cycle (enzymes of which are found in carboxysomes), reserve carbohydrate as glycogen, and nitrogen as cyanophycin
   d. Do not have functional TCA cycle; the pentose phosphate pathway plays a central role in their metabolism
   e. Some can grow slowly in the dark as chemoheterotrophs, and some species can carry out anoxygenic photosynthesis if in an anaerobic environment
2. Vary greatly in shape and appearance
   a. May be unicellular, exist as colonies of many shapes, or form filaments called trichomes (rows of bacterial cells that are in close contact with one another over a large area)
   b. Most appear blue-green because of phycocyanin, but some species are red-brown because of the pigment phycoerythrin; the relative amounts of these pigments can be modulated through chromatic adaptation
   c. Have typical procaryotic structures with a gram-negative cell wall
   d. Often use gas vesicles to move vertically in the water (a form of phototaxis); many filamentous cyanobacteria have a gliding motility; although cyanobacteria lack flagella, some marine species are able to swim by an unknown mechanism
3. Reproduce by binary fission, budding, fragmentation, and multiple fission
   a. Fragmentation of filamentous cyanobacteria generates small, motile filaments called hormogonia
   b. Some species develop akinetes, thick-walled resting cells that are resistant to desiccation; these often germinate to form new filaments
4. Many filamentous cyanobacteria fix atmospheric nitrogen in special cells (heterocysts), which protect the oxygen-sensitive nitrogenase; other cyanobacteria that lack heterocysts also can fix nitrogen
5. The taxonomy of cyanobacteria is unsettled; *Bergey's Manual* divides them into five subsections with 56 genera
   a. The five subsections differ markedly in terms of morphology and reproduction
      1) Subsection I—unicellular rods or cocci; most are nonmotile; reproduce by binary fission or budding
      2) Subsection II—unicellular, though some may be held together in an aggregate by an outer wall; reproduce by multiple fission to form baeocytes
      3) Subsection III—filamentous cyanobacteria with branched trichomes
      4) Subsection IV—unbranched filamentous cyanobacteria that can form heterocysts and akinetes

          5)   Subsection V—branched filamentous cyanobacteria that can form heterocysts and akinetes

        b.   Prochlorophytes include the genera *Prochloron*, *Prochlorococcus*, and *Prochlorothrix*
           1)   Differ from other cyanobacteria by having chlorophyll *b* as well as chlorophyll *a* and by lacking phycobilisomes
           2)   These are the best candidates for the endosymbiont that became chloroplasts
           3)   *Prochlorococcus* may be the most abundant oxygenic photosynthetic organism on Earth

     6.   Are tolerant of environmental extremes; thermophilic species can grow at temperatures up to 75°C; they are also successful at establishing symbiotic relationships (e.g., in lichens; with protozoa, fungi and plants)

IV.   Phylum *Planctomycetes*
    A.   Contains one class, one order, and four genera
    B.   Spherical or oval, budding bacteria that lack peptidoglycan and have distinctive crateriform structures (pits) in their cell walls
    C.   In two genera, *Gemmata* and *Pirullela*, the nuclear body is membrane bounded, something that is not seen in other procaryotes
    D.   A recently discovered form of chemolithotrophy is seen in some species where anaerobic ammonia oxidation in anammoxosomes uses ammonia as an electron source and nitrite as a terminal electron acceptor yielding nitrogen gas
    E.   The genus *Planctomyces* attaches to surfaces through a stalk and holdfast; other genera lack stalks
    F.   Most have life cycles in which sessile cells bud to produce motile swarmer cells

V.   Phylum *Chlamydiae*
    A.   This phylum has one class, one order, four families, and six genera; *Chlamydia* is the most important and best-studied genus
    B.   Obligately intracellular parasites that are 0.2 to 1.5 μm in size
    C.   Genus *Chlamydia*
        1.   Nonmotile, coccoid, gram-negative bacteria
        2.   Reproduce within cytoplasmic vesicles of host cells by a unique developmental cycle involving elementary bodies (EBs) and reticulate bodies (RBs)
           a.   Begins with attachment of an EB to host cell
           b.   Host cell phagocytoses the EB, but fusion of lysosome with the phagosome is prevented by the EB
           c.   EB reorganizes itself into a RB, which is specialized for reproduction
           d.   RB reproduces repeatedly, giving rise to many RBs, all within a vacuole
           e.   RBs change back into EBs, and these are released when the host cell lyses
        3.   Cell wall lacks muramic acid and peptidoglycan; EBs use cross-linking of outer membrane proteins, and possibly, periplasmic proteins to achieve osmotic stability
        4.   Found mostly in mammals and birds, but have been recently isolated from spiders, clams, and freshwater invertebrates
        5.   Have one of the smallest procaryotic genomes
        6.   Usually thought of as being completely dependent on host for ATP (using translocases for uptake); however, recent genomic analysis indicates that some genes for ATP synthesis are present in the genome; RBs have a number of biosynthetic capabilities (e.g., DNA, RNA, glycogen, lipid, protein, some amino acids, and coenzymes); EBs have very little metabolic activity and seem to be dormant forms concerned exclusively with transmission and infection
        7.   Three are recognized human pathogens
           a.   *C. trachomatis*—trachoma, nongonococcal urethritis, and other diseases in humans and mice
           b.   *C. psittaci*—causes psittacosis in humans and infects many other mammals as well; invades the respiratory and genital tracts, the placenta, developing fetuses, the eye, and synovial fluid of the joints
           c.   *C. pneumoniae*—a common cause of human pneumonia

VI.   Phylum *Spirochaetes*

213

A. Consists of one class, one order, three families, and 13 genera
B. Gram-negative, chemoheterotrophic, flexibly helical bacteria that exhibit a creeping (crawling) motility due to a structure called an axial filament
C. The axial filament (a complex of periplasmic flagella) lies in a flexible outer sheath (outer membrane) outside the protoplasmic cylinder (houses the nucleoid and cytoplasm); function of the sheath is essential (spirochetes will die if it is removed) but unknown
D. Flagellar rotation is responsible for motility by an unknown mechanism, presumably by rotating the outer sheath or flexing the cell for a crawling motion
E. Can be anaerobic, facultatively anaerobic, or aerobic and can use a diverse array of organic molecules as carbon and energy sources
F. Ecologically diverse
   1. *Spirochaeta*—free-living and often found in anaerobic, sulfide-rich aquatic environments
   2. *Leptospira*—aerobic water and moist soils
   3. Many, including *Criptispira* and *Treponema* form symbiotic associations with other organisms
   4. Some members of *Treponema*, *Borrelia*, and *Leptospira* cause disease (e.g., *T. pallidum* is the causative agent of syphilis, and *B. burgdorferi* is the causative agent of Lyme disease)

VII. Phylum *Bacteroidetes*
A. Consists of three classes (*Bacteroides*, *Flavobacteria*, and *Sphingobacteria*), 12 families, and 63 genera
B. Class *Bacteroides*
   1. Obligate anaerobes, nonsporing, chemoheterotrophic, fermentative, rods
   2. Found in oral cavity and intestinal tract of humans and other animals and the rumen of ruminants where they often benefit the host by degrading cellulose, pectins, and other complex carbohydrates, thereby providing extra nutrition for the host
   3. Some species can be associated with disease
C. Class *Sphingobacteria*
   1. Often have sphingolipids in their cell walls
   2. Contains several genera including *Flexibacter*, *Cytophaga* and *Sporocytophaga;* they differ in morphology, life cycle, and physiology
      a. *Cytophaga* are slender rods with pointed ends and exhibit gliding motility; *Sporocytophaga* are similar to *Cytophaga* but form spherical resting cells called microcysts; *Flexibacter* form long threads
      b. *Cytophaga* and *Sporocytophaga* are aerobes that actively degrade complex carbohydrates (e.g., cellulose, chitin, keratin); they play a major role in the mineralization of organic matter and can damage exposed wooden structures; they contribute significantly to wastewater treatment; *Flexibacter* are unable to degrade complex carbohydrates
      c. Most cytophagas are free-living, but some pathogenic species are known (e.g., *C. columnaris* causes disease in freshwater and marine fish)
      d. Cytophagas are nonmotile when in suspension, but exhibit gliding motility when in contact with a surface; they leave a slime trail; gliding motility has several advantages; these advantages include enabling them to find and digest insoluble material encountered as they move, allowing motility in drier habitats, and enabling them to position themselves for optimal environmental conditions

214

TERMS AND DEFINITIONS

Place the letter of each term in the space next to the definition or description that best matches it.

_____ 1. A photosynthetic process in which water is used as an electron donor

_____ 2. A photosynthetic process that does not involve water as an electron donor

_____ 3. Ellipsoidal vesicles containing the green sulfur bacteria's photosynthetic pigments; these vesicles are attached to but are not continuous with the plasma membrane

_____ 4. Particles that line the thylakoid membranes of most cyanobacteria

_____ 5. Small, motile filaments produced by fragmentation of filamentous cyanobacteria

_____ 6. Dormant, thick-walled resting cells of cyanobacteria that are resistant to desiccation

_____ 7. Special cells of cyanobacteria that are able to fix atmospheric nitrogen

_____ 8. Small reproductive cells formed by those species of cyanobacteria that reproduce by multiple fission; these escape when the outer wall ruptures

_____ 9. A row of bacterial cells that are in close contact with one another over a large area

_____ 10. The infective form of chlamydiae; it is taken into the host cell by phagocytosis

_____ 11. The form of chlamydiae that is specialized for reproduction

_____ 12. A complex of periplasmic flagella that mediate spirochete movement

_____ 13. The flagella that extend from both ends of the protoplasmic cylinder of spirochetes

a. akinetes
b. anoxygenic photosynthesis
c. axial fibrils (periplasmic flagella)
d. axial filament
e. baeocytes
f. chlorosomes
g. elementary body
h. heterocysts
i. hormogonia
j. oxygenic photosynthesis
k. phycobilisomes
l. reticulate body (initial body)
m. trichome

FILL IN THE BLANK

1. In _____ photosynthesis, water is used as the electron donor and oxygen is produced. In contrast, _____ photosynthesis does not involve water, but instead uses reduced molecules such as hydrogen sulfide as electron sources for the generation of NADH and NADPH.
2. Some filamentous cyanobacteria fix atmospheric nitrogen using special cells called _____.
3. The infectious stage of chlamydiae is called the _____ body, while the reproductive stage is called the _____ body.
4. The bacteriochlorophyll pigments of _____ _____ bacteria, the _____ _____ bacteria, and the purple bacteria enable them to live in deeper, anaerobic zones of aquatic habitats. The chlorophyll pigments characteristic of the _____ limit their growth to upper levels in water columns.
5. Members of the phylum *Chloroflexi* are also known as the _____ _____ bacteria. Many of these bacteria exhibit _____ motility and can carry out _____ photosynthesis.
6. Members of the phylum *Chlorobi* are also known as the _____ _____ bacteria. All carry out _____ photosynthesis using hydrogen sulfide, elemental sulfur, and hydrogen as electron sources. Their photosynthetic pigments are located in vesicles called _____.

THE DEINOCOCCI AND NONPROTEOBACTERIA GRAM NEGATIVES

Phylum	Representative Genera	Important Characteristics (structural, physiological, ecological)
Aquificae		
Thermotogae		
Deinococcus-Thermus		
Chloroflexi		
Chlorobi		
Cyanobacteria		
Planctomycetes		
Chlamydiae		
Spirochaetes		
Bacteroidetes		
Prochlorophytes		

MULTIPLE CHOICE

For each of the questions below select the *one best* answer.

1. Which of the following are capable of carrying out oxygenic photosynthesis?
   a. purple bacteria
   b. green bacteria
   c. cyanobacteria
   d. All of the above are capable of oxygenic photosynthesis.

2. Which of the following accumulates sulfur granules outside the cell?
   a. purple sulfur bacteria
   b. green sulfur bacteria
   c. cyanobacteria
   d. colorless sulfur bacteria

3. Which of the following activities is associated with cytophagas?
   a. damage to fishing gear and wood structures
   b. digestion of sewage in sewage treatment plants
   c. Both (a) and (b) are associated with cytophagas.
   d. Neither (a) nor (b) is associated with cytophagas.

4. Which of the following is thought to be the oldest branch of bacteria?
   a. *Thermotogae*
   b. *Aquificae*
   c. *Cyanobacteria*
   d. *Spirochaetes*

5. Where do members of the phylum *Thermotogae* generally grow?
   a. marine hydrothermal vents
   b. terrestrial solfataric hot springs
   c. Both (a) and (b) are correct.
   d. Neither (a) nor (b) is correct.

6. Which of the following is extremely radiation resistant?
   a. deinococci
   b. *Aquificae*
   c. Both (a) and (b) are correct.
   d. Neither (a) nor (b) is correct.

7. Which of the following contain chlorophyll *a*?
   a. *Chlorobi*
   b. *Cyanobacteria*
   c. *Chloroflexi*
   d. all of the above

8. In which of the following phyla can membrane-bound nuclear bodies be observed?
   a. *Planctomycetes*
   b. *Chlamydiae*
   c. *Bacteroidetes*
   d. *Spirochaetes*

9. Cyanobacterial heterocysts have very thick walls and lack photosystem II. Why are these characteristics important?
   a. They allow the heterocysts to resist desiccation.
   b. They protect the oxygen-sensitive nitrogenase enzyme.
   c. Both (a) and (b) are correct.
   d. Neither (a) nor (b) is correct.

10. Prochlorophytes differ from cyanobacteria by not having:
    a. chlorophyll *a*
    b. chlorophyll *b*
    c. chloroplasts
    d. phycobilisomes

TRUE/FALSE

_____ 1. The cyanobacteria comprise a small group of photosynthetic bacteria and do not differ greatly from one another.

_____ 2. All of the cyanobacteria have a blue-green color that comes from the pigment phycocyanin.

_____ 3. The photosynthetic partner in most lichen associations is generally a cyanobacterium.

_____ 4. A trichome is a bacterial cell with three different photosynthetic pigments.

_____ 5. The deinococci stain gram positive but have a layered cell wall and an outer membrane that is more like the cell wall of a gram-negative organism.

_____ 6. The deinococci have an unusual ability to repair chromosome damage, even fragmentation.

_____ 7. The chlorosomes of *Chlorobi* are continuous with the plasma membrane.

_____ 8. The flexible outer sheath in which the axial filaments of spirochetes lay is nonessential (i.e., the bacteria will survive if it is removed).

_____ 9. Members of the genus *Bacteroides* constitute as much as 30% of the bacteria isolated from human feces.

_____ 10. Many bacteria with gliding motility are able to use as a nutrient source the insoluble material encountered while gliding.

_____ 11. Cytophagas contribute significantly to waste water treatment.

_____ 12. The phyla *Planctomycetes* and *Chlamydiae* lack peptidoglycan in their cell walls.

CRITICAL THINKING

1. In what ways are cyanobacteria similar to eucaryotic phototrophs? Why, then, are they classified in the domain *Bacteria* (i.e., in what ways are they more like bacteria)?

2. Compare and contrast oxygenic and anoxygenic photosynthesis in terms of both substrates and products.

3. The chlamydiae are only slightly larger than poxviruses and are obligately intracellular parasites. These characteristics make them similar to viruses. What features distinguish them from viruses?

ANSWER KEY

Terms and Definitions

1. j, 2. b, 3. f, 4. k, 5. i, 6. a, 7. h, 8. e, 9. m, 10. g, 11. l, 12. d, 13. c

Fill in the Blank

1. oxygenic; anoxygenic 2. heterocysts 3. elementary; reticulate 4. green sulfur; green nonsulfur; cyanobacteria 5. green nonsulfur; gliding; anoxygenic 6. green sulfur; anoxygenic; chlorosomes

Multiple Choice

1. c, 2. b, 3. c, 4. b, 5. c, 6. a, 7. b, 8. a, 9. b, 10. d

True/False

1. F, 2. F, 3. T, 4. F, 5. T, 6. T, 7. F, 8. F, 9. T, 10. T, 11. T, 12. T

22   Bacteria: The Proteobacteria

CHAPTER OVERVIEW

This chapter presents the phylum *Proteobacteria*, the largest and most diverse group of bacteria. The distinguishing characteristics of these gram-negative bacteria are presented. The phylogenetic relationships are discussed and representative species are examined.

CHAPTER OBJECTIVES

After reading this chapter you should be able to:

- discuss the importance of this diverse group of organisms
- describe the diverse lifestyles and metabolism of members of proteobacteria
- discuss the complex structures (prosthecae, stalks, buds, sheaths, or fruiting bodies) produced by some proteobacteria
- discuss the ecological impact of chemolithotrophic bacteria
- discuss the dependence of parasitic bacteria, such as *Bdellovibrio* and the rickettsias, on their hosts for energy and/or cell constituents

CHAPTER OUTLINE

I.   The Proteobacteria
    A.   Largest and most diverse group of bacteria; thought to have arisen from a photosynthetic ancestor
    B.   Divided into five classes: *Alphaproteobacteria*, *Betaproteobacteria*, *Gammaproteobacteria*, *Deltaproteobacteria*, and *Epsilonproteobacteria*
II.   Class *Alphaproteobacteria*
    A.   Consists of seven orders and 20 families; includes most of the oligotrophic proteobacteria
    B.   The purple nonsulfur bacteria
        1.   Like all purple bacteria, the purple nonsulfur bacteria use anoxygenic photosynthesis, possess bacteriochlorophyll *a* or *b*, have their photosystems in membranes that are continuous with the plasma membrane, and are usually motile by polar flagella; with one exception, all purple nonsulfur bacteria are α-proteobacteria
        2.   Are flexible in their choice of an energy source; normally they grow anaerobically as photoorganoheterotrophs, but can grow aerobically as chemoorganotrophs; some can carry out fermentations; because of their metabolism they are found in the mud and water of lakes and ponds with abundant organic matter and low sulfide levels; some marine species are known
        3.   Are morphologically diverse (spirals, rods, circles, and half-circles); some form prosthecae and buds
        4.   *Rhodospirillum* and *Azospirillum* can form cysts (resting cells that differ from endospores) that are rich in poly-β-hydroxybutyrate and resistant to desiccation
    C.   *Rickettsia* and *Coxiella*
        1.   Members of the genus *Rickettsia* are in the alphaproteobacteria (order Rickettsiales) and members of the genus *Coxiella* are in the gammaproteobacteria (order *Legionallales*); however, they are discussed together because of their similar lifestyles

2. Rod-shaped, coccoid, or pleomorphic, with typical gram-negative walls and no flagella; size varies but tends to be small (0.3–2.0 μm); all are parasitic or mutualistic
3. Reproduction
    a. Rickettsias enter the host by phagocytosis, escape the phagosome, and then reproduce in the cytoplasm by binary fission
    b. *Coxiella* remains in the phagosome after fusion with a lysosome and reproduces within the resulting phagolysosome.
    c. For both genera, the host cell eventually bursts, releasing new organisms
4. Rickettsias lack the glycolytic pathway and do not use glucose as an energy source; instead they oxidize glutamate and TCA cycle intermediates; they take up nutrients, coenzymes, and ATP from their host cell
5. These two genera contain many important pathogens
    a. *R. prowazekii* and *R. typhi*—typhus fever
    b. *R. rickettsii*—Rocky Mountain Spotted Fever
    c. *Coxiella burnetii*—Q fever
    d. They are also important pathogens in dogs, horses, sheep, and cattle
D. The *Caulobacteraceae* and *Hyphomicrobiaceae*
    1. These two families of the alphaproteobacteria belong to different orders, but share the characteristic of being appendaged bacteria; many appendaged bacteria reproduce by budding
        a. Prostheca—an extension of the cell, including the plasma membrane, that is narrower than the mature cell
        b. Stalk—a nonliving appendage produced by cells and extending from it
        c. Reproduction by budding—parental cell retains its identity, and progeny are much smaller than the parental cell
    2. Genus *Hyphomicrobium*
        a. Chemoheterotrophic, aerobic, budding bacteria that frequently attach to solid objects in freshwater, marine, and terrestrial environments
        b. During budding process, mature cell produces a hypha or prostheca that elongates; the nucleoid divides and a copy moves into the hypha while a bud forms at its end; the bud matures, produces one to three flagella, and a septum divides the bud from the hypha; the bud is released as an oval- to pear-shaped swarmer cell
        c. Has distinctive nutrition and physiology; grows on ethanol, acetate, and one-carbon molecules such as methanol, formate, and formaldehyde (i.e., it is a facultative methylotroph); may be as much as 25% of the total bacterial population in oligotrophic (nutrient-poor) freshwater habitats
    3. Genus *Caulobacter*
        a. May be polarly flagellated rods or possess prostheca and holdfast by which they attach to solid substrata
        b. Usually found in low-nutrient freshwater and marine habitats, but also present in soil; often adhere to bacteria, algae, and other microorganisms and may absorb nutrients released by their hosts
        c. Prostheca differs from that of *Hyphomicrobium* in that it lacks cytoplasmic components and is composed almost totally of plasma membrane and cell wall
        d. Reproduction involves formation of a single flagellum at the end opposite the prostheca; asymmetric transverse fission forms a swarmer cell that swims off; when the swarmer comes to rest, it forms a new prostheca at the flagellar end and loses the flagellum, and begins to form swarmers; whole cycle takes only 2 hours
E. Family *Rhizobiaceae*
    1. This family is found in the order *Rhizobiales* and includes the gram-negative genera *Rhizobium* and *Agrobacterium*; some suggest that these genera are a single genus
    2. Genus *Rhizobium*
        a. Motile rods (often containing poly-β-hydroxybutyrate granules) that become pleomorphic under adverse conditions

        b.   Grow symbiotically within root nodule cells of legumes as nitrogen-fixing bacteroids

    3.   Genus *Agrobacterium*

        a.   Not capable of nitrogen fixation

        b.   Invades crown, roots, and stems of many plants and transforms infected plant cells into autonomously proliferating tumors

        c.   *A. tumefaciens* (best studied) causes crown gall disease by means of a tumor-inducing (Ti) plasmid

  F.   Nitrifying bacteria

    1.   Nitrifying bacteria fall into three classes (alpha-, beta-, and gammaproteobacteria), and several families, but are considered together here

    2.   All are aerobic, gram-negative organisms without endospores, and are able to oxidize either ammonia or nitrite; they differ in terms of morphology (rod-shaped, ellipsoidal, spherical, spirillar, or lobate with either polar or peritrichous flagella), presence of extensive membrane complexes in the cytoplasm, and reproduction (budding or binary fission)

    3.   Ecologically important in nitrogen cycle

        a.   *Nitrosomonas*, *Nitrosospira*, and *Nitrosococcus* oxidize ammonia to nitrite, and then *Nitrobacter* and *Nitrococcus* oxidize nitrite to nitrate

        b.   If two genera such as *Nitrobacter* and *Nitrosomonas* grow together in a habitat, ammonia is converted to nitrate (nitrification); nitrate is readily used by plants but is also easily leached from the soil or denitrified to nitrogen gas

III. Class *Betaproteobacteria*

  A.   Consists of seven orders and 12 families

  B.   Order *Neisseriales*

    1.   Contains one family with 15 genera, including the genus *Neisseria*

    2.   Genus *Neisseria*

        a.   Are nonmotile, aerobic, gram-negative cocci that most often occur in pairs with adjacent sides flattened; may have capsules and fimbriae

        b.   Chemoorganotrophic, oxidase-positive, and almost always catalase-positive

        c.   Inhabitants of the mucous membranes of animals; some are human pathogens (e.g., *Neisseria gonorrhoeae*, the causative agent of gonorrhea and *Neisseria meningitidis*, one of the causative agents of bacterial meningitis)

  C.   Order *Burkholderiales*

    1.   Contains four families

    2.   Genus *Burkholderia* (family *Burkholderiaceae*)

        a.   Gram-negative, aerobic, nonfermentative, non-spore-forming, mesophilic, straight rods; all but one species are motile with a single flagellum or a tuft of polar flagella

        b.   Catalase-positive and often oxidase-positive; use poly-β-hydroxybutyrate as their carbon reserve

        c.   *B. cepacia* is very active in recycling organic materials; is a plant pathogen; can cause disease in hospital patients (e.g., cystic fibrosis patients)

        d.   Two genera appear to be capable of nitrogen fixation and nodule formation, containing *nod* genes similar to those of the rhizobia

    3.   Genus *Bordetella* (family *Alcaliginaceae*)

        a.   Gram-negative, aerobic coccobacilli

        b.   Chemoorganotrophs with respiratory metabolism; require organic sulfur and nitrogen (in the form of amino acids) for growth

        c.   Mammalian parasites that multiply in respiratory epithelial cells; *B. pertussis* is a nonmotile, encapsulated species that is the causative agent for whooping cough

    4.   Some genera have a sheath—a hollow, tubelike structure surrounding a chain of cells—which helps bacteria attach to surfaces and obtain nutrients from slowly running water; sheath also provides protection against predators

        a.   Members of the genus *Sphaerotilus* form long, sheathed chains of rods, often attach to solid surfaces by a holdfast, reproduce and spread via swarmer cells, prefer slowly running freshwater polluted with sewage or industrial waste, and can form tangled masses that interfere with activated sludge tanks used in sewage treatment

      b. Members of the genus *Leptothrix* deposit large amounts of iron and manganese oxides in the sheath; this provides protection and allows growth in the presence of high concentrations of soluble iron compounds

  D. Order *Nitrosomonadales*
    1. Includes the nitrifying bacteria *Nitrosomonas, Nitrosococcus,* and *Nitrosospira,* discussed earlier
    2. Also includes the genera *Gallionella* (a stalked chemolithotroph) and *Spirillum*

  E. Order *Hydrogenophilales*
    1. Contains the genus *Thiobacillus,* a prominent member of the colorless sulfur bacteria (use sulfur as electron source, but are not photosynthetic)
    2. Genus *Thiobacillus*
      a. Gram-negative rods, lacking extensive internal membranes
      b. Grow aerobically by oxidizing inorganic sulfur compounds; supply carbon needs with carbon dioxide (chemolithoautotrophs); some are heterotrophs; some grow anaerobically, using nitrate as an electron acceptor
      c. Found in soil and aquatic habitats, especially those acidified by sulfuric acid; their production of sulfuric acid and ferric iron allows them to corrode concrete and pipe structures; may also be beneficial by increasing soil fertility and processing low-grade ores (leaching)

IV. Class *Gammaproteobacteria*
  A. Largest class of proteobacteria; divided into 14 orders and 28 families with several deeply branching groups; most important members are chemoorganotrophic and facultatively anaerobic; rRNA superfamily I are facultative anaerobes that use Embden-Myerhof and pentose phosphate pathways; rRNA superfamily II are mostly aerobes that use Entner-Doudoroff and pentose phosphate pathways

  B. The purple sulfur bacteria (order *Chromatiales*)
    1. Divided into two families: *Chromatiaceae* and *Ectothiorhodospiraceae*
      a. Family *Ectothiorhodospiraceae* contains eight genera, including *Ectothiorhodospira,* which has red, polarly flagellated, spiral-shaped cells that deposit sulfur globules externally and internal photosynthetic membranes that are organized as lamellar stacks
      b. The typical purple sulfur bacteria are in the family *Chromatiaceae,* which contains 26 genera
    2. Typical purple sulfur bacteria are strict anaerobes and usually photolithoautotrophs; oxidize hydrogen sulfide to sulfur and deposit it internally as sulfur granules; hydrogen may also serve as an electron donor; genera *Thiospirillum, Thiocapsa,* and *Chromatium* are typical purple sulfur bacteria; they are usually found in anaerobic, sulfide-rich zones of lakes

  C. Order *Thiotrichales*
    1. Contains three families, the largest of which is the *Thiotrichiaceae,* which contains some of the colorless sulfur bacteria (e.g., *Beggiatoa* and *Thiothrix*)
    2. Genus *Beggiatoa*
      a. Microaerophilic; grow in sulfide-rich habitats
      b. Filamentous; lack a sheath
      c. Metabolically versatile; can oxidize hydrogen sulfide to sulfur (deposited internally) and can oxidize sulfur to sulfate; can also grow heterotrophically with acetate as a carbon source; some may incorporate CO_2 autotrophically
    3. Genus *Leucothrix*
      a. Aerobic chemoorganotrophs that form long filaments (trichomes); are marine bacteria that attach to solid substrates by a holdfast
      b. Have complex lifestyle in which dispersal is by formation of gonidia
    4. *Thiothrix* is similar to *Leucothrix* in that it forms gonidia; forms sheathed filaments and is chemolithotrophic; oxidizes hydrogen sulfide and deposits sulfur granules internally; requires

organic compounds for growth (mixotroph); found in sulfide-rich, flowing water and activated sludge sewage systems

D. Order *Methylococcales*
1. Contains one family and seven genera (e.g., *Methylococcus and Methylomonas*)
2. Rods, vibrios, and cocci use methane and methanol as their sole carbon and energy source (methylotrophs) under aerobic or microaerobic conditions
   a. Have complex arrays of intracellular membranes when oxidizing methane
   b. Methane is oxidized to methanol and then to formaldehyde; formaldehyde is then assimilated into cell material
   c. Found in anaerobic habitats, where methane is often abundant

E. Order *Pseudomonadales*
1. Contains two families and numerous genera
2. Genus *Pseudomonas* is the most important genus in this order with about 60 species
   a. Straight or slightly curved rods, motile by polar flagella; lack a sheath or prosthecae
   b. Aerobic respiratory chemoheterotrophs, though sometimes carry out anaerobic respiration using nitrate as the final electron acceptor; have functional TCA cycle and use Entner-Doudoroff pathway rather than the Embden-Myerhof pathway
   c. Have great impact: mineralization (degradation) of a wide variety of organic compounds; important model organisms; some are major animal and plant pathogens; some involved in the spoilage of refrigerated food because they can grow at 4°C and degrade lipids and proteins
3. Members of the genus *Azotobacter* are large, ovoid bacteria that are motile by peritrichous flagella; are aerobic, catalase positive, and fix nitrogen nonsymbiotically; widespread in soil and water

F. Order *Vibrionales*
1. Contains one family (*Vibrionaceae*) and eight genera
   a. Gram-negative, straight or curved rods with polar flagella
   b. Oxidase-positive and use D-glucose as their sole or primary carbon and energy source
   c. Aquatic with widespread distribution in freshwater and marine habitats
2. Pathogens in this order include *V. cholerae* (cholera), *V. parahaemolyticus* (gastroenteritis after eating contaminated seafood), and *V. anguillarum* (a fish pathogen)
3. Some (e.g., *V. fischeri* and at least two species of *Photobacterium*) are among the few marine bacteria capable of bioluminescence; some bioluminescent species live symbiotically in the luminous organs of fish, while others are free-living

G. Order *Enterobacteriales*
1. Consists of one family containing over 44 genera; all are gram-negative, peritrichously flagellated or nonmotile, facultatively anaerobic, straight rods with simple nutritional requirements
2. Their metabolic properties are useful for characterization and identification
   a. Degrade sugars by Embden-Meyerhof pathway and cleave pyruvic acid into formic acid in formic acid fermentations; some produce gas during their fermentations
   b. The majority (e.g., *Escherichia, Proteus, Salmonella,* and *Shigella*) carries out mixed acid fermentation, while others (e.g., *Enterobacter, Serratia, Erwinia,* and *Klebsiella*) carry out butanediol fermentation
   c. Other metabolic characteristics (e.g., lactose utilization, indole production) also are used to differentiate the genera; they are usually identified using rapid commercial identification systems (e.g., Enterotube, API 20-E) that are based on their biochemical characteristics
3. Very common, widespread, and important
   a. *Escherichia coli* is probably the best-studied bacterium and experimental organism of choice for many microbiologists; it is an intestinal tract inhabitant and an indicator organism for water quality (fecal contamination)
   b. *Salmonella*—typhoid fever and gastroenteritis

    c.   *Shigella*—bacillary dysentery
    d.   *Klebsiella*—pneumonia
    e.   *Yersinia*—plague
    f.   *Erwinia*—plant pathogens
  H. Order *Pasteurellales*
    1. Consists of one family and seven genera
    2. Small, nonmotile, normally oxidase-positive with complex nutritional requirements; parasitic in vertebrates
    3. Best known for the diseases they cause
        a.   *P. multilocida*—fowl cholera
        b.   *P. haemolytica*—pneumonia in cattle, sheep, and goats (e.g., "shipping fever" in cattle)
        c.   *H. influenzae*—major human pathogen that causes a variety of diseases, including meningitis in children

V. Class *Deltaproteobacteria*
  A. Consists of eight orders and 20 families
  B. Orders *Desulfovibrionales, Desulfobacterales,* and *Desulfuromonadales*
    1. Gram-negative, sulfate- or sulfur-reducing bacteria; strict anaerobes; use elemental sulfur or sulfur compounds as electron acceptors during anaerobic respiration
    2. Important in sulfur cycling within ecosystems
        a.   Thrive in mud, polluted lake sediments, sewage lagoons and methane digesters, waterlogged soils, and anaerobic marine and estuarine sediments
        b.   Can have negative impact on industry because of their primary role in the anaerobic corrosion of iron in pipelines, heating systems, and other structures
  C. Order *Bdellovibrionales*
    1. Has one family and three genera, including the genus *Bdellovibrio*
    2. Genus *Bdellovibrio*
        a.   Gram-negative curved rods with polar flagella; prey on other gram-negative bacteria and alternate between a nongrowing predatory phase and an intracellular reproductive phase
        b.   Have a complex lifestyle that begins with a high-speed collision with its prey, after which it enters its prey by boring a hole through the host cell wall; this is accomplished by a combination of mechanical and enzymatic action, and the flagellum is lost during penetration; it inhabits the host within the space between cell wall and plasma membrane, where it inhibits host DNA, RNA, and protein synthesis, disrupts the host cell's plasma membrane so that cell constituents leak out, and grows into a long filament that divides (multiple fission) into many smaller flagellated progeny; these escape when the host cell lyses
  D. Order *Myxococcales*
    1. Divided into six families
    2. Commonly called the myxobacteria; they are gram-negative, aerobic soil bacteria with gliding motility, and complex life cycles that involve the formation of fruiting bodies and dormant myxospores; the families are distinguished based on the shape of vegetative cells (rods that may be either slender with tapered ends or stout with rounded, blunt ends), myxospores, and sporangia
    3. Most are micropredators or scavengers that lyse bacteria and yeasts by secretion of an array of digestive enzymes; many also secrete antibiotics to kill prey; they use the released peptides and amino acids as primary carbon, nitrogen, and energy source; all are chemoheterotrophs with respiratory metabolism
    4. Lifestyle resembles that of cellular slime molds
        a.   When food is plentiful, they migrate along solid surfaces, feeding and leaving a slime trail
        b.   When their nutrient supply is exhausted, they aggregate and differentiate into fruiting bodies

      c.   Some cells in the fruiting body develop into myxospores; these are frequently enclosed in walled structures called sporangioles or sporangia; myxospores are dormant and desiccation-resistant; fruiting bodies protect and aid dispersal of myxospores; a colony develops automatically when myxospores are released and this aids digestion by providing higher enzyme concentration than any individual bacterium could

     5.   Found in neutral soils, decaying plant material, and animal dung; are most abundant in warm areas but will grow in the arctic tundra

VI. Class *Epsilonproteobacteria*

    A.   Smallest of the proteobacteria groups; contains one order with three families

       1.   All are slender, gram-negative rods that can be straight, curved, or helical

       2.   *Campylobacter* and *Helicobacter* are the most important genera; both are microaerophiles, motile, helical or vibroid, gram-negative rods

       3.   Some can be found in sulfide-rich cave waters and as thermophilic chemolithoautotrophs (oxidize H_2 while reducing sulfur) at deep-sea hydrothermal vents

    B.   Genus *Campylobacter*—contains both pathogenic and nonpathogenic species

       1.   *C. fetus*—causes reproductive disease and abortions in cattle and sheep; can cause septicemia and enteritis in humans

       2.   *C. jejuni*—causes abortion in sheep and enteritis diarrhea in humans

    C.   Genus *Helicobacter*

       1.   Isolated from stomachs and upper intestines of humans, dogs, cats, and other mammals

       2.   *H. pylori*—cause of gastritis and peptic ulcer disease; produces large quantities of urease, and urea hydrolysis appears to be associated with their virulence

TERMS AND DEFINITIONS

Place the letter of each term in the space next to the definition or description that best matches it.

____ 1.   A nonliving appendage produced by a cell and extending from it

____ 2.   An extension of the cell, including the plasma membrane and cell wall, that is narrower than the cell

____ 3.   A hollow, tubelike structure that surrounds a chain of cells

____ 4.   An organism that uses methane or methanol as its sole carbon and energy source

____ 5.   The generation of light by biological processes

____ 6.   Inflammation of the intestinal tract

____ 7.   The presence of pathogens or their toxins in the blood

____ 8.   The microbial breakdown of organic materials to inorganic substances

a.   bioluminescence
b.   enteritis
c.   methylotroph
d.   mineralization process
e.   prostheca
f.   septicemia
g.   sheath
h.   stalk

THE PROTEOBACTERIA

Class	Important Taxa or Subgroups	Representative Genera	Important Characteristics (structural, physiological, ecological)
Alphaproteobacteria	Purple nonsulfur bacteria		
	Rickettsiales		
	Caulobacteriaceae and *Hyphomicrobiaceae*		
	Rhizobiaceae		
	Bradyrhizobiaceae		
Betaproteobacteria	*Coxiellaceae*		
	Neisseriales		
	Burkhoderiales		
	Nitrosomonadales		
	Hydrogenophilales		
Gammaproteobacteria	Purple sulfur bacteria		
	Thiotrichales		
	Methylococcales		
	Pseudomonadales		
	Vibrionales		
	Enterobacteriales		
	Pasteurellales		
Deltaproteobacteria	*Desulfovibrionales, Desulfobacterales,* and *Desulfuromonadales*		
	Bdellovibrionales		
	Myxococcales		
Epsilonproteobacteria	*Campylobacteraceae*		
	Helicobacteraceae		

FILL IN THE BLANK

1. The _____ _____ bacteria (e.g., *Rhodospirillum*) can grow anaerobically as photoorganoheterotrophs and often aerobically as chemoorganoheterotrophs. The _____ _____ bacteria are anaerobes and usually photoautolithotrophs. They oxidize hydrogen sulfide to sulfur and deposit the granules internally.

2. Members of the genus *Hyphomicrobium* have several distinctive characteristics. For instance, they are facultative _____; that is, they can derive both energy and carbon from reduced one-carbon compounds. Another interesting feature of this genus is that they form a hypha (a type of _____) that produces a bud at its tip. The bud develops flagella as it matures and eventually detaches from the hypha and swims off. This type of reproduction is called _____, and it differs significantly from _____ _____, where the parent cell elongates and then splits in half.

3. Some proteobacteria (e.g., members of the genus *Caulobacter*) attach to solid surfaces by way of a structure called a _____.

4. *Sphaerotilus, Leptothrix,* and members of several other genera have_____, hollow, tubelike structures that surround chains of cells.

5. The _____ _____ bacteria (e.g., members of the genus *Thiobacillus*) are chemolithotrophic bacteria that oxidize elemental sulfur, hydrogen sulfide, and thiosulfate to sulfate.

6. The _____ are gram-negative, aerobic soil bacteria with a complex life cycle. In the presence of food, they migrate along solid surfaces, feeding and leaving slime trails. When their food is exhausted, cells aggregate to form a _____ _____. Some cells within this structure develop into _____, which are dormant, desiccation-resistant, and can survive up to 10 years under adverse conditions.

7. The _____ proteobacteria contain anaerobic bacteria that use elemental sulfur and oxidized sulfur compounds as electron acceptors during anaerobic respiration. They are important in sulfur cycles in the ecosystem.

8. The _____ bacteria include a number of important pathogens of both plants (e.g., *Erwinia*) and animals (e.g., *Salmonella*).

9. Bacteria capable of oxidizing ammonia or nitrite (a process called _____) are called _____ bacteria. These chemolithotrophic bacteria are found in _____, _____, and _____-proteobacteria.

MULTIPLE CHOICE

For each of the questions below select the best answer.

1. Members of the genus *Pseudomonas* do which of the following?
   a. mineralization of organic compounds
   b. spoilage of refrigerated milk
   c. Both (a) and (b) are correct.
   d. Neither (a) nor (b) is correct.

2. What does mineralization mean?
   a. breakdown of organic materials to inorganic materials
   b. release of various minerals from various ores
   c. use of minerals as energy sources
   d. None of the above is correct.

3. What is the function of the sheath observed surrounding some bacterial species?
   a. helps bacteria attach to surfaces
   b. helps obtain nutrients from slowly running water as it flows past
   c. helps protect against predators
   d. All of the above are correct.

4. What is the largest class of the proteobacteria?
   a. *Alphaproteobacteria*
   b. *Betaproteobacteria*
   c *Gammaproteobacteria*
   d. *Deltaproteobacteria*

5. Which of the following is true of *Beggiatoa?*
   a. They can oxidize hydrogen sulfide to sulfur.
   b. They can oxidize sulfur to sulfate.
   c. They can grow heterotrophically with acetate as their carbon source.
   d. All of the above are true of *Beggiatoa.*

6. Which of the following genera contain bacteria that are capable of bioluminescence?
   a. *Vibrio*
   c. *Photobacteria*
   c. Both (a) and (b) are correct.
   d. Neither (a) nor (b) is correct.

TRUE/FALSE

____ 1. The enterobacteria can be easily distinguished from one another by morphological criteria.
____ 2. The purple nonsulfur bacteria are so named because they usually do not oxidize elemental sulfur to sulfide.
____ 3. Rickettsias do not use glucose as an energy source.
____ 4. *Escherichia coli* is not a good indicator of fecal contamination of water supplies because it cannot be readily detected.
____ 5. Members of the *Desulfovibrionales* have a positive impact on industry because they protect iron-containing pipelines from corrosion.
____ 6. Myxobacteria lyse bacteria and yeasts by secretion of digestive enzymes. They then use the resulting peptides and amino acids as their primary carbon, nitrogen, and energy source.

CRITICAL THINKING

1. At one time rickettsias, like the chlamydiae, were thought to be viruses rather than bacteria. What was the basis for this misconception? Explain why it is a misconception.

2 Explain the basis of rapid commercial identification systems such as the Enterotube or the API 20-E system. Why was it necessary and desirable to develop these types of rapid identification systems?

ANSWER KEY

Terms and Definitions

1. h, 2. e, 3. g, 4.c 5. a, 6. b, 7. f, 8. d

Fill in the Blank

1. purple nonsulfur; purple sulfur 2. methylotrophs; prostheca; budding; binary fission 3. holdfast 4. sheaths 5. colorless sulfur 6. myxobacteria; fruiting body; myxospores 7. δ- (delta-) 8. enteric 9. nitrification; nitrifying; α- (alpha-), β- (beta-), γ- (gamma-)

Multiple Choice

1. c, 2. a, 3. d, 4. c, 5. d, 6. c

True/False

1. F, 2. T, 3. T, 4. F, 5. F, 6. T

23   Bacteria: The Low G + C Gram Positives

CHAPTER OVERVIEW

This chapter describes the low G + C gram-positive bacteria. These bacteria are placed in the phylum *Firmicutes*, which is divided into three classes: *Mollicutes, Clostridia*, and *Bacilli*. For each class, important genera are discussed.

CHAPTER OBJECTIVES

After reading this chapter you should be able to:

*   discuss the variation in peptidoglycan structure that traditionally has been useful to the taxonomy of bacteria
*   discuss the various roles of these organisms
*   describe the important taxa of the *Firmicutes*

CHAPTER OUTLINE

I.   The *Firmicutes*
   A.   Is divided into three classes (*Mollicutes, Clostridia,* and *Bacilli*), 10 orders, 34 families, 255 genera, and over 1,300 species
   B.   All have low G + C content, and with the exception of the Mollicutes, which lack a cell wall; all have gram-positive type cell walls
II.   Class *Mollicutes* (The Mycoplasmas)
   A.   Has five orders and six families having the following characteristics:
      1.   Lack cell walls and cannot synthesize peptidoglycan precursors; therefore are penicillin-resistant, pleomorphic, and susceptible to lysis by osmotic shock and detergent treatment
      2.   Most are nonmotile, but some can glide along liquid-covered surfaces
      3.   Most species require sterols (unusual for bacteria)
      4.   Usually facultative anaerobes, but a few are strict anaerobes
      5.   Are smallest bacteria capable of self-reproduction; have some of the smallest genomes observed in free-living procaryotes; G + C content ranges from 23 to 41%
      6.   Can be saprophytes, commensals, or parasites
   B.   Metabolism is not particularly unusual
      1.   Deficient in several biosynthetic pathways
      2.   Some produce ATP by the Embden-Meyerhof pathway and lactic acid fermentation; others catabolize arginine to urea; the pentose phosphate pathway functions in some; none has a complete TCA cycle
   C.   Widespread
      1.   Can be isolated from plants, animals, soil, and compost piles
      2.   Serious contaminants of mammalian cell cultures, where they are difficult to detect and difficult to eliminate
      3.   In animals, they colonize mucous membranes and joints and are often associated with diseases of the respiratory and urogenital tracts; pathogenic species include:
         a.   *M. mycoides*—bovine pleuropneumonia in cattle
         b.   *M. gallisepticum*—chronic respiratory disease in chickens
         c.   *M. pneumoniae*—primary atypical pneumonia in humans

        d.   *M. hominis* and *Ureaplasma urealyticum*—pathogenic in humans

        e.   Spiroplasmas—pathogenic in insects, ticks, and a variety of plants

III. Peptidoglycan and Endospore Structure

    A.   Classified on the basis of cell shape, clustering and arrangement of cells, oxygen relationships, and fermentation patterns; however, the most important characteristics are the presence or absence of endospores and peptidoglycan chemistry

    B.   Peptidoglycan structure varies considerably; these variations are characteristic of particular groups and are taxonomically useful in phenetic classification schemes

        1.   Some contain *meso*-diaminopimelic acid cross-linked through its free amino group to the carboxyl group of the terminal D-alanine of the adjacent chain

        2.   Others contain lysine cross-linked by interpeptide bridges

        3.   Others contain L,L-diaminopimelic acid and have one glycine as the interpeptide bridge

        4.   Others use ornithine to cross-link between positions 2 and 4 of the peptide chains rather than positions 3 and 4

        5.   Other cross-links and differences in cross-link frequency also contribute to variation in structure

    C.   Bacterial endospores are complex structures that allow survival under adverse conditions; endospore-forming bacteria are distributed widely, but found mainly in soil; the presence or absence of endospores also is taxonomically useful in phenetic classification schemes

IV. Class *Clostridia*

    A.   Contains three orders and 11 families

    B.   The largest genus is *Clostridium*, having over 100 species in distinct phylogenetic clusters

        1.   Obligate anaerobes; form endospores; do not carry out dissimilatory sulfate reduction

        2.   Practical impact

            a.   Responsible for many cases of food spoilage, even in canned foods (e.g., *C. botulinum*, botulism)

            b.   *C. perfringens*—gas gangrene

            c.   *C. tetani*—tetanus

            d.   Some are of industrial value (e.g., *C. acetobutylicum*, which is used to manufacture butanol)

    C.   Genus *Desulfotomaculum*

        1.   Anaerobic, endospore-forming bacteria that reduce sulfate and sulfite to hydrogen sulfide during anaerobic respiration

        2.   Stains gram negative but has a gram-positive type cell wall with a lower than normal peptidoglycan content

    D.   Genera *Heliobacterium* and *Heliophilum*

        1.   Anaerobic, photosynthetic bacteria that use bacteriochlorophyll *g*; have a photosystem like the green sulfur bacteria, but lack intracytoplasmic photosynthetic membranes (pigments are in the plasma membrane)

        2.   Stain gram negative but have gram-positive type cell wall with lower than normal peptidoglycan content

    E.   Genus *Veillonella* (family *Veillonellaceae*)

        1.   Anaerobic, chemoheterotrophic cocci; usually diplococci

        2.   Have complex nutritional requirements; ferment carbohydrates, lactate and other organic acids, and amino acids; produce gas and a mixture of volatile fatty acids; unable to ferment glucose or other carbohydrates

        3.   Parasites of homeothermic animals; part of the normal microflora of the mouth, the gastrointestinal tract, and urogenital tract of humans and other animals

V. Class *Bacilli*

    A.   Consists of two orders, 17 families, and over 70 genera

    B.   Order *Bacillales*

        1.   Genus *Bacillus* (family *Bacillaceae*)

            a.   Largest genus in the order

        b.   Gram-positive, endospore-forming, chemoheterotrophic rods that are usually motile with peritrichous flagella

        c.   Usually aerobic, sometimes facultative, and catalase positive

        d.   Many species are of considerable importance: some produce antibiotics, some cause disease (e.g., *B. cereus* causes food poisoning, and *B. anthracis* causes anthrax), and some are used as insecticides (e.g., *B. thuringiensis* and *B. sphaericus*)

  2.   Genus *Thermoactinomyces* (family *Thermoactinomycetaceae*)

        a.   Thermophilic; form single spores within the hyphae of both aerial and substrate mycelia; the spores are very heat-resistant and thus are true bacterial endospores

        b.   Commonly found in damp haystacks, compost piles, and other high-temperature habitats

        c.   *T. vulgaris*—causative agent for farmer's lung disease, an allergic respiratory disease in agricultural workers

  3.   Genus *Caryophanon*—strict aerobe, catalase positive, motile by peritrichous flagella; lives in cow dung; has disk-shaped cells that join together to form rods

  4.   Genus *Staphylococcus* (family *Staphylococcaceae*)

        a.   Facultatively anaerobic, nonmotile cocci that form irregular clusters; have techoic acids in their cell walls

        b.   Catalase positive; oxidase negative; ferment glucose anaerobically

        c.   Normally associated with skin, skin glands, and mucous membranes of warm-blooded animals

        d.   Cause many human diseases (e.g., endocarditis, wound infections, surgical infections, urinary tract infections, various skin infections, pneumonia, toxic shock syndrome, and food poisoning); those with coagulase or α-hemolysins most virulent

        e.   Major health concern since they are often antibiotic resistant (including methicillin and vancomycin)

  5.   Genus *Listeria* (family *Listeriaceae*)—short rods that are peritrichously flagellated; aerobic or facultative, and catalase positive; *L. monocytogenes* is a human pathogen that causes listeriosis, an important food-borne infection

C.   Order *Lactobacillales*

  1.   Lactic acid bacteria—nonsporing; nonmotile; lack cytochromes; fermentative (lactic acid fermentation); nutritionally fastidious; facultative or aerotolerant anaerobes

  2.   Largest genus is *Lactobacillus* with around 100 species

        a.   Can be rods and sometimes coccobacilli; lack catalase and cytochromes, and are facultative or microaerophilic

        b.   Can carry out heterolactic or homolactic acid fermentation

        c.   Grow optimally between pH 4.5 and pH 6.4

        d.   Found on plant surfaces and in dairy products, meat, water, sewage, beer, fruits, and many other materials; are also normal microflora of mouth, intestinal tract, and vagina; usually not pathogenic

        e.   Used in the production of fermented vegetable foods, beverages, sour dough, hard cheeses, yogurt, and sausages

        f.   Responsible for spoilage of beer, milk, and meat

  3.   Genus *Leuconostoc* (family *Leuconostocaceae*)

        a.   Facultative cocci that may be elongated or elliptical shape; arranged in pairs or chains

        b.   Lack catalase; carry out heterolactic fermentation

        c.   Isolated from plants, silage, and milk

        d.   Important in wine production, fermentation of vegetables such as cabbage and cucumbers, manufacture of buttermilk, butter, cheese, and dextrans; involved in food spoilage

  4.   Genus *Streptococcus* (family *Streptococcaceae*)

        a.   Most are facultative anaerobes; catalase negative; a few are obligate anaerobes

        b.   Form pairs or chains in liquid media; do not form endospores; nonmotile

        c.   Homolactic fermentation; produces lactic acid but no gas

d. The many species of this genus are distinguished by hemolysis reactions (α-hemolysis—incomplete with greenish zone or β-hemolysis—complete with clear zone but no greening), serologically (e.g., Lancefield grouping system), and a variety of biochemical, genetic, and physiological tests

5. Members of the genera *Enterococcus*, *Streptococcus*, and *Lactococcus* have great practical importance
   a. *S. pyogenes*—causes streptococcal sore throat, acute glomerulonephritis, and rheumatic fever
   b. *S. pneumonia*—causes lobar pneumonia
   c. *S. mutans*—associated with dental caries
   d. *E. faecalis*—opportunistic pathogen that can cause urinary tract infections and endocarditis
   e. *L. lactis*—used in the production of buttermilk and cheese

TERMS AND DEFINITIONS

Place the letter of each term in the space next to the definition or description that best matches it.

_____ 1. Contains protein toxins that are insecticidal
_____ 2. An enzyme that causes blood plasma to clot
_____ 3. Incomplete lysis of red blood cells with formation of a greenish ring around the bacterial colony
_____ 4. Complete lysis of red blood cells with no greenish zone formed

a. coagulase
b. α-hemolysis
c. β-hemolysis
d. parasporal body

FILL IN THE BLANK

1. *Staphylococcus aureus* is the most important staphylococcal human pathogen. It differs from other staphylococcal species in that it is positive for the enzyme _____, which causes blood plasma to clot.

2. One way of distinguishing streptococci is by determining their effect on red blood cells. Some cause _____ , which is only partially lysis of red blood cells, leading to the formation of a _____ ring around the growing colony. Others cause _____, which is characterized by a zone of complete killing or lysis. Streptococci are often identified using the _____ _____, which is based on differences in the teichoic acids and polysaccharides of streptococcal species.

3. The lactobacilli are _____ _____ bacteria and are of importance in the food industry. They carry out either a homolactic fermentation using the Embden-Meyerhof pathway or a _____ fermentation with the pentose phosphate pathway.

4. Some *Firmicutes* (called the _____) actually stain gram negative because they lack a cell wall. They are the smallest bacteria capable of self-reproduction

MULTIPLE CHOICE

For each of the questions below select the *one best* answer.

1. Lactobacilli are normal inhabitants of the human body. Where are they found?
   a. mouth
   b. intestines
   c. vagina
   d. all of the above

2. Which of the following can be attributed to members of the genus *Bacillus?*

a. They produce certain useful antibiotics.
b. They are the causative agents of some food poisonings.
c. They have been used as a biological insecticide.
d. All of the above are properties attributed to members of the genus *Bacillus.*

3. Which of the following is caused by *Clostridium perfringens?*
   a. gas gangrene
   b. tetanus
   c. botulism
   d. None of the above is caused by this organism.
4. Which genus produces endospores?
   a. *Staphylococcus*
   b. *Streptococcus*
   c. *Bacillus*
   d. *Lactobacillus*
5. Which of the following is produced by the action of lactobacilli?
   a. sauerkraut and pickles
   b. hard cheeses
   c. yogurt
   d. All of the above involve lactobacilli in their production.
6. Which of the following processes does NOT involve members of the genus *Leuconostoc?*
   a. fermentation of vegetables like cabbage (sauerkraut) and cucumbers (pickles)
   b. manufacturing of buttermilk, butter, and cheese
   c. manufacturing of beer
   d. All of the above involve *Leuconostoc.*
7. Which of the following is a characteristic of the heliobacteria (genera *Heliobacterium* and *Heliophilum*)?
   a. They contain bacteriochlorophyll *g.*
   b. They have two photosystems.
   c. They have a higher than normal peptidoglycan content in their cell walls.
   d. all of the above
8. Which genus contains gram-positive cocci in pairs or chains that are usually facultative and carry out homolactic fermentation?
   a. *Streptococcus*
   b. *Enterococcus*
   c. *Lactococcus*
   d. all of the above

9. Which of the following causes primary atypical pneumonia in humans?
   a. *Streptococcus pneumoniae*
   b. *Mycoplasma pneumoniae*
   c. *Clostridium pneumoniae*
   d. none of the above
10. Which genus contains facultatively anaerobic, nonmotile, gram-positive cocci that form irregular clusters?
   a. *Streptococcus*
   b. *Staphylococcus*
   c. *Enterococcus*
   d. all of the above
11. Which of the following is a reason why mycoplasmas are a major problem when working with mammalian cell cultures?
   a. They are difficult to detect.
   b. They are difficult to eliminate.
   c. Both (a) and (b) are correct.
   d. Neither (a) nor (b) is correct.
12. Which of the following is characteristic of the genus *Desulfotomaculum?*
   a. Its members are aerobic.
   b. Its members do not form endospores.
   c. Its members use sulfate as a terminal electron acceptor for anaerobic respiration.
   d. Its members lack cell walls.
13. Which of the following is characteristic of the family *Veillonellaceae?*
   a. Its members have a higher than normal peptidoglycan content in their cell walls.
   b. Many of its members are parasites of homeothermic animals.
   c. Its members stain gram positive.
   d. Its members are aerobic.

Firmicutes

Class	Representative Genera	Important Characteristics (structural, physiological, ecological)
Mollicutes		
Clostridia		
Bacilli		
Bacillales		
Lactobacillales		

TRUE/FALSE

_____ 1. When the members of the gram-positive cocci are examined both phenetically and phylogenetically, it is found that the two ways of classifying these organisms show a close match.

_____ 2. The streptococci carry out homolactic fermentation, producing lactic acid but no gas.

_____ 3. *Leuconostoc* will tolerate a higher sugar content (unlike most organisms) and is, therefore, a major problem in sugar refineries. It can also cause food spoilage of foods that are baked in heavy syrup.

_____ 4. Lactobacilli are alkalophilic and prefer conditions that are slightly alkaline for optimal growth.

_____ 5. Mycoplasmas are serious contaminants of mammalian cell cultures.

_____ 6. *Clostridium botulinum* is responsible for cases of food spoilage, even in canned foods.

_____ 7. The division of gram-positive bacteria into groups having low G + C and high G + C content is not supported by 16S rRNA analysis.

_____ 8. Most taxonomically significant variations in peptidoglycan are in amino acid 3 of the peptide subunit or in the interpeptide bridge.

_____ 9. Members of the genus *Thermoactinomyces* are thermophilic and produce mycelia and endospores. One species causes a disease called farmer's lung.

CRITICAL THINKING

1. In the production of yogurt, two organisms are used. The first to grow is *Streptococcus thermophilus* and the second is *Lactobacillus bulgaricus.* This is true even though the two are inoculated into the milk simultaneously. Explain. In your explanation be sure to consider the properties of lactobacilli and how these may contribute to the sequence of events.

2. Tetanus, which is caused by *Clostridium tetani,* is of serious concern in deep puncture wounds. However, it is seldom a problem with surface lacerations. Using your knowledge of the properties of the genus *Clostridium,* explain these observations.

3. Many endospore-forming bacteria are inhabitants of soil. Speculate why this characteristic is advantageous in soil environments.

ANSWER KEY

Terms and Definitions

1. d, 2. a, 3. b, 4. c

Fill in the Blank

1. coagulase 2. α-hemolysis; greenish; β-hemolysis; Lancefield grouping system 3. lactic acid; heterolactic 4. mycoplasmas

Multiple Choice

1. d, 2. d, 3. a, 4. c, 5. d, 6. c, 7. a, 8. d, 9. b, 10. b, 11. c, 12.c, 13. b

True/False

1. F, 2. T, 3. T, 4. F, 5. T, 6. T, 7. F, 8. T, 9. T

24 Bacteria: The High G + C Gram Positives

CHAPTER OVERVIEW

This chapter surveys the general characteristics of members of the phylum *Actinobacteria*. All are high G + C gram positives and many are actinomycetes. The actinomycetes are filamentous bacteria that form branching hyphae and asexual spores.

CHAPTER OBJECTIVES

After reading this chapter you should be able to:

- describe actinomycetes
- list and describe the phenotypes that are important to the taxonomy of actinobacteria
- discuss the roles of actinomycetes in the mineralization of organic compounds and in the production of antibiotics
- describe the important human pathogens in *Actinobacteria*

CHAPTER OUTLINE

I. General Properties of the Actinomycetes
    A. Many of the high G + C gram-positive bacteria are actinomycetes; they are aerobic bacteria that exhibit filamentous growth
        1. Nearly all form substrate mycelia made of hyphae; septa divide the mycelia into long cells (20 μm and longer), each containing several nucleoids
        2. They may have aerial mycelia that form conidospores at the ends of filaments or sporangiospores within a sporangium; spores are not heat resistant but withstand desiccation
    B. Actinomycetes are generally nonmotile, but spores may be flagellated
    C. Cell wall types vary and can be distinguished by the amino acid in position 3 of the tetrapeptide, the presence of glycine in the interpeptide bridge, and the sugar content; four major types are known
    D. Cell wall type, sugars in extracts, morphology and color of mycelia and sporangia, G + C content, membrane phospholipid composition, and heat resistance of the spores are all important in classifying these organisms, as is comparison of 16S rRNA sequences
    E. Considerable practical importance
        1. Those in soil degrade a number of organic compounds and are important in the mineralization process; also produce many of the medically important, natural antibiotics and drugs
        2. A few species are pathogenic in humans, other animals, and plants
    F. The high G + C gram positives were characterized primarily based on cell wall type, conidia arrangement, and the presence or absence of a sporangium, but now 16S rRNA sequences have been used to create a large phylum, *Actinobacteria*, containing one class, five subclasses, six orders, 14 suborders, and 44 families
II. Suborder *Actinomycineae*
    A. Most genera (one family, five genera) are irregularly shaped, nonsporing rods (straight or slightly curved; usually with swellings or other deviations from rod shape), with aerobic or facultative metabolism
    B. Genus *Actinomyces*

1. Straight or slightly curved rods and slender filaments with true branching; may have swollen, clubbed, or clavate ends
2. Facultative or obligate anaerobes; require CO_2 for best growth
3. Cell walls contain lysine but not *meso*-diaminopimelic acid
4. Normal inhabitants of mucosal surfaces (often oral) of warm-blooded animals; some cause disease in their hosts

III. Suborder *Micrococcineae*
   A. Contains 14 families and many genera
   B. Genus *Micrococcus*
      1. Aerobic, catalase-positive cocci that occur in pairs, tetrads, or irregular clusters; usually nonmotile; often have yellow, red, or orange pigmentation
      2. Widespread in soil, water, and on mammalian skin; usually not pathogenic
   C. Genus *Arthrobacter*
      1. Aerobic, catalase-positive rods with respiratory metabolism and lysine in peptidoglycan
      2. Exhibit a rod-coccus growth cycle
         a. When growing in exponential phase, they are rods that reproduce by snapping division
         b. In stationary phase, they change to a coccoid form
         c. Upon transfer to fresh medium, coccoid cells produce outgrowths and resume active reproduction as rods
      3. Most important habitat is soil; also is isolated from fish, sewage, and plant surfaces
         a. Resistant to desiccation and nutrient deprivation
         b. Very flexible nutritionally; able to degrade some herbicides and pesticides
   D. Genus *Dermatophilus*
      1. Forms packets of motile spores with tufts of flagella
      2. Facultative anaerobe
      3. Mammalian parasite responsible for a skin infection called streptothrichosis

IV. Suborder *Corynebacterineae*
   A. Contains seven families with several important genera
   B. Genus *Corynebacterium* (family *Corynebacteriaceae*)
      1. Aerobic or facultative; catalase-positive; straight to slightly curved rods, often with tapered ends; club-shaped forms also are seen
      2. Remain partially attached after snapping division resulting in angular arrangements
      3. Form metachromatic granules and their cell walls contain *meso*-diaminopimelic acid
      4. Some species are harmless soil and water saprophytes; many are animal and human pathogens (e.g., *C. diphtheriae* is the causative agent of diphtheria in humans)
   C. Genus *Mycobacterium* (family *Mycobacteriaceae*)
      1. Straight or slightly curved rods that sometimes branch or form filaments that readily fragment
      2. Aerobic and catalase positive; grow very slowly
      3. Cell walls contain waxes with 60 to 90 carbon mycolic acids; these make them acid-fast (i.e., basic fuchsin dye cannot be removed with acid-alcohol treatment)
      4. Some are free-living saprophytes; but many are animal pathogens
         a. *M. bovis*—tuberculosis in cattle and other ruminants
         b. *M. tuberculosis*—tuberculosis in humans
         c. *M. leprae*—leprosy in humans
   D. Genera *Nocardia* and *Rhodococcus* (family *Nocardiaceae*)
      1. These and related genera are collectively called nocardioforms
      2. Develop a substrate mycelium that readily breaks into rods and coccoid elements; some develop aerial mycelia
      3. Most are strict aerobes
      4. Most have peptidoglycan with *meso*-diaminopimelic acid and no peptide interbridge; mycolic acids are present
      5. They are found in soil and aquatic habitats
         a. Members of *Nocardia* degrade hydrocarbons and waxes and are involved in biodegradation of rubber joints in water and sewage pipes; most are free-living

saprophytes, but some species (e.g., *N. asteroides*) are opportunistic pathogens causing nocardiosis

   b. *Rhodococcus* can degrade a wide variety of molecules, including those found in toxic wastes

V. Suborder *Micromonosporineae*
- A. Contains one family and many genera; are often referred to as actinoplanetes
  1. Extensive substrate mycelia; aerial mycelia are absent or rudimentary; have type IID cell walls
  2. Form spores within a sporangium that extends above the surface of the substratum; spores can be motile or nonmotile
  3. Genera vary in arrangement and development of spores
- B. Found in soil and freshwater habitats and occasionally in the ocean
  1. Soil dwellers play an important role in plant and animal decomposition
  2. Some produce antibiotics such as gentamicin

VI. Suborder *Propionibacterineae*
- A. Contains two families and 14 genera
- B. Genus *Propionibacterium*
  1. Pleomorphic, nonmotile rods that are often club shaped; cells also may be coccoid or even branched; arranged as single cells, short chains, or in clumps
  2. Facultatively anaerobic or aerotolerant; can ferment sugars to produce propionic acid
  3. Found on skin and in the digestive tract of animals; also in dairy products such as cheese; contribute to the production of Swiss cheese; *P. acne* is involved in the development of body odor and acne vulgaris

VII. Suborder *Streptomycineae*
- A. Contains one family, *Streptomycetaceae,* and three genera
  1. Have aerial mycelia that divide in a single plane to form chains of nonmotile conidiospores
  2. Commonly called streptomycetes
- B. Genus *Streptomyces*
  1. An enormous genus with around 150 species, but all are strict aerobes, have wall type I, and form nonmotile spores within a thin sheath; one of the largest procaryotic genomes with many regulatory genes to control its complex life cycle
  2. Species are distinguished based on morphological and physiological characteristics
- C. Streptomycetes are ecologically and medically important
  1. Natural habitat is soil where they represent from 1–20% of the organisms present (impart the characteristic odor of moist earth by producing volatile substances such as geosmin)
  2. Metabolically flexible; major contributors to mineralization
  3. Best known for the synthesis of a vast array of antibiotics useful in medicine and research, including streptomycin, neomycin, tetracycline, erythromycin, amphotericin B, chloramphenicol, and nystatin
  4. Only *S. somaliensis* is known to be pathogenic in humans; it causes actinomycetoma, an infection of subcutaneous tissues that produces swelling, abscesses, and bone destruction

VIII. Suborder *Streptosporangineae*
- A. Contains three families and 16 genera
- B. Many of the genera are referred to as maduromycetes because the sugar madurose (3-O-methyl-D-galactose) is found in their cell extracts; have aerial mycelia that produce pairs or short chains of spores; substrate mycelia are branched; some genera form sporangia
- C. Genus *Thermomonospora* produces single spores on the aerial mycelium or on both the aerial and the substrate mycelium; isolated from high-temperature habitats such as compost piles and hay

IX. Suborder *Frankineae*
- A. Genera *Frankia* and *Geodermatophilus*
  1. Form clusters of spores and have type III cell walls
  2. The genus *Geodermatophilus* has motile spores and is an aerobic soil organism
  3. Genus *Frankia*
       a. Forms nonmotile sporangiospores in a sporogenous body

b. Grows in symbiotic relationship with at least eight families of higher nonleguminous plants

c. Microaerophilic and able to fix atmospheric nitrogen within nodules formed in host plant roots

B. Genus *Sporichthya* lacks a substrate mycelium but uses holdfasts to anchor to the substratum; grows upward to form aerial mycelia that release motile, flagellated conidia in the presence of water

X. Order *Bifidobacteriales*

A. Contains one family and 10 genera

B. Genera *Falcivibrio* and *Gardnerella* are found in the human genitourinary tract; *Gardnerella* may be a major cause of vaginitis

C. Genus *Bifidobacterium* is best studied

1. Nonmotile, nonsporing, gram-positive rods of varied shapes that are slightly curved and clubbed; often they are branched; rods can be single cells, in clusters or in V-shaped pairs

2. Anaerobic and ferment carbohydrates to produce acetic and lactic acids but no carbon dioxide

3. Found in the mouth and intestinal tract of warm-blooded animals, in sewage, and in insects

a. *B. bifidus* is a pioneer colonizer of the human intestinal tract, particularly when babies are breast-fed

b. Some infections of humans have been reported, but genus does not appear to be a major cause of disease

TERMS AND DEFINITIONS

Place the letter of each term in the space next to the definition or description that best matches it.

_____ 1. A tissuelike mass of cells formed by actinomycetes

_____ 2. Asexual spores held on the ends of filaments

_____ 3. Asexual spores located within a structure at the end of the filament

_____ 4. A volatile substance produced by members of the genus *Streptomyces* that imparts the characteristic odor of moist earth

_____ 5. An infection of subcutaneous tissues that produces lesions and leads to swelling, abscesses, and bone destruction if untreated.

_____ 6. 3-*O*-methyl-D-galactose

_____ 7. Complex fatty acids with a hydroxyl group on the β-carbon and an aliphatic chain attached to the α-carbon.

a. actinomycetoma
b. conidiospores (conidia)
c. geosmin
d. madurose
e. mycolic acids
f. sporangiospores
g. thallus

FILL IN THE BLANK

1. Aerobic, gram-positive bacteria that form branching, usually nonfragmenting, hyphae and asexual spores are called _____.

2. Actinomycetes form asexual spores called _____ or _____, if they are at the tips of hyphae, or _____ if they are located in a sporangium.

3. The actinomycetes called _____ have hyphae that readily fragment into rods and coccoid elements. Some also have _____ in their cell walls, as do mycobacteria.

4. Members of the genus *Arthrobacter* exhibit a type of reproduction in which the inner layer of the cell grows inward to generate a transverse wall dividing the new cells. As the transverse wall thickens, it puts pressure on the outer layer of the original cell wall, eventually causing the outer layer to rupture at its weakest point. This type of reproduction is called _____ _____.

5. The _____ are composed of _____ and their high G + C relatives.

242

6. Members of the genus *Mycobacterium* are said to be _____ (i.e., basic fuchsin dye cannot be removed with acid-alcohol treatment). This is because their cell walls contain waxes with 60 to 90 carbon _____ acids.

7. The maduromycetes, members of the suborder *Streptosporangineae*, all have type III cell walls containing the sugar _____.

8. Members of the suborder *Streptomycineae* are often called _____.

MULTIPLE CHOICE

For each of the questions below select the *one best* answer.

1. Which of the following is NOT a reason for studying the actinomycetes?
   a. They contribute to the mineralization of organic material.
   b. They produce the majority of the medically useful, natural antibiotics.
   c. Some members are pathogenic.
   d. All of the above are reasons for studying actinomycetes.

2. Which of the following is characteristic of the members of the suborder *Micromonosporineae*?
   a. They lack a substrate mycelium.
   b. Some species produce antibiotics such as gentamicin.
   c. both (a) and (b)
   d. neither (a) nor (b)

3. Which of the following colors is frequently observed in the colonies of members of the genus *Micrococcus*?
   a. red
   b. yellow
   c. orange
   d. all of the above.

4. Which of the following is correct about the genus *Arthrobacter*?
   a. They are irregular, branched rods during exponential growth.
   b. They are coccoid in stationary phase.
   c. Both (a) and (b) are correct.
   d. Neither (a) nor (b) is correct.

5. A member of which of the following genera causes diphtheria?
   a. *Corynebacterium*
   b. *Mycobacterium*
   c. *Propionibacterium*
   d. none of the above

6. Which of the following diseases is caused by members of the genus *Mycobacterium*?
   a. tuberculosis
   b. leprosy
   c. Both (a) and (b) are correct.
   d. Neither (a) nor (b) is correct.

7. Which of the following is correct about members of the genus *Propionibacterium*?
   a. They contribute to the production of Swiss cheese.
   b. They cause acne vulgaris and contribute to the development of body odor.
   c. Both (a) and (b) are correct.
   d. Neither (a) nor (b) is correct.

8. Which of the following is characteristic of the streptomycetes?
   a. They have type I cell walls.
   b. They synthesize a vast array of antibiotics.
   c. They play a major role in mineralization processes.
   d. All of the above are characteristic of the streptomycetes.

9. Which of the following is NOT characteristic of the *Frankineae*?
   a. They produce clusters of spores at hyphal tips.
   b. They have type III cell walls.
   c. Some form symbiotic associations with leguminous plants.
   d. Some can fix nitrogen.

ACTINOBACTERIA

Order	Suborder	Representative Genera	Important Characteristics (structural, physiological, ecological)
Actinomycetales	*Actinomycineae*		
	Micrococcineae		
	Corynebacterineae *Corynebacteriaceae*		
	Mycobacteriaceae		
	Nocardiaceae		
	Micromonosporineae		
	Propionibacterineae		
	Streptomycineae		
	Streptosporangineae		
	Frankineae		
Bifidobacteriales			

TRUE/FALSE

_____ 1. Most actinomycetes are not motile, but when present, motility is confined to flagellated spores.

_____ 2. Unlike most other actinomycetes, the nocardioforms have filaments that readily fragment into rods and coccoid elements.

_____ 3. No streptomycetes are known to be pathogenic for humans.

_____ 4. In addition to being opportunistic pathogens, some species of *Nocardia* cause biodeterioration of rubber joints in water and sewer pipes.

_____ 5. Streptomycetes can constitute up to 20% of the culturable microorganisms found in the soil.

_____ 6. Of the nearly 500 species in the genus *Streptomyces*, only *S. somaliensis* is known to be pathogenic in humans.

_____ 7. In the actinomycetes, six major types of cell walls can be distinguished.

_____ 8. Members of the genus *Bifidobacterium* are among the first colonizers of the intestinal tract of nursing infants.

CRITICAL THINKING

1. Discuss the major ecological and medical contributions of the genus *Streptomyces*.

ANSWER KEY

Terms and Definitions

1. g, 2. b, 3. f, 4. c, 5. a, 6. d, 7. e

Fill in the Blank

1. actinomycetes 2. conidia; conidiospores; sporangiospores 3. nocardioforms; mycolic acids 4. snapping division 5. actinobacteria; actinomycetes 6. acid-fast; mycolic 7. madurose 8. streptomycetes

Multiple Choice

1. d, 2. b, 3. d, 4. c, 5. a, 6. c, 7. c, 8. d, 9. c

True/False

1. T, 2. T, 3. F, 4. T, 5. T, 6. T, 7. F, 8. T

25 The Protists

CHAPTER OVERVIEW

This chapter discusses the characteristics of the diverse, polyphyletic group of organisms commonly known as the protists, which includes the algae and the protozoans. They range from single cells to multicellular organisms over 75 meters in length. They are found in oceans and freshwater environments and are the major producers of oxygen and organic material. In addition to a discussion of their general features and the vast array of their niches and habitats, individual coverage of some representative protists is given.

CHAPTER OBJECTIVES

After reading this chapter you should be able to:

• discuss the various habitats, types of locomotion, and specialized organelles of protists
• discuss the various morphological characteristics of protists
• discuss the taxonomic relationships of this diverse, polyphyletic group of organisms
• discuss asexual and sexual reproduction of protists
• discuss the various types of nuclei found in protists
• describe the various feeding mechanisms used by protozoans
• place the protists within current classification schemes

CHAPTER OUTLINE

I. Introduction
    A. Protists are a polyphyletic collection of organisms. Most are unicellular and lack the level of tissue organization found in higher organisms
    B. The protists include protozoans (chemoorganotrophic protists) and algae (photosynthetic protists); protozoology is the study of protozoans; phycology is the study of algae; protistology is the general study of protists
    C. Older classification schemes separated protozoans based on locomotion; the current system of classification uses morphological, biochemical, and phylogenetic analyses
II. Distribution
    A. Found in moist environments, mainly free living (planktonic) in freshwater or marine environments; in terrestrial systems associated with decaying organic matter
    B. Every group of protists contains species that live in association with other organisms such as in lichens (symbiosis of algae and fungi)
III. Nutrition
    A. Photosynthetic protists are aerobes that perform plant-like oxygenic photosynthesis; some are photoheterotrophic
    B. Chemoheterotrophic protists are holozoic (phagocytize solid foods such as bacteria) or saprozoic (take up soluble nutrients)
    C. Some protists are mixotrophic, having flexible metabolisms (e.g., simultaneous photosynthesis and holozoic feeding)
IV. Morphology
    A. Complex single-celled eucaryotes with specialized organelles; typically maximize surface-to-volume ratio to enhance diffusion

B. Protists have a cell membrane called a plasmalemma like other eucaryotes; however, some protists have a gelatinous layer of cytoplasm, the ectoderm, just inside the plasmalemma and a fluid region of cytoplasm, the endoplasm, toward the center of the cell; the pellicle includes the plasmalemma and structures immediately inside it

C. Specialized organelles include contractile vacuoles (osmoregulation), phagocytic vacuoles (food digestion in holozoic and parasitic species), cytosome (cell mouth), cytoproct (excretory channel), pyrenoids (starch storage in chloroplasts)

D. Aerobic chemoorganotrophic species have mitochondria while most anaerobic protists lack mitochondria, cytochromes, and the TCA cycle; some anaerobes have membrane-bound hydrogenosomes that convert pyruvate into acetyl-CoA, CO_2, and H_2, generating ATP

E. Many protists have flagella or cilia with basal-body-like kinetosomes, not only for movement but also for making waves to aid feeding and respiration

V. Encystment and Excystment

A. Many protists can de-differentiate (become simpler) into cysts, a dormant form with a cell wall and little metabolic activity; encystation protects the organism from adverse environments, can be reproductive, and can be used for transfer to new hosts

B. Excystment (escape from cysts) is usually triggered by a change to favorable conditions

VI. Reproduction

A. Most protists reproduce asexually by binary fission or budding; some filamentous forms fragment

B. Most protists also can reproduce sexually by producing haploid gametes called gamonts and gamete fusion called syngamy with two morphologically similar gametes (isogamy) or different gametes (anisogamy); nuclear material may be exchanged between individuals (conjugation) or between genetically distinct nuclei in a single individual (autogamy)

C. There is considerable diversity in nuclear structure

1. Vesicular nucleus—most common, spherical 1 to 10 μm with distinct nucleolus and uncondensed chromosomes

2. Ovular nucleus—larger bodies (10 to 100 μm) with many peripheral nucleoli

3. Chromosomal nucleus—single nucleolus associated with one chromosome; chromosomes remain condensed during interphase

4. *Ciliophora* have two nuclei, a large macronucleus with distinct nucleoli and condensed chromatin, and a diploid micronucelus with dispersed chromatin and no nucleoli

VII. Protist Classification

A. Super group *Excavata*

1. Most primitive and deeply branching eucaryotes; most possess a suspension-feeding groove (cytostome) with a flagellum to create feeding currents

2. *Fornicata*

a. Have flagella but lack mitochondria; reproduce asexually by binary fission

b. Most are harmless symbionts with some free-living forms in polluted waters

c. Can be parasites in fish, pathogens in turkeys, and *Giardia* in humans causes diarrhea and is a public health concern

3. *Parabasilia*

a. Flagellated, but without a distinct cytostome; use phagocytosis for feeding; most are endosymbionts of animals

b. *Trichonymphida*

1) Obligate mutualist in the digestive tract of wood-eating insects (termites)

2) Secrete cellulase to aid animal digestion

3) Usually asexual but sometimes reproduce sexually in response to host hormone called ecdysone

c. *Trichomonadida*

1) Anaerobic, do not contain mitochondria, but have hydrogenosomes, and reproduce asexually

2) Symbionts of digestive, reproductive, and respiratory tracts of vertebrates

(i) *Dientamoeba fragilis* causes diarrhea in humans

(ii) *Trichomonas vaginalis* can cause disease in human genitourinary tract

248

(iii) Tritrichomonas foetus can cause spontaneous abortion in cattle

4. *Euglenozoa*
   a. Commonly found in freshwater, but can be marine
   b. Some are photoautotrophic, while the majority are mainly saprotrophic chemoorganotrophs
   c. *Euglena* is the representative photoautotroph; contains chlorophylls *a* and *b* and carotenoids; primary storage is in paramylon (polysaccharide of β-1,3 linked glucose); has giant eye spot (stigma) for orientation to light, two flagella, and a large contractile vacuole
   d. Some *Euglenozoa* are pathogens, particularly the trypanosomes
      1) *Leishmania* cause systemic and skin/mucous afflictions (leishmaniasis)
      2) *Trypanosoma cruzi* cause Chagas's disease in Central and South America characterized by peripheral nervous system dysfunction
      3) *T. gamiense* and *T. rhodesiense* cause African sleeping sickness
      4) Pathogenic trypanosomes rapidly change the glycoprotein coating of their cell walls (antigenic variation), which helps them evade the host's immune system

B. Super group *Amoebozoa*
   1. Amoeboid motility involves pseudopodia (cell extensions) that are rounded (lobopodia), long and narrow (filopodia), or form a mesh (reticulopodia)
   2. Amoebae can lack a cell wall (naked amoebae) or be covered (testate amoebae)
   3. Asexual reproduction is be binary fission, although some form cysts
   4. *Tubulinea*
      a. Widely found in moist environments; *Amoeba proteus* is commonly used in laboratories
      b. Some are endosymbionts, commensals, or parasites of animals
      c. Some harbor intracellular symbionts including algae, bacteria, and viruses
   5. *Entamoebida*
      a. *Entamoeba histolytica* causes amoebic dysentary, a leading cause of parasitic death
      b. Cysts ingested from feces-contaminated food or water pass through stomach and multiply in intestines, producing digestive enzymes that damage gut epithelial cells; can spread to other areas of the body
   6. *Eumycetozoa*
      a. Slime molds have complex morphology, life cycle, and behavior
      b. *Myxogastria* are acellular slime molds that live on rotting organic matter, where multinucleate masses of protoplasm (plasmodium) move in an amoeboid fashion and feeding by endocytosis; under adverse conditions, fruiting bodies form with stalks and spores; spores germinate to produce haploid amoeboflagellates and eventually zygotes and new plasmodium
      c. *Dictyostelium discoideum* is the classic cellular slime mold
         1) Masses called pseudoplasmodium retain their cell walls; under adverse conditions cells aggregate into motile multicellular slugs that form fruiting bodies with spores that can germinate into amoeboid cells
         2) Sexual reproduction involves special spores called macrocysts that arise via conjugation

C. Super group *Rhizaria*
   1. Amoeboid cells with filopodia; some filopodia called axopodia are supported by microtubules that protrude from an axoplast and are used for feeding
   2. *Radiolaria*
      a. Most Radiolaria have internal skeletons of siliceous material or strontium sulfate; some have siliceous exoskeletons, while others lack skeletons
      b. The skeletons of ancient radiolarians are found in seafloor sediments (siliceous ooze)
      c. Feed by endocytosis via entrapment in mucous; some have algal endosymbionts
      d. Reproduction can be sexual with release of many biciliated isogametic cells or asexual by binary or multiple fission or budding

3. *Foraminifera*
   a. Amoeboid with reticulopedia and form calcerous shells; can have algal symbionts
   b. Complex life cycle that may laternate between asexual and sexual phases; the sexual phase includes flagellated gametes and haploid gamonts; autogamy is known
   c. *Foraminifera* are found in marine and estuarine waters and most are benthic; their calcerous shells accumulate on the sea floor creating chalk, limestone, and marble layers hundreds of meters deep
D. Super group *Chromalveolata*
   1. A diverse group that includes autotrophic, mixotrophic, and heterotrophic protists, deriving their chloroplasts from archaeplastids
   2. *Alveolata*—a large group of protists that includes apicomplexans, dinoflagellates, and ciliates
      a. Apicomplexans
         1) *Acomplexa* are intra- and intercellular parasites of animals with a unique array of cytoskeleton, vacuoles, and other organelles at one end of the cell called the apical complex designed to penetrate host cells; some have motile gametes or zygotes
         2) Complex life cycle has sexual and asexual stages that sometimes occur in different hosts; the motile haploid infective stage is called the sporozoite; in the host, gamonts are generated, which fuse to form zygotes that become spores; these undergo meiosis to generate more haploid sporozoites
         3) Apicomplexans are important infectious agents
         (i) *Plasmodium*—malaria, a major killer worldwide
         (ii) *Eimera*—cecal coccidiosis in chickens
         (iii) *Cryptosporidia*—intestinal infection
         (iv) *Theilaria*—tick-borne blood diseases in cattle
      b. Dinoflagellates
         1) *Dinoflagellata* large group of common marine plankton; involved in coral formation the causative agents of toxic red tides
         2) Many are photosynthetic but not autotrophic; most are saprotrophic, but some use endocytosis
         3) Two flagella cause the cells to whirl; thecate cells are armored with cellulose plates
         4) Most are free-living, but some are endosymbionts that release motile cells called zooxanthellae
      c. Ciliates
         1) *Ciliophora* are large group (12,000 species) of chemoorganotrophs that inhabit benthic and planktonic aquatic systems and moist soils; *Paramecium* and *Stentor* are well-known members
         2) Employ many cilia for movement and feeding; the action of cilia is coordinated; some have tentacles or expel poison darts (toxicysts) to capture prey
         3) Most have macro- and micronuclei; the micronucleus contains the diploid chromosomes for sexual reproduction; the macronucleus is polyploid and produces mRNA to maintain cellular metabolism and function
         4) Some ciliates reproduce asexually by binary fusion, while many can reproduce sexually by conjugation; during sexual reproduction, the macronuclei degrade and gametic nuclei are produced from the micronucleus in a complex process
         5) Some ciliates are symbiotic and may be parasites; cause ick in fish
      d. Stramenophiles
         1) Large, diverse group that includes photosynthetic diatoms, brown and golden algae, brown seaweeds and kelp, and chemoorganotrophic öomcytes and labyrinthulids; all possess heterokont flagella (straw hair flagella) in a life cycle phase
         2) Diatoms
         (i) Photoautotrophs with chlorophylls *a* and c_1/c_2 and fucoxanthin; major carbohydrate is chrysolaminarin (polysaccharide of β-1,3 linked glucose)

       (ii) Diatoms form silica shells (frustules) made of two overlapping halves; frustule morphology is diverse and used for identification; can be unicellular, colonial, or filamentous; lack flagella

       (iii) Asexual reproduction occurs until frustule becomes too small and then sexual reproduction occurs where the zygote develops into an auxospore that divides and forms a new wall to produce vegetative cells with frustules

       (iv) Found in aquatic systems; major contributors to carbon fixation in the ocean and vital to global carbon cycles

     3) Öomycetes

       (i) Once considered fungi but have cellulose and β-glucan cell walls instead of chitin

       (ii) Form large egg cell during sexual reproduction that is fertilized by a sperm cell or antheridium, generating zoospores with heterokont flagellation

       (iii) Important members include saprophytes on decaying organic matter, fish parasites, downy mildews on plants, and *Phytophthora*, the causative agent of the Irish potato blight in the 1840s

     4) Labyrinthulids

       (i) Once considered fungi, they form heterokont flagellated spores

       (ii) Move along a external ectoplasmic network of actinlike filaments

       (iii) Mainly feed by osmotrophy, releasing enzymes to degrade organic matter

   e. *Haptophyta*—planktonic photosynthetic protists that are encrusted in calcite scales and influence global carbon cycles

E. Super group *Archaeplastida*

   1. Includes all organisms with a photosynthetic plasmid arising from endosymbiosis with an ancient cyanobacterium; the group includes higher plants and protists

   2. *Chloroplastida*

     a. Commonly called green algae, these phototrophs are found in aquatic and soil systems; many cell types including unicellular, filamentous, and colonial

     b. Have chlorophylls a and b and carotenoids; store carbohydrates as starch

     c. Asexual reproduction is through zoospores and sexual reproduction involves four flagellated zygotes and resting phase with meiosis, producing haploid organisms

     d. Some important Chloroplastida include:

       1) *Chlamydomonas* is highly studied; has two flagella, a haploid nucleus, a giant chloroplast, contractile vacuoles, and a stigma (eyespot)

       2) *Chlorella* is widespread in the environment; lacks flagella, but has eyespot and contractile vacuoles; only asexual reproduction

       3) *Volvox* forms colonies of hollow spheres that coordinate flagella beating; only a few cells of the colony are reproductive (sexually or asexually)

       4) *Protheca* is common in soil and can cause severe subcutaneous and systemic blood infections (protothecosis) in animals and humans

TERMS AND DEFINITIONS

Place the letter of each term in the space next to the definition or description that best matches it.

_____ 1. Term that describes algae that are suspended in the aqueous environment

_____ 2. A disease caused by a green alga; the disease usually starts as a small subcutaneous lesion and spreads until it covers large areas of the body

_____ 3. A dense, proteinaceous area of the chloroplast that is associated with the synthesis and storage of starch

_____ 4. Asexual spores that are flagellated and, therefore, motile

_____ 5. Vegetative cells that are the site of egg formation

_____ 6. A structure that aids in phototactic responses

_____ 7. An articulated, proteinaceous structure inside the plasma membrane that is rigid but somewhat flexible

_____ 8. Intricately patterned coverings on the plasma membrane of some algae

_____ 9. The distinctive two-piece wall (valve) of silica found on diatoms

_____ 10. The larger of the two pieces of the diatom valve

_____ 11. The smaller of the two pieces of the diatom valve

_____ 12. Symbiotic dinoflagellates found in marine organisms

_____ 13. The vegetative form of a protozoan

_____ 14. The dormant form of a protozoan

_____ 15. Semisolid or gelatinous cytoplasm just beneath the plasma membrane

_____ 16. Fluid cytoplasm in the interior of protozoa

_____ 17. The single site of phagocytosis in some ciliated protozoans

_____ 18. Formation of the cyst stage from the vegetative stage

_____ 19. Formation of the vegetative stage from the cyst stage

_____ 20. Cytoplasmic extensions used for locomotion and food capture

_____ 21. Nuclear mitotic division followed by division of the cytoplasm

_____ 22. A rapid series of mitotic events producing many small, infective organisms through the formation of uninuclear buds

_____ 23. The process whereby there is an exchange of gametes between paired protozoans of complementary mating types

_____ 24. The loose-fitting shell around some amoeboid organisms

_____ 25. An arrangement of fibrils, microtubules, vacuoles, and other organelles at one end of the cell

_____ 26. A region of the pellicle of ciliates where phagocytic vacuoles empty their contents after food digestion has taken place

_____ 27. Specialized region found in some zooflagellates where their mitochondrial DNA is stored

a. apical complex
b. binary fission
c. conjugation
d. cyst
e. cytoproct
f. cytostome
g. ectoplasm
h. encystation
i. endoplasm
j. epitheca
k. excystation
l. frustule
m. hypotheca
n. kinetoplast
o. oogonia
p. pellicle
q. planktonic
r. protothecosis
s. pseudopodia
t. pyrenoid
u. scales
v. schizogony
w. stigma (eyespot)
x. test
y. trophozoite
z. zoospores
aa. zooxanthellae

FILL IN THE BLANK

1. The term _____ was originally used to define simple "aquatic plants." The study of these organisms is called _____.

2. Many of the algae have a structure called an eyespot, or _____, that aids the organism in phototactic responses.

3. Brown algae are multicellular and can be very large. The largest are called _____.

4. The *Stramenopiles* includes golden-brown algae, yellow-green algae, and _____. The latter have a two-piece cell wall of silica, called a _____, that accumulates at the bottom of aquatic environments. This material is used in detergents, polishes, paint removers, and many other products.

5. The _____, though photosynthetic, are most closely related to ciliated protozoans. Algal blooms of these organisms, called _____ _____, can be very destructive.

6. The *Euglenozoa* lack cell walls but have a flexible proteinaceous structure inside the plasma membrane. This structure is called a _____. The *Euglenozoa* that are pathogenic to humans fall in class called _____ and includes two systemic diseases of global importance, _____, and _____.

7. In some species of protozoans, the cytoplasm immediately under the plasma membrane is gelatinous and is termed _____. The plasma membrane and structures immediately beneath it are referred to as the _____. The more fluid portion of the cytoplasm in the interior of the cell is referred to as the _____.

8. One or more vacuoles are usually found in the cytoplasm of protozoans. Vacuoles that function as osmoregulatory organelles are called _____ vacuoles and are usually found in organisms that live in a hypotonic environment, such as a freshwater lake. Holozoic and parasitic organisms have _____ vacuoles, which serve as the site of food digestion. Vacuoles with enzymes that function during the excystation process are referred to as _____ vacuoles.

9. Two types of heterotrophic nutrition are observed in protozoans. In _____ nutrition, nutrients are acquired by phagocytosis. In ciliated organisms, phagocytosis occurs at a single location called the _____. In _____ nutrition, nutrients are acquired by pinocytosis or other forms of direct transport.

10. Members of the phylum *Ciliophora* have cilia and two types of nuclei. The _____ is associated with trophic activities, and the _____ is diploid and involved in reproductive processes.

11. Ciliated protozoans reproduce sexually by a process called _____. Two ciliates, called _____, fuse their pellicles at a contact point. Then the _____ in each is degraded and the _____ divide meiotically to produce haploid pronuclei. These are exchanged, fuse, undergo mitotic divisions, and eventually develop into new nuclei.

12. The _____ slime molds feed as a multinucleate mass of protoplasm called a _____. When food or water is in short supply, they form a fruiting body that produces and releases spores. When the conditions are again favorable, the spores germinate to release either nonflagellated amoeboid _____ or flagellated _____ cells that feed and are haploid. The _____ slime molds feed as individual amoeboid cells called _____. When their food supply is exhausted, the cells swarm together to form a sluglike _____, which eventually forms a fruiting body called a _____ (_____).

13. The _____, or water molds, even though phylogenetically distinct, resemble fungi in appearance because they consist of finely branched filaments called _____.

MULTIPLE CHOICE

For each of the questions below select the *one best* answer.

1. Which of the following is NOT a mechanism used by algae for asexual reproduction?
   a. fragmentation
   b. spore formation
   c. binary fission
   d. All of the above are mechanisms used by algae for asexual reproduction.

2. Diatom shells have been used for which of the following?
   a. abrasives in detergents and polishes
   b. filtering agents
   c. soundproofing and insulating material
   d. All of the above are correct.

3. Which of the following is NOT a reason for studying protozoans?
   a. They are important links in many food chains and food webs.
   b. They are good models for studying eucaryotic metabolism.
   c. They cause important diseases in humans and other animals.
   d. All of the above are reasons for studying protozoa.

4. Which of the following is NOT a major function of cysts?
   a. They have locomotory organelles for movement through an aqueous environment.
   b. They protect against adverse changes in the environment.
   c. They play a role in reproduction.
   d. They serve as a means of transfer from one host to another in parasitic species.

5. Which of the following is a method of locomotion used by protozoans?
   a. flagella
   b. cilia
   c. pseudopodia
   d. All of the above are correct.

6. Which of the following is NOT a flagellated parasite to humans?
   a. *Giardia lamblia*
   b. *Entamoeba histolytica*
   c. *Trichomonas vaginalis*
   d. *Trypanosoma cruzi*

7. Protozoans that have spindle-shaped or spherical nonamoeboid vegetative cells, and that move within a network of mucous tracks by a typical gliding motion, belong to which of the following phyla?
   a. *Apicomplexa*
   b. *Ascetospora*
   c. *Labyrinthomorpha*
   d. *Myxozoa*

8. Which of the following is NOT true about the apicomplexans*?*
   a. The motile infective stage is called the sporozoite.
   b. Their complex life cycles involve an alternation of diploid and haploid generations.
   c. The male and female gamonts fuse to form the zygote.
   d. All of the above are true about the phylum *Apicomplexa.*

9. Which of the following diseases is NOT caused by members of the apicomplexans?
   a. malaria
   b. toxoplasmosis
   c. giardiasis
   d. coccidiosis

TRUE/FALSE

_____ 1. There are no diseases of humans caused by photosynthetic protists.

_____ 2. Algae are not known to be endosymbionts, parasites, or members of mutualistic relationships.

_____ 3. *Euglena* uses a contractile vacuole and an anterior reservoir for osmotic regulation.

_____ 4. While most photosynthetic protists are photoautotrophic, some chemoheterotrophs have been identified.

_____ 5. The photosynthetic protists are polyphyletic, meaning that they are associated with multiple lineages with independent origins and do not, therefore, represent a single evolutionary branch of development.

_____ 6. Photosynthetic protists can be unicellular, colonial, filamentous, membranous, or tubular.

_____ 7. Photosynthetic protists have well-developed vascular systems and complex reproductive structures.

_____ 8. The distribution of protists in nature is limited to marine habitats.

_____ 9. In conjugation, the micronuclei divide by meiosis to form the haploid gamete nuclei.

_____10. In conjugation, the parental macronuclei disappear and the daughter cells develop new macronuclei after the diploid zygote has formed.

_____11. In both classical and more recent molecular classification schemes, some protozoans and some algae belong to the same taxa.

_____12. All protozoans exhibit motility as adult organisms.

_____13. Protozoans have been shown to be polyphyletic and therefore not an evolutionary distinct taxon.

_____14. In many respects, the morphology and physiology of protozoans resembles those of multicellular animals.

_____15. Protozoans do not exhibit sexual reproduction.

PROTOZOAL TAXONOMY

Super Group	Evolutionary Lineage	Representative Genera	Important Characteristics (structural, physiological, ecological)
Excavata	Fornicata		
	Euglenozoa		
	Parabasilia		
Amoebozoa	Tubulinea		
	Entamoebida		
	Eumycetozoa		
Rhizaria	Radiolaria		
	Foraminifera		
Chromalveolata	Alveolata		
	Stramenopiles		
	Haptophyta		
Archaeplastida	Chloroplastida		

CRITICAL THINKING

1. The cyanobacteria were once included in the algae. Why were they once included? Why aren't they now included?

2. Many protozoans use contractile vacuoles as osmoregulatory organelles. How do these organelles function to maintain osmotic balance? Why are they found in freshwater protozoans but not in marine protozoans?

3   Many algae and all the protozoans are referred to as protists. Compare and contrast algal protists and protozoans. Do recent molecular studies support their being grouped together in a single taxon? Explain your answer.

ANSWER KEY

Terms and Definitions

1. q, 2. r, 3. t, 4. z, 5. o, 6. w, 7. p, 8. u, 9. l, 10. j, 11. m, 12. aa, 13. y, 14. d, 15. g, 16. i, 17. f, 18. h, 19. k, 20. s, 21. b, 22. v, 23. c, 24. x, 25. a, 26. e, 27. n

Fill in the Blank

1. algae; phycology 2. stigma 3. kelps 4. diatoms; frustule 5. dinoflagellates; red tides 6. pellicle; trypanosomes; Chaga's disease; African sleeping sickness 7. ectoplasm; pellicle; endoplasm 8. contractile; phagocytic; secretory 9. holozoic; cytostome; saprozoic 10. macronucleus; micronucleus 11. conjugation; conjugants; macronucleus; micronuclei 12. acellular (plasmodial); plasmodium; myxamoebae; swarm; cellular; myxamoebae; pseudoplasmodium; sorus; sorocarp 13. oomycetes; hyphae

Multiple Choice

1. d, 2. d, 3. d, 4. a, 5. d, 6. b, 7. c, 8. c, 9. c

True/False

1. F, 2. F, 3. T, 4. T, 5. T, 6. T, 7. F, 8. F, 9. T, 10. T, 11. T, 12. F, 13. T, 14. T, 15. F

26 The Fungi (Eumycota)

CHAPTER OVERVIEW

This chapter discusses the characteristics of the members of the kingdom *Fungi*. The diversity of these organisms is described, and their ecological and economic impact is discussed. In addition, certain protists—the slime molds and water molds, which resemble fungi—are also presented in this chapter.

CHAPTER OBJECTIVES

After reading this chapter, you should be able to:

- discuss the distribution of fungi and their roles in the environment
- discuss the morphological characteristics of fungi
- describe the external digestion of organic matter by fungi
- explain the formation of both asexual and sexual spores for reproduction
- discuss the five major types of true fungi: zygomycetes, ascomycetes, basidiomycetes, deuteromycetes, and chytrids
- discuss the criteria upon which fungi are classified
- compare and contrast slime molds, water molds, and true fungi

CHAPTER OUTLINE

I.   Introduction
    A.   Fungi—eucaryotic, spore-bearing organisms with absorptive metabolism and no chlorophyll; reproduce sexually and asexually
    B.   Mycologists—scientists who study fungi
    C.   Mycology—the study of fungi
    D.   Mycotoxicology—the study of fungal toxins and their effects on various organisms
    E.   Mycoses—diseases caused by fungi in animals
    F.   Belong to the kingdom *Fungi* (*Eumycota*) within the domain *Eucarya*; is a monophyletic group also known as the true fungi

II.   Distribution
    A.   Primarily terrestrial with a few freshwater and marine organisms
    B.   Many are pathogenic in plants or animals
    C.   Form beneficial associations with plant roots (mycorrhizae) or with algae or cyanobacteria (lichens)

III. Importance
    A.   Decomposers—break down organic material and return it to environment
    B.   Major cause of plant disease; also cause disease in animals, including humans
    C.   Industrial fermentation—bread, wine, beer, cheese, tofu, soy sauce, steroid manufacture, antibiotic production, and the production of the immunosuppressive drug cyclosporin
    D.   Research—fundamental biological processes can be studied in these simple eucaryotic organisms with *Saccharomyces cereviseae* being particularly important

IV.   Structure
    A.   Thallus—body or vegetative structure of a fungus; fungal cell walls are usually composed of chitin, a nitrogen-containing polysaccharide consisting of N-acetyl glucosamine residues
    B.   Yeast—unicellular fungus with single nucleus; reproduces asexually by budding or sexually by spore formation; daughter cells may separate after budding or may aggregate to form colonies

C. Mold—a fungus with long, branched, threadlike filaments
1. Hyphae—the filaments of a mold; may be coenocytic (no cross walls within the hyphae) or septate (having cross walls)
2. Mycelia—bundles or tangled masses of hyphae
D. Dimorphism—a property of some fungi that change from the yeast (Y) form (within an animal host) to the mold (M) form (in the environment); this is referred to as the YM shift; the reverse relationship exists in plant-associated fungi
V. Nutrition and Metabolism
A. Most fungi are saprophytes, securing nutrients from dead organic material (chemoorganoheterotrophs); fungi secrete hydrolytic enzymes that promote external digestion
B. Glycogen is the primary storage polysaccharide
C. Most are aerobic (some yeasts are facultatively anaerobic); obligate anaerobic fungi are found in the rumen of cattle
VI. Reproduction
A. Asexual reproduction—occurs by several mechanisms
1. Transverse fission
2. Budding of somatic vegetative cells
3. Spore production
a. Hyphal fragmentation—resulting cells behave as arthrospores or chlamydospores (if enveloped in thick cell wall before separation)
b. Sporangiospores are produced in sporangium (sac) at the end of an aerial hypha (sporangiophore)
c. Conidiospores are unenclosed spores produced at the tip or on the sides of an aerial hypha
d. Blastospores are produced from a vegetative mother cell by budding
B. Sexual reproduction
1. Involves the union of compatible nuclei
2. Some fungi are self-fertilizing, producing male and female gametes on the same mycelium (homothallic), while others require outcrossing between different but sexually compatible mycelial mating types (heterothallic)
3. Zygote formation proceeds by one of several mechanisms
a. Fusion of gametes
b. Fusion of gamete-producing bodies (gametangia)
c. Fusion of hyphae
d. Sometimes there is immediate fusion of nuclei and cytoplasm; however, more common is a delayed fusion of nuclei, resulting in the formation of a cell with two haploid nuclei (dikaryotic stage)
4. Zygotes can develop into spores (zygospores, ascospores, or basidiospores)
C. Both sexual and asexual spores are used for identification purposes and aid fungal dissemination
VII. Characteristics of the Fungal Divisions
A. Fungi have an evolutionary history rich in convergent and divergent evolution where similar structures have developed independently; the fungi are a monophyletic group with eight subdivisions based on gene sequence analysis
B. *Chytridiomycetes*
1. The simplest of the true fungi, often called chytrids
2. Terrestrial and aquatic fungi that reproduce asexually by forming motile zoospores
3. Microscopic in size; may consist of single cells; a small, multinucleate mass; or a true mycelium
4. Reproduce asexually or sexually (with sporangiospores)
5. Some saprophytic; others are parasites of plants and animals
C. *Zygomycota*—zygomycetes
1. Most are saprophytes; a few are plant and animal parasites
2. Have coenocytic hyphae (no cross walls), with many haploid nuclei
3. Asexual reproduction leads to the formation of sporangiospores

4. Sexual reproduction leads to the formation of zygospores; these are tough, thick-walled zygotes that can remain dormant when the environment is too harsh for growth
5. *Rhizopus stolonifer* (commonly known as bread mold, but also grows on fruits and vegetables)
   a. Normally reproduces asexually
   b. Reproduces sexually by fusion of gametangia if food is scarce or environment is unfavorable; zygospores (diploid) are produced and remain dormant until conditions are favorable; meiosis often occurs at time of germination
   c. *Rhizopus* is an important pathogen in rice, causing seedling blight
6. Zygomycetes are used in the production of foods, anesthetics, coloring agents, and other useful products

D. *Ascomycota*—ascomycetes
1. Members of this division cause food spoilage, and a number of plant diseases (e.g., powdery mildew, chestnut blight, ergot, and Dutch elm disease); they have also been important research organisms
2. Include many types of yeast, edible morels, and truffles, as well as the bread molds *Neurospora crassa* and *Aspergillus nidulans*
3. Mycelia have septate hyphae
4. Many produce conidiospores when reproducing asexually
5. Ascospores (haploid spores located in a sac called an ascus) are formed when reproducing sexually; thousands of asci may be packed together in a cup-shaped ascocarp
6. Many yeast genera are classified with ascomycetes with *Saccharomyces cereviseae* being highly studied; it alternates between haploid and diploid states in response to nutrients
7. *Aspergillus* is widespread and can be pathogenic in immunocomprimised individuals; some species are also used for fermenting saki and soy sauce; several *Aspergillus* genomes have been sequenced; insights into eucaryotic evolution and the immune system are obtained from comparative analyses

E. *Basidiomycota*—basidiomycetes (club fungi)
1. Includes jelly fungi, rusts, shelf fungi, stinkhorns, puffballs, toadstools, mushrooms, and bird's nest fungi
2. Basidia are produced at the tips of the hyphae, in which the basidiospores will develop; basidiospores are held in fruiting bodies called basidiocarps
3. Usefulness—many basidiomycetes are decomposers; some mushrooms serve as food (some are poisonous); one is the causative agent of cryptococcosis, asystemic infection involving the lungs and central nervous system

F. *Urediniomycetes* and *Ustilaginomycetes*
1. Previously placed in the basidiomycetes, even though they don't produce basidiocarps; this group includes important plant pathogens (smuts and rusts) and some human pathogens
2. *Ustilago maydis* causes corn smut; it is a dimorphic, yeastlike saprophyte; it penetrates the leaf with a specialized structure called an appresorium and triggers tumor formation, eventually releasing diploid spores (teliospores) that germinate to form haploid sporidia

G. *Glomeromycota*
1. Previously placed in the zygomycetes, this group includes important endomycorrhizal symbionts in plants
2. Mycorrhizal fungi form mutualistic associations with plant roots, delivering nutrients to the plant and deriving nutrients from the plant
3. Reproduction is asexual with spores that germinate on plant roots, creating mycelia that penetrate the roots; fragmentation is another mode of propagation

H. *Microsporidia*
1. Includes odd fungi that have been considered as protists; includes some human pathogens
2. Lack mitochondria, peroxisomes, and centrioles
3. Upon germination on host cells, the spores release an organelle called the polar tube with such force that it pierces the host cell membrane allowing entrance; more spores then develop in the host
4. Common human pathology with infection includes diarrhea, pneumonia, encephalitis, and nephritis; can be used a biocontrol agent for some insect pests

TERMS AND DEFINITIONS

Place the letter of each term in the space next to the definition or description that best matches it.

_____	1.	The study of fungi
_____	2.	The study of the poisonous substances released by fungi and their effects on various organisms
_____	3.	The vegetative structure of a fungus
_____	4.	The cross walls observed in some fungal hyphae
_____	5.	Filaments that are uninterrupted by cross walls
_____	6.	Filaments with cross walls
_____	7.	Asexual spores produced within a sac
_____	8.	Asexual spores produced without a sac
_____	9.	Asexual spores produced by budding from a vegetative cell
_____	10.	Cells that behave like spores, are produced by fragmentation of hyphae, but are not surrounded by a thick wall
_____	11.	Cells that behave like spores, are produced by fragmentation of hyphae, and are surrounded by a thick wall
_____	12.	Sexual spores that are diploid, dormant, and surrounded by a tough coating
_____	13.	Sexual spores that are haploid and produced within a sac
_____	14.	Sexual spores that are diploid and produced within a club-shaped sac
_____	15.	A cup- or flask-shaped structure that can hold thousands of asci
_____	16.	Large, multinucleated mass of protoplasm that exhibits amoeboid movement, leaving a slime trail
_____	17.	Large aggregate of amoeboid cells that moves as a unit, leaving a slime trail
_____	18.	Gamete-producing bodies that can fuse to form a zygote during sexual reproduction of some fungi

a. arthroconidia (arthrospores)
b. ascocarp
c. ascospores
d. basidiospores
e. blastospores
f. chlamydospores
g. coenocytic hyphae
h. conidiospores
i. gametangia
j. mycology
k. mycotoxicology
l. plasmodium
m. pseudoplasmodium
n. septa
o. septate hyphae
p. sporangiospores
q. thallus
r. zygospores

FILL IN THE BLANK

1. Unicellular fungi that have a single nucleus and reproduce either asexually by budding and transverse division or sexually by producing spores are called _____, while _____ consist of long, threadlike filaments called _____ that aggregate in bundles to form _____. Some fungi can alternate between the two forms and are said to be_____.

2. *Rhizopus stolonifer* is a _____ fungus, which is characterized by having _____ hyphae. *R. stolonifer* is a _____; it secures its nutrients from dead organic material. Sexual reproduction occurs when two mating types form hyphal projections called _____. These mature to form _____ and eventually fuse, allowing formation of a zygote and later a dormant _____.

3. The _____, also known as club fungi, form a club-shaped structure called a _____ during sexual reproduction. Two or more sexual spores, called _____, are formed by this structure. These fungi may also form fruiting bodies called _____, within which numerous basidia are held.

4. Asexual spores that are produced in saclike structures known as _____ are called _____. Asexual spores that are not enclosed in a sac, but that are produced at the tips or sides of hyphae are called _____, while spores that are produced by budding from a vegetative cell are called _____.

5. The _____ are the simplest terrestrial and aquatic fungi.

6. Diseases caused by fungi are called _____. One interesting plant fungal disease is _____, a disease of rye and other grasses. If humans or other animals eat grain from these diseased plants, they can develop a toxic condition called _____. Another fungal disease of note is _____, a systemic infection of the lungs and central nervous system.

7. The _____ occurs when dimorphic fungi switch from the _____ form (single cells) to the _____ form (hyphae).

8. The _____ are named for a characteristic structure, the saclike _____, which they make during sexual reproduction. In the more complex members of this group, the formation of this structure is preceded by the development of special hyphae called _____ hyphae. Pairs of nuclei migrate into these hyphae—one from the male mycelium (_____) or cell and one from a female organ or cell (_____).

9. The cell walls of most fungi contain _____.

10. The body or vegetative structure of a fungus is called the _____. Fungi may be grouped into _____ (filamentous fungi) or _____ (unicellular fungi) based on the development of this structure.

FUNGAL TAXONOMY

Division	Representative Genera	Type of Asexual Spores	Type of Sexual Spores	Yeast, Septate Hyphae, or Coenocytic Hyphae
Zygomycota				
Ascomycota				
Basidiomycota				
Chytridiomycetes				
Urediniomycetes and Ustilaginomycetes				

Glomeromycota				
Microsporidia				

MULTIPLE CHOICE

For each of the questions below select the *one best* answer.

1. In addition to the production of asexual spores, fungi can reproduce asexually by which of the following mechanisms?
   a. transverse fission of a parental cell
   b. fragmentation of hyphae whereby the component cells behave as spores
   c. budding of either somatic vegetative cells or vegetative mother cells
   d. All of the above are correct.

2. In most cases when sexual gametes of fungi fuse, the cytoplasm fuses first, and the fusion of the nuclei is delayed. This leads to a stage in which there is one cell containing two haploid nuclei. What is this stage called?
   a. dikaryotic stage
   b. dinucleated stage
   c. monocytoplasmic stage
   d. None of the above is correct.

3. In which of the following divisions are coenocytic hyphae found?
   a. zygomycetes
   b. ascomycetes
   c. basidiomycetes
   d. None of the above is correct.

4. In which of the following are the sexual spores haploid?
   a. zygomycetes
   b. ascomycetes
   c. basidiomycetes
   d. all of the above

5. In which of the following will you find the common bread mold *Rhizopus stolonifer*?
   a. zygomycetes
   b. ascomycetes
   c. basidiomycetes
   d. none of the above

6. The fungi are very important. Why?
   a. They are important decomposers.
   b. They cause diseases in plants and animals.
   c. They are used in many industrial processes.
   d. all of the above

TRUE/FALSE

_____ 1. Most fungi are aerobic or facultatively anaerobic. However, a few are obligate anaerobes.

_____ 2. Many fungal spores are responsible for the bright colors and fluffy texture of the molds that produce them.

_____ 3. Fungal spores are light and can therefore remain suspended in the air for long periods of time.

_____ 4. In all fungi that have been observed to have a sexual reproduction cycle, male and female gametes are produced on separate mycelia and must then find each other for fertilization to occur.

_____ 5. Fungi secrete enzymes outside their body structure and absorb the digested food.

_____ 6. Apressoria are used for reproduction in *Ustilago*.

_____ 7. Microsporidia are eucaryotic fungi that lack mitochondria and centrioles.

CRITICAL THINKING

1.   Dimorphic fungi have traditionally been grouped together as deuteromycetes. Today mycologists have spread them among different fungal divisions. For what reasons would deuteromycete grouping make sense? Why would the field have shifted away form that taxonomic scheme?

ANSWER KEY

Terms and Definitions

1. j, 2. k, 3. q, 4. n, 5. g, 6. o, 7. p, 8. h, 9. e, 10. a, 11. f, 12. r, 13. c, 14. d, 15. b, 16. l, 17. m, 18. i

True/False

1. T, 2. T, 3. T, 4. F, 5. T, 6. F, 7. T

Fill in the Blank

1. yeasts; molds; hyphae; mycelia; dimorphic 2. zygomycete; coenocytic; saprophyte; progametangia; gametangia; zygospore 3. basidiomycetes; basidium; basidiospores; basidiocarps 4. sporangia; sporangiospores; conidiospores; blastospores 5. chytrids 6. mycoses; ergot, ergotism; cryptococcosis 7. YM shift; yeast, mold (mycelial) 8. ascomycetes; ascus; ascogenous; antheridium; ascogonium 9. chitin 10. thallus; molds; yeasts

Multiple Choice

1. d, 2. a, 3. a, 4. b, 5. a, 6. d

27   Biogeochemical Cycling and Introductory Microbial Ecology

CHAPTER OVERVIEW

This chapter focuses on microbial ecology—the study of the relationships microorganisms have with other organisms, each other, and their physical environment. This chapter considers nutrient cycling and the roles of microorganisms in ecosystems. The chapter also includes a discussion of extreme environments and concludes with a discussion of the methods used in microbial ecology.

CHAPTER OBJECTIVES

After reading this chapter you should be able to:

- describe the roles microbes play in ecosystems
- discuss the interactions of microorganisms with one another and with nonmicrobial members of ecosystems
- describe the microorganisms found in extreme environments
- discuss the various aspects of microbial assemblages that are of interest to microbial ecologists
- discuss the methods used by microbial ecologists to study microbial populations and communities

CHAPTER OUTLINE

I.   Foundations of Microbial Diversity and Ecology
    A.   Microbial ecology is the study of the interactions of microbes with other organisms (symbioses) and with their physical environment; symbiosis is an association of two or more different species
    B.   Interactions of organisms with each other and with their physical environment contribute to the functioning of ecosystems
        1.   Populations—assemblages of similar organisms within an ecosystem
        2.   Communities—mixtures of different populations within an ecosystem
        3.   Ecosystems—self-regulating biological communities and their physical environment
    C.   A major problem in understanding microbial ecology is that most microscopically observable microorganisms cannot be cultured in the lab; however, recent advances in molecular techniques are providing information on the uncultured microorganisms in ecosystems
II.   Biogeochemical Cycling
    A.   Biogeochemical cycling (nutrient cycling) involves both biological and chemical processes; oxidation-reduction reactions change the chemical and physical properties of the nutrient
    B.   Carbon cycle
        1.   Carbon can be interconverted among methane, complex organic matter, and carbon dioxide; conversions involve reductants (sources of electrons) and oxidants (sinks for electrons)

a. Organic compounds are produced primarily by carbon fixation carried out by the activities of plants, cyanobacteria, green algae, photosynthetic bacteria, and chemolithoautotrophs

b. Organic carbon is returned to the atmosphere by oxidation to carbon dioxide (inorganic carbon) through respiration and fermentation

c. Inorganic and organic carbon can be reduced anaerobically to methane by archaea in anoxic zones (e.g., marches, rumen guts, rice paddies)

2. Carbon dioxide is produced by the degradation of organic matter through the process of mineralization

a. Organic matter varies in terms of elemental composition, structure of basic repeating units, linkages between repeating units, and physical and chemical characteristics

b. Degradation of organic matter is influenced by nutrients present in the environment, abiotic conditions (pH, oxidation-reduction potential, O_2, osmotic conditions), and the microbial community present

c. Microbial degradation of complex organic material occurs when microbes use these molecules for growth

1) Chitin, protein, and nucleic acids contain large amounts of nitrogen; when these are degraded, any excess nitrogen is released by mineralization

2) Molecules containing only hydrogen, carbon, and oxygen cannot support the growth of microbes; the inability to assimilate sufficient levels of nutrients limits growth

3) Immobilization occurs when nutrients held in biomass are unavailable for nutrient cycling

d. Most organic substrates can be degraded in the presence or absence of oxygen; however, degradation of hydrocarbons and lignin usually occurs aerobically

1) Hydrocarbon degradation usually requires oxygen because the first step adds molecular oxygen to the molecule; recently, however, slow anaerobic digestion in the presence of sulfate or nitrate has been observed

2) Filamentous fungi are major lignin degraders and they require oxygen

e. The presence or absence of oxygen affects the final products that accumulate when organic substances are degraded

1) Aerobic conditions—oxidized products are made (carbon dioxide, nitrate, sulfate)

2) Anaerobic conditions—reduced end products are formed (ammonium, sulfide, methane)

f. Carbon dioxide and methane are greenhouse gases that appear to be accumulating in the atmosphere, trapping heat and raising global temperatures

C. Nitrogen cycle

1. Nitrogen is cycled among organic nitrogen, ammonia, nitrogen gas, nitrite, nitrous oxide, and nitrate

2. Nitrogen fixation

a. Sequential reduction of gaseous nitrogen to ammonia; it requires an expenditure of energy; the ammonia is immediately incorporated into organic matter as an amine

b. Can be carried out by procaryotic aerobes or anaerobes; the actual reduction process must be done anaerobically, even by aerobic microorganisms; physical barriers, O_2-scavenging molecules, and high rates of metabolic activity are used to maintain the anaerobic conditions required for nitrogen fixation; cyanobacteria form specialized cells (heterocysts) for nitrogen fixation

3. Nitrification—aerobic oxidation of ammonium ion to nitrite and ultimately to nitrate; it is carried out by chemolithoautotrophs and by some heterotrophs (heterotrophic nitrification)

4. Nitrogen assimilation

a. Assimilatory reduction—reduction of nitrate and incorporation into new microbial biomass

b. Dissimilatory reduction—reduction of nitrate to nitrite, nitrous oxide, and gaseous N_2; the sequence of reactions is known as denitrification

5. Anoxic ammonia oxidation (anammox reaction)—oxidation of ammonia is coupled with the reduction of nitrite to nitrogen gas by chemolithotrophs

D. Phosphorus cycle
1. This cycle has no gaseous component; found in soils in inorganic and organic forms as phosphates or polyphosphates; it is always in valence state +5 (not redox active)
2. Inorganic phosphorus compounds are relatively insoluble and not bioavailable, often limiting microbial growth in the environment

E. Sulfur cycle
1. Sulfur can be interconverted among elemental sulfur, sulfide, and sulfate by the actions of various microorganisms; under certain conditions, transformations of the sulfur cycle can occur in the absence of microorganisms
2. Dissimilatory sulfate reduction (anaerobic respiration) produces sulfide, which accumulates in the environment; assimilatory sulfate reduction results in the reduction of sulfate for use in amino acid biosynthesis
3. Sulfur oxidation is carried out by both aerobic and anaerobic anoxygenic phototrophs

F. Iron cycle
1. Iron can be interconverted among ferric iron, ferrous iron, and magnetite
2. Iron oxidation from ferrous iron to ferric iron is carried out by a number of genera under aerobic conditions; some microorganisms can carry out the process under anaerobic conditions using nitrate as the electron acceptor
3. Iron reduction from ferric iron to ferrous iron occurs under anaerobic conditions and is carried out by bacteria that use ferric iron as a final electron acceptor
4. Magneto-aerotactic bacteria reduce iron to magnetite, which is used to construct intracellular magnetic compasses; these bacteria use magnetic fields to migrate to a position in a bog or swamp where the oxygen level is optimal for their functioning

G. Manganese cycle—microbial transformation of manganous ion (Mn^{+2}) to MnO_2 (with Mn^{+4}); occurs in hydrothermal vents and bogs

H. Microorganisms and metal toxicity
1. Metals have varied toxic effects on microorganisms and homeothermic animals; microorganisms modify this toxicity
2. Noble metals (silver, gold, platinum, etc.)—cannot cross the blood-brain barrier of vertebrates, but have distinct effects on microorganisms
3. Metals and metalloids that can be methylated (e.g., mercury, arsenic, lead, selenium, and tin)—methylation enables them to affect microorganisms and to cross the blood-brain barrier and affect the central nervous system of higher organisms; methylated metals can be concentrated in the food chain (a process known as biomagnification)
4. Metals that occur in ionic forms (e.g., copper, zinc, cobalt) are often required as trace elements, but can be directly toxic to microorganisms and more complex organisms if in excess

III. The Physical Environment
A. The microenvironment and niche
1. Microenvironment—specific physical location of a microorganism
2. Niche—the function of an organism in a complex system, including place of the organism, the resources used in a given location, and the time of use
3. Microorganisms can create their own environments and niches

B. Biofilms and microbial mats
1. Biofilms—organized microbial systems consisting of layers and aggregates of microbial cells associated with surfaces
2. Biofilms can be observed in three dimensions using confocal scanning laser microscopy
3. Biofilms on living surfaces usually play a role in causing disease; biofilms can protect pathogens from disinfectants, create a focus for later occurrence of disease, or release microbial cells or products that may affect the immune system of the host
4. Microbial mats—thick biofilms that have macroscopic dimensions; found in many freshwater and marine habitats

C. Microorganisms and ecosystems
   1. Ecosystems—communities of organisms and their physical and chemical environments that function as self-regulating units
   2. The main roles of microorganisms are as primary producers and decomposers; the microbial loop is the complex set of interactions between microbes that contribute to and use a common pool of dissolved organic matter (DOM)
   3. They impact ecosystems in many ways
      a. Synthesize organic matter through photosynthetic and chemosynthetic processes
      b. Decompose organic matter, often with the release of inorganic compounds (mineralization)
      c. Serve as nutrient-rich food sources for chemoheterotrophic microorganisms and animals
      d. Modify substrates and nutrients used in symbiotic growth processes and interactions, thus contributing to biogeochemical cycling
      e. Change the amounts of materials in soluble and gaseous forms
      f. Produce inhibitory compounds that decrease microbial activity or limit the survival and functioning of plants and animals
      g. Contribute to the functioning of plants and animals through positive and negative symbiotic interactions
D. Microorganism movement between ecosystems
   1. Microorganisms are constantly moving and being moved between ecosystems by a variety of mechanisms
   2. The fate of microorganisms when moved to another ecosystem is of theoretical and practical importance (e.g., what happens to pathogens when moved to an aquatic environment)
   3. When microorganisms are moved out of their normal environment, they usually die or enter a viable but nonculturable (VBNC) state
E. Extreme environments
   1. Factors such as pH, temperature, pressure, salinity, water availability, and ionizing radiation can act as stress factors
      a. If one or more of these factors is extremely high or low, it creates an extreme environment
      b. Extremophiles are organisms that survive in extreme environments
   2. Salinity—favors extreme halophiles
   3. High barometric pressure (e.g., deep-sea environments) favors barotolerant, moderately barophilic, and extremely barophilic bacteria
   4. Extremes of pH favor acidophiles (maintain a high internal pH relative to the environment) and alkalophiles (maintain a low internal pH relative to the environment)
   5. High temperature (up to 121°C) favors thermophiles and extreme thermophiles
IV. Microbial Ecology and Its Methods: An Overview
   A. Microbial ecologists study natural microbial communities that exist in a wide variety of habitats and their interactions
   B. Examination of microbial populations
      1. Microscopic, cultural, physical, chemical, and molecular techniques have all been useful
      2. Direct viable count procedures and enrichment techniques are useful both in determining population size and finding new and undescribed microbes
      3. One problem in determining numbers and types of microbes is that the vast majority of microorganisms observed in nature cannot yet be grown, including VBNC microorganisms
      4. Another problem is that many microorganisms (e.g., protozoa, algae, and cyanobacteria) are actually microbial assemblages; these microbes can have surface-associated commensal partners or endosymbionts, from which they cannot be separated
      5. All methods have inherent limitations; a critical challenge for microbial ecologists is to recognize the limitations and to understand what information a particular method will and will not provide; generally it is best to use more than one method to obtain complementary information
      6. Only about 1% of microbial species can be cultured today

7. Direct DNA extraction from communities and genetic analysis have become invaluable to microbial ecologists; however, there are inconsistencies in extraction methods and concerns as to whether the DNA is from living organisms, and whether the DNA is from nonfunctional propagules (e.g., spores)

8. Typically PCR is used to examine small subunit (SSU) ribosomal RNA (16S for procaryotes; 18S for eucaryotes) gene sequences in direct DNA extracts
   a. rRNA gene amplicons from different species can be separated based on melting characteristics (reflects primary sequence and G + C content) by denaturing gradient gel electrophoresis (DGGE)
   b. Gene sequences can be compared using statistical methods and related organisms grouped into phylotypes
   c. Genes that code for enzymes can be used to determine functional characteristics of a community

9. Microbial species diversity can be estimated by counting different genomes using DNA reassociation kinetics

C. Examination of microbial community structure
   1. Direct observation coupled with molecular techniques provides valuable information about community structure
   2. Community structure can also be described in terms of the nutrients contained in the community (e.g., carbon, nitrogen); such single-value estimates are of limited value when dealing with nondiscrete microorganisms such as filamentous fungi
   3. Metagenomics uses total community DNA (or RNA) extracts to create clone libraries to identify functional elements in the gene pool and their dynamics

D. Microbial activity and turnover
   1. Specific processes (e.g., nitrification, denitrification) are studied by the use of direct chemical measurements; microarrays are used to measure gene expression; real-time reverse-transcriptase PCR is used to determine community gene expression; stable isotope measurements can indicate whether elements (e.g., carbon) have been processed by an organism
   2. Microbial growth rates can be measured directly (microscopic examination of culture) or indirectly (e.g., incorporation of radiolabeled components such as thymidine)

E. Recovery or addition of individual microbes
   1. Single-cell isolations can be carried out using optical tweezers (a laser beam is used to drag a microbe away from its neighbors) and by micromanipulation (a fine capillary is used to separate a microbe from its neighbors); once isolated the cell can be studied individually (i.e., PCR amplification, sequencing, and phylogenetic analysis)
   2. Phenotypic or population heterogeneity—dissimilar phenotypes are observed in a genetically uniform population
   3. Reporter microbes can be used to characterize the physical microenvironment; this involves constructing reporter genes that have altered expression or produce gene products with altered behavior (e.g., green fluorescent protein) in response to environmental and physiological changes

TERMS AND DEFINITIONS

Place the letter of each term in the space next to the definition or description that best matches it.

_____ 1.   Reduction of sulfate for use in amino acid biosynthesis
_____ 2.   Self-regulating biological communities and their physical environment
_____ 3.   The incorporation of simple, soluble substances into the body of an organism, making it unavailable for use by other organisms
_____ 4.   An organism that synthesizes and accumulates new organic matter
_____ 5.   Chemoheterotrophs that feed on organisms that accumulate and decompose organic matter
_____ 6.   An organism that cycles material from dead organisms and changes it into forms that can be utilized by other organisms
_____ 7.   Organized microbial systems consisting of layers of microbial cells associated with surfaces.
_____ 8.   Layered communities of microorganisms that have macroscopic dimensions and are associated with surfaces
_____ 9.   The synthesis of new organic matter from carbon dioxide and other inorganic compounds
_____ 10. The decomposition of organic matter that results in the production of inorganic nutrients
_____ 11. An assemblage of similar organisms living within an ecosystem
_____ 12. A mixture of populations living and interacting within an ecosystem
_____ 13. The increase in concentration of a substance (e.g., toxic metals) in higher-level consumer organisms
_____ 14. Produces an accumulation of sulfide in the environment
_____ 15. The process by which atmospheric nitrogen gas is converted into ammonia
_____ 16. The aerobic process by which ammonia is converted into nitrite and nitrate
_____ 17. The process by which nitrate is converted into nitrogen gas
_____ 18. The anaerobic process by which ammonia is oxidized
_____ 19. The function of an organism in a complex system
_____ 20. Common pool of organic matter found in aquatic systems

j.   dissolved organic matter (DOM)
k.   ecosystems
l.   immobilization
m.   microbial mats
n.   mineralization
o.   niche
p.   nitrification
q.   nitrogen fixation
r.   population
s.   primary producer
t.   primary production

a.   anammox
b.   assimilatory sulfate reduction
c.   biofilms
d.   biomagnification
e.   community
f.   consumers
g.   decomposer
h.   denitrification
i.   dissimilatory sulfate reduction

FILL IN THE BLANK

1. Microorganisms function in physical locations called their _____. The resources available at this location and the time of their use define the microorganism's _____.

2. An assemblage of similar organisms (i.e., the same species) is called a _____. A mixture of different assemblages is called a _____.

3. *Chromatium* and *Chlorobium* are anoxygenic phototrophs that function under strict anaerobic conditions in deep water columns. However, a large group of bacteria, often found in _____ _____ (biofilms of macroscopic dimensions), carry out _____ _____ _____.

4. Bacteria such as *Nitrosococcus* are chemolithoautotrophs that carry out _____. In acidic environments, certain bacteria and fungi oxidize ammonia in a process called _____ _____. Still other microorganisms are capable of coupling the oxidation of ammonia with the reduction of nitrite to nitrogen gas. This is called the _____ (anoxic ammonia oxidation) process.

5. Microbial communities play major roles in _____ cycling, in which both biological and chemical processes are involved in the cycling of nutrients. These cycles often involve _____ and _____ reactions, and change the concentrations of gaseous components (e.g., carbon, nitrogen, and sulfur).

6. Sulfate, when used as an external electron acceptor, is reduced to sulfide and accumulates in the environment. This is an example of _____ reduction. In contrast, if sulfate is reduced for use in the synthesis of proteins and other molecules, it is referred to as _____ reduction.

7. The reduction of nitrate to nitrite, nitrous oxide, and gaseous molecular nitrogen is referred to as _____. The reduction of nitrogen gas to ammonia is called _____ _____.

8. Most of the microorganisms observed in complex natural communities are currently _____ microorganisms. This complicates the understanding of microbial ecology. Fortunately, recent advances such as the use of _____ _____ and _____ enable scientists to recover individual cells from their communities and obtain genomic and phylogenetic information about the recovered cells.

9. Many microorganisms serve as _____ _____, accumulating organic matter, which becomes available to other organisms in the ecosystem. Some of these microbes are photoautotrophs, but others are chemoautotrophs that use hydrogen, sulfide, and methane as energy sources.

10. Bacteria called _____ bacteria transform extracellular iron to magnetite and use it to construct intracellular compasses that are used to orient the bacteria in bogs or swamps at the appropriate oxygen level.

11. Stress factors such as high temperature and high salinity can create _____ environments that are not hospitable to many organisms. Organisms that survive in such environments are called _____. Some examples include _____ bacteria or _____ bacteria, which grow under conditions ranging from 1–500 atm; _____ _____ bacteria, which have a growth optimum at 5,000 atm but can also grow at 1 atm; and _____ _____ bacteria, which require 400 atm or higher for growth.

MULTIPLE CHOICE

For each of the questions below select the *one best* answer.

1. What role is played by vascular plants in terrestrial environments?
    a. They are the major primary producers.
    b. They are the major consumers.
    c. They are the major decomposers.
    d. They are the major top-level consumers.

2. In which of the following environments do chemoautotrophs serve as primary producers?
    a. terrestrial soils
    b. shallow freshwater environments
    c. deep marine seeps of hydrogen sulfide

d. In none of the above environments do chemoautotrophs serve as primary producers.

3. Which role is NOT one of the main functions of microorganisms in ecosystems?
   a. primary producers
   b. consumers
   c. decomposers
   d. top level consumers

4. Which of the following techniques is useful for measuring gene expression in complex microbial communities?
   a. microarrays
   b. optical tweezers
   c. micromanipulation
   d. All of the above can be used to measure gene expression in complex microbial communities.

5. Which of the following metals can be methylated and thereby rendered capable of crossing the vertebrate blood-brain barrier and causing central nervous system damage?
   a. silver
   b. mercury
   c. zinc
   d. platinum

6. What is the term that describes the process in which toxic compounds accumulate in organisms of a food chain such that the top-level consumers ingest concentrations higher than those that are normally found in the environment?
   a. bioconcentration
   b. biomagnification
   c. bioamplification
   d. bioaccumulation

7. Which of the following organic compounds cannot be degraded in the absence of oxygen?
   a. lignin
   b. hemicellulose
   c. chitin

d. lipids

8. Which of the following is used by aerobic nitrogen-fixing bacteria to maintain the anaerobic conditions needed for the reduction of nitrogen?
   a. physical barriers to keep oxygen out of certain compartments
   b. O_2-scavenging molecules to remove oxygen
   c. high metabolic rates to utilize oxygen as quickly as it becomes available
   d. All of the above are correct.

9. Which group of metals is harmful to humans and other homeothermic animals?
   a. noble metals (e.g., silver)
   b. metals that can be transformed to organometals (e.g., mercury)
   c. metals that occur in ionic forms (e.g., copper)
   d. None of the above are harmful to homeothermic animals.

10. In which cycle are there inorganic and organic forms that are not redox active and no gaseous component?
    a. iron
    b. phosphorus
    c. manganese
    d. sulfur

11. In which technique are DNA fragments separated based on melting point?
    a. DGGE
    b. SSU
    c. VBNC
    d. microarrays

12. In which technique is total community DNA extracts used to create clone libraries for analysis of functional elements in the gene pool?
    a. microarrays
    b. direct observation
    c. metagenomics
    d. numerical taxonomy

TRUE/FALSE

_____ 1. Mineralization is the process of obtaining various minerals from ores by the action of microorganisms.

_____ 2. In addition to their usual roles as producers, consumers, and decomposers, microorganisms also play a role as food sources for other chemoheterotrophic microorganisms and for predatory and parasitic organisms.

_____ 3. The aspects of microbial communities that are of interest to microbial ecologists include nutrient cycling, population size, activity of the microbes, and community structure.

_____ 4. Although most organisms observable in natural habitats cannot be grown, molecular techniques are making it possible to obtain information on these noncultured organisms.

_____ 5. Nitrogen fixation can be carried out by organisms growing aerobically or anaerobically, but the process itself must take place in an anaerobic microenvironment—even in aerobic microorganisms.

_____ 6. Metals such as copper are toxic at extremely low levels and serve no useful purpose for microorganisms.

_____ 7. What we would consider an extreme environment may be considered optimum and may be required by the microorganisms found in them.

_____ 8. Degradation of organic material by microorganisms is influenced by several factors, including the composition and structure of the organic material and the microbial community present.

_____ 9. Genes that code for enzymes used to determine functional characteristics are called phylotypes.

CRITICAL THINKING

1. Explain why wood structures can exist for longer periods of time in swamps and bogs with little or no structural degradation. What would happen if the water table fell or the swamp was drained?

2. Discuss the necessity of using molecular techniques for microbial ecology. What kinds of analyses can be performed after direct DNA extraction? Would these techniques be useful for assigning functions to individual phylotypes?

ANSWER KEY

Terms and Definitions

1. b, 2. k, 3. l, 4. s, 5. f, 6. g, 7. c, 8. m, 9. t, 10. n, 11. r, 12. e, 13. d, 14. i, 15. q, 16. p, 17. h, 18. a, 19. o, 20. j

Fill in the Blank

1. microenvironment; niche 2. population; community 3. microbial mats; aerobic anoxygenic photosynthesis 4. nitrification; heterotrophic nitrification; anammox 5. biogeochemical; oxidation; reduction 6. dissimilatory; assimilatory 7. denitrification; nitrogen fixation 8. nonculturable; optical tweezers; micromanipulators 9. primary producers 10. magneto-aerotactic 11. extreme; extremophiles; barotolerant; piezotolerant; moderately barophilic; extreme barophilic

Multiple Choice

1. a, 2. c, 3. d, 4. a, 5. b, 6. b, 7. a, 8. d, 9. b, 10. b, 11. a, 12. c

True/False

1. F, 2. T, 3. T, 4. T, 5. T, 6. F, 7. T, 8. T, 9. F

28   Microorganisms in Marine and Freshwater Environments

CHAPTER OVERVIEW

This chapter discusses the general characteristics of marine and freshwater environments and describes some of the microorganisms that live there. Life in aquatic environments is most impacted by the levels of light, oxygen, and nutrients present. These are determined by physical properties and by the activities of microorganisms. The chapter discusses life in estuaries and salt marshes, the photic zone of the open ocean, and the benthic zone at the seafloor. Freshwater environments discussed include ices, rivers, and lakes.

CHAPTER OBJECTIVES

After reading this chapter you should be able to:

•   list the gases of major importance to aquatic microbial communities
•   discuss the nutrient levels observed in various types of bodies of water
•   describe the vertical stratification of microbial populations, nutrients, and oxygen that are observed in a Winogradsky column
•   discuss the cycling of carbon, nitrogen, phosphorus, and sulfur in aquatic environments
•   describe the microbial loop and its role in the cycling of nutrients in aquatic environments
•   give examples of the types of adaptations observed in aquatic microorganisms
•   compare and contrast marine and freshwater environments
•   compare and contrast lake environments and stream and river environments
•   discuss frozen aquatic environments

CHAPTER OUTLINE

I.   Marine and Freshwater Environments
   A.   Gases and aquatic environments
      1.   Oxygen
         a.   Aquatic environments are low-oxygen diffusion environments; this can lead to the formation of hypoxic or anoxic zones, which are inhabited by anaerobic microbes
         b.   At higher temperatures and with lower pressure the solubility of oxygen in water is further decreased
      2.   Carbon dioxide
         a.   Important in many chemical and biological processes
         b.   The carbonate equilibrium system buffers the pH of water, with seawater highly buffered by the interplay of carbon dioxide, bicarbonate, and carbonate
         c.   This equilibrium is impacted by the activity of aquatic microorganisms
      3.   Other gases such as nitrogen, hydrogen, and methane are important; they vary in terms of their solubility in water, with methane being the least soluble
      4.   Light is critical for primary production by autotrophs in marine and freshwater ecosystems; the zone where light penetrates is the photoic zone; solar radiation can warm surface waters and create a thermocline where warm waters float on cooler waters
   B.   Nutrient cycling in marine and freshwater environments

1. Major source of organic matter in illuminated water is phytoplankton (including tiny picoplankton), consisting of photoautotrophic organisms that acquire needed nitrogen and phosphorous from surrounding water
2. Redfield ratio—ratio of carbon-nitrogen-phosphorus (C:N:P) in phytoplankton; is important for following nutrient dynamics and for studying factors that limit microbial growth
3. The microbial loop recycles much of the organic matter produced by phytoplankton (photosynthate in the form of dissolved organic matter); the chemoheterotrophic bacteria that function in the microbial loop (thought of as particulate organic matter) are consumed by a series of larger predators (protists)

II. Microbial Adaptations to Marine and Freshwater Environments
   A. A wide variety of microorganisms are found in waters; most have been discussed in previous chapters; this section focuses on specific microorganisms of particular interest
   B. Ultramicrobacteria or nanobacteria—very small bacteria (volume <0.08 μm^3); dominant bacteria in marine environments
   C. *Thiomargarita namibiensis*—very large bacterium (100 to 300 μm in diameter); contains a huge vacuole that is used to store nitrate; also stores sulfur granules; these characteristics allow it to survive during times when nitrate and sulfur are not readily available
   D. *Thioploca* spp.—individual cells, that can be many centimeters long, live in filamentous sheathed structures containing numerous cells; individual cells move up and down within the sheath, moving from oxygen-poor, nitrate-rich waters into sulfide-rich sediments; by doing so it fulfills its nutrient needs
   E. Microorganisms associated with surfaces include prosthecate bacteria and gliding bacteria; they use organic matter on surfaces
   F. Zoosporic fungi and chytrids have motile reproductive spores; chytrids are important decomposers and can prey on algae
   G. Some filamentous fungi (e.g., Ingoldian fungi) can sporulate under water; they play an important role in processing organic matter

III. Marine Environments
   A. Represent major portion of biosphere; contain 97% of the Earth's water; vital to global biogeochemical cycles
   B. Coastal marine systems: Estuaries and salt marshes
      1. In estuaries, tidal mixing of freshwater and saltwater creates a salinity profile characterized by salt wedges, where heavier saltwater forms a layer below freshwater
      2. The salinity varies in the estuary in time and space; microbial inhabitants need to be halotolerant (withstands salinity changes)
      3. These nutrient-rich waters are often polluted and this can create dead zones or greatly increase microbial growth; harmful algal blooms (HABs), including red tides caused by dinoflagellates, occur when algae grow to high numbers and then produce toxic compounds
   C. Winogradsky columns can be used to model salt marshes and illustrate the interactions and gradients that occur in aquatic environments
      1. Made by mixing together mud, water, and sources of nutrients (e.g., cellulose and other materials), then incubating the column in light
      2. In the bottom, anaerobic conditions foster the activity of fermentative microorganisms, which leads to the accumulation of fermentation products
      3. Other anaerobic microorganisms use the fermentation products to carry out anaerobic respiration using sulfate as the electron acceptor and producing sulfide
      4. Sulfide diffuses upward, creating an anaerobic, sulfide-rich zone where anoxygenic photosynthetic bacteria reside
      5. Further up in the column, chemolithotrophic and mixotrophic organisms may use hydrogen sulfide as an energy source and oxygen as the electron acceptor
      6. At the top are oxygenic photosynthetic organisms such as diatoms and cyanobacteria
   D. The photic zone of the open ocean
      1. The open ocean regions (pelagic) account for half of the world's photosynthesis; it is an oligotrophic environment (very low nutrient levels)

278

2. Organic matter found in the photic zone (upper 200–300 meters) where light penetrates falls as marine snow into lower depths; much of the marine snow is not easily degraded; it is colonized by microbes that consume it, such that only about 1% reaches the seafloor; the collection of organic matter on the seafloor is important in moderating the effects of global warming

3. Cyanobacteria (including *Trichodesmium*) are important in fixing nitrogen in the pelagic zone; some bacteria can perform the anammox reaction wherein ammonia is oxidized to nitrogen gas anaerobically below the photic zone, and lost from the seas

4. The most abundant monophyletic group of organisms is an α-proteobacterial lineage called SAR11; these small bacteria have recently been isolated and have small, efficient genomes; they are responsible for about 50% of bacterial biomass production and DOM flux in some marine environments

E. Archaea are important members of oceanic picoplankton (cells < 2 µm); can also be found in freshwater and deep in the ocean; Crenarchaeota represent about 20% of oceanic picoplankton; the relative distribution of bacteria and archaea differs widely among ecosystems

F. Viruses—present in 10-fold higher concentration than bacteria; the virioplankton may influence microbial loop, may be involved in horizontal gene transfer, and may control microbial community diversity

G. Benthic marine environments
   1. Benthos—oceanic sediments and the seafloor including hydrothermal vents; mainly high pressure, low light, and cold (1 to 4 °C) environments
   2. Microbes found not only on seafloor but to a depth of at least 0.6 km where pressures are high and the organisms must be barophilic
   3. Massive deposits of methane hydrate are found at the ocean floor below 500 meters; some marine bacteria consume methane hydrates and serve as food for ice worms; some archaea metabolize methane in conjunction with sulfate, in a process called reverse methanogenesis

IV. Microorganisms in Freshwater Environments
   A. Microorganisms in glaciers and permanently frozen lakes
      1. These ancient deposits hold microorganisms that may provide information about the biogeographic distribution of microorganisms and life on icy, extraterrestrial worlds (e.g., Europa)
      2. Metabolically active environments that support a variety of microbes; frozen lakes may have thick ice caps that block solar radiation, leading to chemosynthetic communities
   B. Microorganisms in streams and rivers
      1. Differ from lakes in that horizontal movement minimizes vertical stratification and most of the functional biomass is attached to surfaces (not planktonic)
      2. Nutrients available in streams and rivers can be from in-stream production (autochthonous) or from out-stream sources (allochthonous) such as leaves, and runoff from riparian areas; under most conditions, such added organic material does not exceed the oxidative capacity of the stream and it remains productive and aesthetically pleasing
      3. Ability to process organic matter is limited
         a. If the amount of organic material is excessive (eutrophic), oxygen is used faster than it can be replenished, which causes the water to become anaerobic
            1) Point sources of pollution include inadequately treated municipal wastes and other materials from specific locations
            2) Nonpoint sources of pollution include field and feedlot runoffs
         b. If the amount of organic material is not excessive, algae will grow, which leads to oxygen production in the daytime and respiration at night (diurnal oxygen shifts)
      4. Dams, which cause the loss of silicon, also impact streams and rivers; this loss reduces the diatom population and increases the activity of toxic nitrate-utilizing algae
   C. Microorganisms in lakes
      1. Lakes vary in nutrient status
         a. Oligotrophic lakes are nutrient-poor; typically oxic throughout
         b. Eutrophic lakes are nutrient-rich; typically anoxic at the bottom

2. Lakes can be thermally stratified; stratified waters undergo seasonal turnovers because of temperature and specific gravity changes
   a. Epilimnion—warm, aerobic, upper layer
   b. Thermocline—region of rapid temperature decrease
   c. Hypolimnion—cold, often anaerobic (particularly in nutrient-rich lakes) lower layer
3. Eutrophication—stimulation of growth of plants, algae, and bacteria by addition of nutrients (enrichment) to a body of water; cyanobacteria and algae contribute to massive algal blooms in strongly eutrophied lakes; increased respiration reduces oxygen levels and can cause fish kills and odor production, degrading surface waters

TERMS AND DEFINITIONS

Place the letter of each term in the space next to the definition or description that best matches it.

_____ 1. A nutrient-enriched environment

_____ 2. A nutrient-poor environment   a. allochthonous

_____ 3. Environments with high flux rates for oxygen   b. autochthonous

_____ 4. A columnar set-up designed to resemble a naturally   c. carbonate occurring aquatic ecosystem   equilibrium system

_____ 5. Ratio of C:N:P; it is impacted by nutrient dynamics   d. epilimnion and is used to monitor these processes   e. eutrophic

_____ 6. Planktonic photosynthetic microorganisms and small   f. high-oxygen aquatic plants   diffusion environments

_____ 7. Planktonic viruses   g. hypolimnion

_____ 8. Depleted in oxygen   h. hypoxic

_____ 9. The zone where light penetrates an aquatic system   i. marine snow

_____ 10. Clumps of organic matter that escapes the photic zone   j. oligotrophic and falls to the seafloor   k. photic

_____ 11. Nutrients that enter from outside of streams   l. phytoplankton

_____ 12. Nutrients produced by in-stream processes   m. Redfield ratio

_____ 13. The upper warmer layer in thermally stratified lakes   n. virioplankton

_____ 14. The colder bottom layer of thermally stratified lakes   o. Winogradsky column

_____ 15. The process primarily responsible for buffering the pH of oceans

FILL IN THE BLANK

1. Partially enclosed coastal regions where a river meets the sea called _____ have a characteristic salinity profile called a _____ _____ where freshwater floats on denser _____.

2. When nutrients are added to water, _____ (nutrient enrichment) can take place.

3. Lakes can be thermally stratified. The warm upper layer is called the _____, and the deeper colder layer is called the _____. These two are separated by a zone of rapid temperature decrease called the _____.

4. If the amount of organic material added to a river or stream is not excessive, algae grow and produce oxygen during the day. At night, the oxygen is consumed, resulting in _____ _____ _____.

5. Aquatic environments are termed _____ _____ _____ environments because oxygen diffuses slowly through water. Furthermore, the solubility of oxygen in water is limited. This can lead to the formation of _____(low oxygen levels) and _____ (oxygen-free) zones, which allow specialized microorganisms to grow.

6. Photosynthetic microbes and small plants suspended in water are called _____. They create organic matter that enters the _____ _____, a complex process that cycles organic matter from dissolved forms to particulate forms and then back to carbon dioxide and other materials.

7. Methane hydrates are found at low temperatures and high pressures in many marine regions. Methane hydrates are used by certain archaea as an energy source in a process called _____ _____. In this process, sulfate is used as the final electron acceptor.

8. Lakes vary in nutrient status. Some are _____ (nutrient poor) and some are _____ (nutrient rich).

9. The release of inadequately treated wastes and other materials at a single location on a river or stream is a type of _____ _____ pollution. Runoff from fields and feedlots is a type of _____ _____ pollution.

10. The marine microbial community is dominated, in terms of numbers and biomass, by very small bacteria called _____ or _____.

MULTIPLE CHOICE

For each of the questions below select the *one best* answer.

1. Which of the following is NOT a characteristic of *Thiomargarita namibiensis*?
   a. It is considered to be the world's largest bacterium.
   b. It takes advantage of short-term mixing of nitrate (used as an electron acceptor) and hydrogen sulfide (used as an electron donor) by using a huge internal vacuole to concentrate nitrate.
   c. It exhibits gliding motility and moves up and down within a sheath to areas containing the nutrients it needs.
   d. All of the above are characteristics of this interesting microbe.

2. Which of the following is a mechanism by which microbes can be added to aquatic environments?
   a. from the atmosphere
   b. within animal carcasses and plant detritus
   c. within dust
   d. All of the above are mechanisms by which microbes are added to aquatic environments.

3. In which of the following types of waters do seasonal climatic temperature changes result in distinct chemical and microbial stratification?
   a. oceans
   b. eutrophic lakes
   c. oligotrophic lakes
   d. fast-flowing rivers

4. If phosphorus is added to an oligotrophic lake, what organisms will play a major role in nutrient accumulation?
   a. cyanobacteria
   b. nonphotosynthetic bacteria
   c. archaea
   d. eucaryotic algae

5. Which of the following occurs in large numbers in marine environments?
   a. ultramicrobacteria (nanobacteria)
   b. viruses
   c. archaea
   d. All of the above occur in large numbers in marine environments.

6. Which of the following describes when microbes interact at several trophic levels to recycle nutrients?
   a. microbial loop
   b. nutrient cycle
   c. organic dissolution
   d. benthic precipitation

7. Which is an important group of filamentous fungi that can sporulate underwater?
   a. *Chytrids*
   b. Ingoldian fungi
   c. *Hyphomicrobium*
   d. *Thioploca* spp.

8. Which of the following often cause harmful algal blooms?
   a. *Thiothrix* spp.
   b. pan algae
   c. chlorobia
   d. dinoflagellates

9. Which of the following is a term used for open ocean regions?
   a. pelagic
   b. benthic
   c. photic
   d. anoxic

10. Microbial growth in the open ocean is typically limited by which nutrient?
    a. iron
    b. phosphorus
    c. nitrogen
    d. calcium

TRUE/FALSE

_____ 1. Water covered by ice year-round does not support the growth of microorganisms.
_____ 2. The Winogradsky column is a microcosm of an aquatic environment that is used to demonstrate the microbial stratification that can occur in nonagitated waters.
_____ 3. Microbial grazing, or the use of microbes as a food source by protozoans, plays a role in nutrient cycling in aquatic environments.
_____ 4. The horizontal water flow in rivers and streams minimizes vertical stratification.
_____ 5. The largest portion of the Earth's waters is freshwater.
_____ 6. Silicon removal by dams decreases the level of toxic algae in the associated reservoir.

_____ 7. Despite their small size, microorganisms in the ocean can have a significant impact on the atmosphere on a global scale.

_____ 8. The Redfield ratio is impacted by the addition of minerals to the ocean and is important for monitoring nutrient cycling.

_____ 9. Because marine viruses are obligate parasites, they are not important to nutrient cycling in the ocean.

_____ 10. Mixing and diffusion of materials in aquatic environments are critical processes in creating unique environments for different microorganisms.

CRITICAL THINKING

1. Mixing and diffusion of gases and nutrients create unique environments that support the activity of unusual microorganisms. Two examples are *Thiomargarita namibiensis* and *Thioploca* spp. Why are they unusual? Describe their lifestyles and demonstrate the importance of mixing to these microorganisms.

2. Fungi, which are usually thought to be terrestrial, also grow in aquatic environments. Two groups of fungi of particular note are the chytrids and the Ingoldian fungi. How are these fungi adapted to an aquatic existence? What roles do they play in their aquatic habitats?

ANSWER KEY

Terms and Definitions

1. e, 2. j, 3. f, 4. o, 5. m, 6. l, 7. n, 8. h, 9. k, 10. i, 11. a, 12. b, 13. d, 14. g, 15. c

Fill in the Blank

1. estuaries; salt wedge; seawater 2. eutrophication 3. epilimnion; hypolimnion; thermocline 4. diurnal oxygen shifts 5. low-oxygen diffusion; hypoxic; anoxic 6. phytoplankton; microbial loop 7. reverse methanogenesis 8. oligotrophic; eutrophic 9. point source; nonpoint source 10. nanobacteria; ultramicrobacteria

Multiple Choice

1. c, 2. d, 3. b, 4. a, 5. d, 6. a, 7. b, 8. d, 9. a, 10. c

True/False

1. F, 2. T, 3. T, 4. T, 5. F, 6. F, 7. T, 8. T, 9. F, 10. T

29   Microorganisms in Terrestrial Environments

CHAPTER OVERVIEW

This chapter discusses the different types of soils and the microorganisms associated with them. The interactions of microorganisms with plants are discussed, as are the impacts of these interactions on soil fertility and the atmosphere. Other topics discussed include the roles of microorganisms in plant decomposition, in the subsurface biosphere, and in human health.

CHAPTER OBJECTIVES

After reading this chapter you should be able to:

- describe the various environmental conditions found in soils
- describe how soil is formed
- describe the different types of soils and the microorganisms found in them
- discuss the various types of associations between plants and microorganisms
- discuss the interactions of microorganisms and plants and their impact on soil nutrients and on the atmosphere
- discuss organic matter accumulation and decomposition in soils
- discuss the subsurface biosphere
- discuss the impact of soil microbes on human health

CHAPTER OUTLINE

I. Soil as an Environment for Microorganisms
    A. Soil consists primarily of inorganic geological materials, which are modified by the biotic community
    B. Soil varies in terms of the amount of oxygen available to microorganisms
        1. Highly oxygenated areas are found on particle surfaces; microorganisms are often found in thin films of water on these surfaces
        2. The interior of a soil particle can be anaerobic
        3. Under certain conditions, soil may contain isolated pockets of water, which serve as mini-aquatic environments; oxygen concentrations are generally lower in these pockets of water
        4. Waterlogged soil is very similar to anaerobic lake sediments
    C. Changes in water content, gas fluxes, and the growth of plant roots can affect the concentration of carbon dioxide and other gases in the soil
II. Soils, Plants, and Nutrients
    A. Mineral soil contains less than 20% organic carbon, while organic soil contains more
    B. Soil organic matter (SOM)
        1. Helps retain nutrients, maintain soil structure, and hold water for plant use; plowing, irrigation, and other soil disturbances can increase microbial degradation of this organic matter, thereby decreasing soil fertility
        2. Nonhumic SOM has not undergone significant degradation; humic SOM (humus) has been converted into a complex blend of organic materials that are recalcitrant (resist degradation)
        3. Breakdown of plant material to SOM occurs in three steps:
            a. Soluble carbohydrates and proteins are easily converted to CO_2 and biomass

b. Complex carbohydrates (e.g., cellulose) are degraded with extracellular enzymes to smaller sugars that are readily assimilated

c. Resistant materials (e.g., lignin) degrade slowly and accumulate in soils

C. Soils are increasingly being impacted by mineral nitrogen releases resulting from human activities

1. This nitrogen is derived from two major sources: agricultural fertilizers and fossil fuel combustion

2. The nitrogen releases have had a number of effects

a. When the nitrogen added is greater than the amount that can be used by plants, it remains in mobile form and can enter surface water and groundwater; higher nitrate levels in water are related to a number of animal health problems

b. Nitrogen fertilizers alter microbial community structure and function

c. Nitrogen accretion may also contribute to the increasing occurrence of invasive plants

D. Phosphorus fertilizers cause eutrophication of surface waters when excess phosphorus moves into those waters

III. Microorganisms in the Soil Environment

A. Bacteria are found primarily on the surfaces of soil particles, most frequently on surfaces of pores in these particles; this probably protects them from predation by protozoa and gives them access to soluble nutrients

B. Filamentous fungi form bridges between separated particles or aggregates called peds; this exposes the fungi to high levels of oxygen; filamentous fungi move nutrients and water over great distances in the soil

C. Microbial populations can be very high; yet only a small portion of them (1%) has been cultured; actinobacteria and polyprosthecate bacteria are important, but less studied, members of terrestrial ecosystems

D. The microbial community makes important contributions to biogeochemical cycling

E. Soil insects and other animals serve as decomposer-reducers; they not only decompose but also physically reduce the size of organic aggregates; this increases the surface area and makes nutrients more available for utilization by bacteria and fungi

F. Microbial loop in soils differs from that of the open ocean; in soils plants (not microbes) are primary producers; microbes still rapidly turn over organic matter, making the nutrients available to higher trophic levels (predatory protozoa); free hydrolytic enzymes released from animals, plants, insects, and microbes contribute to soil biochemistry

IV. Microorganisms and the Formation of Different Soils

A. Soil formation

1. Colonization takes place after newly exposed geological material begins to weather

2. Nutrients such as phosphorus and iron may be present, but carbon and nitrogen must be imported; cyanobacteria are active in the pioneer stage of nutrient accumulation

3. Successive waves of microbial communities establish a stable climax ecosystem

B. Tropical and temperate region soils

1. In tropical soils, organic matter is decomposed rapidly and mobile inorganic nutrients can be leached out, causing rapid loss of fertility; fertility can be further decreased by slash-and-burn agriculture

2. In temperate region soils, decomposition rates are less than the rate of primary production

a. In temperate grasslands, deep-root penetration leads to formation of fertile soils

b. In cooler coniferous environments, there can be excessive accumulations of organic matter; fire is a major means to control nutrient cycles

c. In bog soils, decomposition is slowed by waterlogged, anoxic conditions; this leads to peat accumulation

C. Cold, moist area soils—colder temperatures decrease both the rate of decomposition and the rate of plant growth; below the plant growth zone is permafrost; these soils are very sensitive to physical disturbances and pollution

D. Desert soils—microbial communities called desert crusts, which consist of cyanobacteria and associated commensals, can retain rainfall and supply nitrogen to other soil microbes and plants

E. Geologically heated hyperthermal soils—populated by bacteria and archaea, many of which are chemolithoautotrophs; chemoorganotrophs also are present

V. Microorganism Associations with Vascular Plants
A. Many types of plant-microbe interactions exist; commensalism is a relationship wherein one partner is benefited, while mutualism benefits both partners; epiphytes live on plant surfaces; endophytes colonize internal tissues
B. Phyllosphere microorganisms—a wide variety of microorganisms (*Sphingomonas*, *Pseudomonas*, *Erwinia*) are found on and in the aerial surfaces of plants (phyllosphere), where they can utilize organic compounds released by the leaves and stems
C. Rhizosphere and rhizoplane microorganisms
    a. The rhizosphere is the volume of soil around plant roots influenced by materials released by the plants; the rhizoplane is the root surface; both provide unique environments for microorganisms; microorganisms in the rhizosphere serve as a labile source of nutrients and play a critical role in organic matter synthesis and degradation
    b. Numerous rhizosphere bacteria influence plant growth through the release of auxins, gibberellins, cytokinins, and other molecules
    c. Associative nitrogen fixation—nitrogen fixation carried out by bacteria on the rhizoplanes and in the rhizosphere
    d. Methanogenic archaea in the rhizosphere of rice have been examined by stable isotope probing; hydrogen produced by fermenters is used by the methanogens to reduce carbon dioxide to methane
D. Mycorrhizae
    1. Mutualistic fungus-root associations in which the fungi are not saprophytic, but use photosynthetically derived host carbohydrates
    2. Six associations have been described; they fall into two broad categories:
        a. Ectomycorrhizae
            1) Ascomycete or basidiomycete fungi that grow as an external sheath around the root and may grow between (but not within) cortical root cells, forming the Hartig net
            2) Fungal hyphae can aggregate in soils, forming rhizomorphs that bring nutrients to the plant
        b. Endomycorrhizae
            1) Fungi that penetrate the outer cortical cells of the plant root
            2) Arbuscular mycorrhizae form characteristic structures known as arbuscules within invaginations in the plasma membrane of cortical root cells; they protect from disease and drought, and bring nutrients to the plant
            3) Orchid endomycorrhizae are saprophytic and bring nutrients to the plant
    3. Mycelia extend far into the soil, forming a mycorrhizosphere, and mediate nutrient transfer to the plant; mycorrhizae increase the competitiveness of the plant and increase water uptake by the plant in arid environments; they also make it possible to share resources (e.g., carbon, minerals, and water); mycorrhizal helper bacteria (MHB) aid in the development of the mycorrhizal relationships
E. Nitrogen fixation—symbiotic nitrogen fixation (conversion of nitrogen gas to ammonia) by bacteria associated with plants is crucial for global nitrogen cycles and agriculture; often performed by *Rhizobium* in root nodules on legumes
F. The Rhizobia
    1. *Rhizobium*—a prominent member of the rhizosphere community; it can also establish an endosymbiotic association with legumes and fix nitrogen for use by the plant
    2. The initial reaction of the plant is an oxidative burst; the *Rhizobium* survives this because of its antioxidant abilities and it is subsequently stimulated (by plant flavonoid inducer molecules) to produce Nod factors, which activate the host responses necessary for root hair infection and nodule development
    3. The bacterium induces formation of an infection thread by the plant; the infection thread grows down into the root hair

4. *Rhizobium* spreads within the infection thread into the underlying root cells, eventually giving rise to the nodule
5. Bacteria terminally differentiate into bacteroids that are enclosed by a plant-derived membrane called the peribacteroid membrane
6. Further growth and differentiation lead to the formation of nitrogen-fixing forms called symbiosomes
7. Nitrogen fixation occurs within symbiosomes within the root nodules which are protected from oxygen by leghemoglobin; the nitrogen is then assimilated into various organic compounds and distributed throughout the plant

G. Actinorrhizae-actinomycete-root associations between members of the genus *Frankia* and woody plants; *Frankia* can fix nitrogen and are important in the life of woody, shrublike plants

H. Stem-nodulating rhizobia—bacteria that form nodules at the base of adventitious roots branching out of the stem above the soil surface; observed primarily in tropical legumes

I. Fungal and bacterial endophytes—a variety of fungal endophytes live within plant tissues, protecting the plant from insect attack, providing fixed nitrogen, or acting as parasites

J. *Agrobacterium*
1. Members of this genus that contain the Ti (tumor-inducing) plasmid cause the formation of crown galls (tumors) on the plant
2. Gall formation is a complex process that involves transfer of Ti DNA into the plant host and expression of a set of *vir* (virulence) genes
3. The Ti plasmid is an important biotechnology tool for transfer of new genetic characteristics into plants

K. Fungi, protists, and bacteria as plant pathogens—many fungi and bacteria are plant pathogens, causing rusts, blights, rots, and other plant diseases

L. Viruses—many viruses infect plants and cause disease and economic losses (e.g., tobacco mosaic virus)

M. Tripartite and tetrapartite associations—involve some combination of plant, rhizobia, mycorrhizae, and actinorrhizae

VI. Soil Microorganisms and the Atmosphere
A. Soil microorganisms can influence the global fluxes of both relatively stable and reactive gases
1. Relatively stable gases include carbon dioxide, nitrous oxide, nitric oxide, and methane
2. Reactive gases include ammonia, hydrogen sulfide, and dimethylsulfide
3. Methane is an important greenhouse gas; methane is produced in landfills, ruminant guts, and water-saturated soils; methane release can be moderated by methanotrophic bacteria, which oxidize methane to obtain energy

B. Use of nitrogen-containing fertilizers affects atmospheric nitrogen levels with production of NO and N_2O, important greenhouse gases

C. Soil microorganisms can influence the atmosphere by degrading airborne pollutants such as benzene, formaldehyde, and trichloroethylene

D. Cyanide is produced in the process of woody plant decomposition; cyanide is poisonous and can inhibit respiration

VII. The Subsurface Biosphere
A. Studied by examining outcrops, surface excavations, petroleum hydrocarbons, well corings, and materials from deep mine sites

B. The subsurface biosphere may contain about one-third of the Earth's living biomass

C. Microbial processes take place in different regions
1. Shallow subsurface aquifers where water flowing from the surface moves below the plant root zone
2. Subsurface regions where organic matter has been transformed by chemical and biological processes to yield coal, kerogens, and oil and gas; mobile materials move up into more porous geological structures where microorganisms can be active
3. Deeper biogenic zones where methane is being synthesized using geologic hydrogen as an energy source

VIII. Soil Microorganisms and Human Health
   A.   Soils contain a wide variety of pathogenic organisms, that can cause disease if they have an entry point into the human body and find favorable conditions there (e.g., *Clostridium*, *Acanthamoeba*, and *Cyclospora*)
   B.   The growth of soil-related microorganisms in buildings can cause mold problems and impact human health

TERMS AND DEFINITIONS

Place the letter of each term in the space next to the definition or description that best matches it.

_____ 1. Desert microbial communities consisting of cyanobacteria and associated commensalistic microbes

_____ 2. The maximum amount of nitrogen that can be incorporated into organic matter by organisms in soil

_____ 3. The use of microorganisms as a food source by protozoans

_____ 4. An initial response by plants to root nodule-forming bacteria; it involves the production of a mixture that can contain superoxide radicals, hydrogen peroxide, and N_2O

_____ 5. Porous geological structures below the plant root zone, through which water often flows

_____ 6. The volume of soil around a plant root that is influenced by materials released from the root

_____ 7. Tumorigenic plasmid carried by *Agrobacterium* strains

_____ 8. Bacteria and fungi that infect and live within plants

_____ 9. Heterogeneous soil aggregates formed partly by bacterial and fungal growth in soil

_____ 10. Substances that activate the host symbiotic processes necessary for infection of root hairs by *Rhizobium*

_____ 11. Substance produced by actinomycetes that gives soil its characteristic earthy odor

_____ 12. Rhizobial cells found in root nodules; when fully differentiated they are capable of nitrogen fixation

_____ 13. The addition of organic matter lacking nitrogen to soil in an attempt to eliminate invasive plant and re-establish native plants

_____ 14. Accumulation of nitrogen in soil from atmospheric sources

a.   aquifers
b.   bacteroids
c.   desert crust
d.   endophytes
e.   geosmin
f.   microbivory
g.   nitrogen saturation point
h.   Nod factors
i.   occult nitrogen accretion
j.   oxidative burst
k.   peds
l.   rhizosphere
m.   soil impoverishment
n.   Ti (Ri) plasmid

FILL IN THE BLANK

1. Plants can develop relationships with two or three different types of microorganisms. Associations of a single plant with two different types of microorganisms are called _____ associations. Several examples are known. One involves a host plant, _____, and rhizobia. Another involves a host plant, endomycorrhizae, and _____ (association with actinomycetes). A third association involves a host plant, _____, and actinorrhizae. Associations of a single plant with three different types of microorganisms are called _____ associations. These consist of the host plant, *Frankia*, _____, and _____.

2. Certain bacterial genera (e.g., *Azotobacter* and *Azospirillum*) carry out nitrogen fixation on the surface of plant roots (i.e., the _____). This type of nitrogen fixation is called _____ nitrogen fixation.

3. Associations between fungi and plant roots are referred to as mycorrhizae: if this association involves the growth of the fungi as a sheath around the root tip with limited penetration of the hyphae into the cortical region of the root, it is called _____; however, if extensive penetration of the fungal hyphae into the cortical regions of the plant root occurs, then it is called _____. Certain of the latter type of mycorrhizae have a characteristic structure called the _____, and fungi with this structure are called _____ _____ fungi.

4. Certain tropical legumes are infected by _____ _____, which form nodules at the base of adventitious roots branching out of the stem just above the soil surface.

5. Members of the genus *Rhizobium* infect the root hairs of legumes, stimulating the plant to form an _____ _____ down which the bacteria travel. Eventually the bacteria are released into a host cell and are surrounded by a plant membrane to become _____. After further differentiation they become nitrogen-fixing _____ and are located in swollen structures called _____ _____.

6. Carbon dioxide, nitrous oxide, nitric oxide, and methane are important _____ _____ (gases that decrease the loss of radiation from the earth and may produce atmospheric warming). Microorganisms in well-drained, aerobic soils can decrease the levels of methane in the atmosphere. These microorganisms, called _____, oxidize methane to obtain energy.

7. A soil with less than 20% organic carbon is called a _____ soil. To retain nutrients, water, and structure, _____ _____ _____ is critical. Organic matter that has undergone significant biochemical degradation is called _____.

8. An analytical technique called _____ _____ _____ can be used to determine the source of organic carbon in studies such as those focused on methanogenesis in rice fields.

MULTIPLE CHOICE

For each of the questions below select the *one best* answer.

1. Which of the following is NOT used by soil microorganisms to support their growth?
   a. moisture contained in soil pores
   b. dissolved nutrients associated with soil particles or in pores
   c. organic matter released from the plant roots
   d. All of the above are used to support growth of soil microorganisms.

2. Which of the following plays an active role in pioneer-stage nutrient accumulation, particularly of carbon and nitrogen?
   a. cyanobacteria
   b. chemoautotrophic bacteria
   c. vascular plants
   d. soil nematodes

3. How do soil insects and other animals contribute to terrestrial ecosystems?
   a. They help decompose organic material in the soil.
   b. They help reduce the size of plant litter in soil.
   c. They release hydrolytic enzymes that degrade proteins and other macromolecules.
   d. All of the above are contributions of soil insects and animals.

4. Which of the following is NOT a benefit to the plant involved in a mycorrhizal association?
   a. increased nutrient availability in moist environments
   b. increased water availability in arid environments
   c. increased utilization of photosynthate
   d. All of the above are benefits to the plant of mycorrhizal associations.

5. Which of the following increases the decomposition of soil organic matter?
   a. composting
   b. irrigation
   c. minimum-till agriculture
   d. All of the above increase soil organic matter decomposition.

6. Which of the following influences soil nutrient cycling?
   a. protozoan microbivory
   b. released enzymes from lysed microorganisms
   c. Both (a) and (b) are correct.
   d. Neither (a) nor (b) is correct.

7. How do soil microorganisms impact the atmosphere?
   a. Some serve as nucleation centers for snow formation, so may impact weather patterns.
   b. Some degrade airborne pollutants such as methane and trichloroethylene.
   c. They impact the global fluxes of stable gases (e.g., carbon dioxide) and reactive gases (e.g., ammonia).

   d. All of the above are ways soil microorganisms impact the atmosphere.

8. Mycorrhizal helper bacteria (MHBs) help with which of the following?
   a. ectomycorrhizal development
   b. endomycorrhizal development
   c. Both (a) and (b) are correct.
   d. Neither (a) nor (b) is correct.

9. Which of the following types of compounds are released by microorganisms and influence plant growth?
   a. auxins
   b. giberellins
   c. cytokinins
   d. all of the above

10. Which term refers to the aerial surfaces of plants?
    a. mycorrhizosphere
    b. phyllosphere
    c. residuesphere
    d. subsurface biosphere

11. Bacterial symbionts have been identified in the cytoplasm of arbuscular mycorrhizal fungi. What is the role of these bacteria thought to be?
    a. They act as an energy source for the fungus.
    b. They synthesize amino acids needed by the plant and fungus.
    c. They inhibit the growth of the fungus and disrupt the mycorrhizae.
    d. none of the above

12. Which microorganisms play a predominant role in the decomposition of lignin-containing materials of woody plants?
    a. bacteria
    b. archaea
    c. fungi
    d. viruses

13. In which of the following mycorrhizal associations does the fungus invade plant cells?
    a. orchid endomycorrhizae
    b. ericoid endomycorrhizae
    c. arbuscular mycorrhizae
    d. all of the above

TRUE/FALSE

_____ 1. Soils that are very wet tend to be anaerobic.

_____ 2. When geological weathering forms new soils, carbon and nitrogen must be imported by biological processes.

_____ 3. The compound geosmin, which gives soil its characteristic earthy odor, is produced by soil nematodes.

_____ 4. Microbivory, or the use of microbes as a food source by protozoans, removes nutrients that would otherwise be available for plant growth.

_____ 5. Excess water in the soil inhibits decomposition of organic matter by decreasing the amount of available oxygen.

_____ 6. Rainfall can change an aerobic soil with anaerobic microsites into a primarily anaerobic soil.

_____ 7. Endomycorrhizae are detrimental to desert plants because they limit water uptake by the plant, which is growing in an already arid environment.

_____ 8. *Agrobacterium,* a tumor-producing bacterium that infects plants, has been of recent interest because of its possible use as a vector for genetically engineering plants.

_____ 9. Microbial consumption of methane easily keeps up with production so that this gas poses no problem on a global scale.

_____ 10. The vast majority of microorganisms found in soils have been cultivated in the laboratory.

_____ 11. Soil consists of both organic and inorganic materials.

_____ 12. The decomposition of woody plants by fungi leads to the production of chloromethane and cyanide—compounds usually associated with industrial pollution.

_____ 13. Microorganisms inhabit five major subsurface zones.

_____ 14. *Stachybotrys chartarum* and other fungi can grow in moist areas in houses and cause major health problems in humans.

_____ 15. In an ideal soil, microorganisms function in thin water films that have close contact with air.

_____ 16. Most microbes in soil are embedded in the anoxic zones of soil particles.

_____ 17. Accumulation of organic matter in soils results from the activities of primary producers and by the import of preformed organic materials.

CRITICAL THINKING

1. Compare and contrast the following soils with respect to rate of decomposition of organic matter, nutrient levels, and predominant microorganisms: tropical soils; temperate grasslands; cool coniferous forest soils; temperate bogs; cold, moist soils; desert soils, and geologically heated, hyperthermal soils.

2. Explain how soil temperature and water content interact to produce various high-fertility and low-fertility soils. Include in your explanation a specific discussion of tropical, grassland, forest, and bog regions.

3. Agricultural practices impact soil fertility and structure, which in turn impact the activity of soil microorganisms. Describe the short-term and log-term impacts of slash-and-burn agriculture, plowing, irrigation, and the use of nitrogen-containing fertilizes on:
    *   The activity of soil microorganisms (be as specific as possible)

    *   The level of nutrients, especially organic matter and nitrogen in the soil

    *   The amount of oxygen in the soil

Also describe those agricultural practices that improve or maintain soil fertility.

4.   Draw a cross section of a soil aggregate (ped), showing the location of bacteria, fungi, and protozoa, and showing the relative availability of oxygen throughout the aggregate.

ANSWER KEY

Terms and Definitions

1. c, 2. g, 3. f, 4. j, 5. a, 6. l, 7. n, 8. d, 9. k, 10. h, 11. e, 12. b, 13. m, 14. i

Fill in the Blank

1. tripartite; endomycorrhizae; actinorrhizae; ectomycorrhizae; tetrapartite; endomycorrhizae, ectomycorrhizae
2. rhizoplane; associative 3. ectomycorrhizae; endomycorrhizae; arbuscule; arbuscular mycorrhizal
4. stem-nodulating rhizobia 5. infection thread; bacteroids; symbiosomes; root nodules 6. greenhouse gases;
methanotrophs 7. mineral; soil organic matter; humus 8. stable isotope probing

Multiple Choice

1. d, 2. a, 3. d, 4. c, 5. b, 6. c, 7. d, 8. a, 9. d, 10. b, 11. b, 12. c, 13. d

True/False

1. T, 2. T, 3. F, 4. F, 5. T, 6. T, 7. F, 8. T, 9. F, 10. F, 11. T, 12. T, 13. F, 14. T, 15. T, 16. F, 17. T

30 Microbial Interactions

CHAPTER OVERVIEW

This chapter focuses on the relationships microorganisms have with other organisms and each other. The chapter begins with a discussion of symbiotic relationships, and examples of each type of relationship are presented. This chapter then focuses on the interactions of the human body with microorganisms. It begins by discussing the normal microbiota—those microorganisms that have established residence in or on the body. The relationship between humans and their normal microbiota is usually either mutualistic or commensal. On occasion, the interaction can shift to parasitism (a pathogenic relationship). In addition, microorganisms that are not part of the normal microbiota can be pathogenic.

CHAPTER OBJECTIVES

After reading this chapter you should be able to:

- discuss symbiotic relationships
- give examples of each type of symbiosis
- describe gnotobiotic animals and their importance
- describe the body sites where normal microbiota are found and give examples of the microorganisms found there

CHAPTER OUTLINE

I.   Microbial Interactions
   A.   Microorganisms can be physically associated with other organisms in a number of ways; the physical associations can be intermittent and cyclic, or they can be permanent
      1.   Ectosymbiosis—microorganism remains outside the other organism
      2.   Endosymbiosis—microorganism is found within the other organism
      3.   Consortium—physical interactions of different organism of similar size; the interaction is usually mutually beneficial
   B.   Mutualism
      1.   An association with some degree of obligation that provides some reciprocal benefit to both partners; when separated, the individual organisms will not survive (some examples are given below)
      2.   Microorganism-insect mutualisms—bacterial endosymbionts often provide essential vitamins and amino acids to host insects, while the insect provides a secure physical habitat and ample nutrients to the bacteria
         a.   *Buchnera*-aphid relationship—bacteria live within insect host cells as obligate mutualists that provide the aphids with essential amino acids not found in plant sap; studied as an example of coevolution
         b.   Protozoan-termite relationship—protozoa live in the guts of insects that ingest but cannot metabolize cellulose; the protozoa secrete cellulases, which metabolize cellulose, releasing nutrients that the insects can use
      3.   Zooxanthellae—algae harbored by marine invertebrates; reef-building (hermatypic) corals use zooxanthellae to satisfy most of their energy needs; the coral pigments protect the algae from ultraviolet radiation
      4.   Sulfide-based mutualisms

       a.   Tube worm-bacterial relationships occur in hydrothermal vent communities where vent fluids are anoxic, have high concentrations of hydrogen sulfide, and can reach temperatures of 350°C

       b.   Endosymbiotic chemolithotrophic bacteria provide the main energy source in the community through the oxidation of hydrogen sulfide

       c.   The endosymbiotic bacteria are maintained in specialized cells (trophosome) of the tube worm

       d.   The tube worm binds hydrogen sulfide to hemoglobin and transports it to the bacteria; the bacteria use the energy from hydrogen sulfide oxidation to synthesize reduced organic material that is supplied to the tube worm

   5.   Methane-based mutualisms—methanotrophs are intracellular symbionts of methane-vent mussels and sponges, which use the bacteria to support their nutritional needs; methanotrophs within sphagnum moss stems oxidize bog methane to carbon dioxide, which is then fixed by the plant, providing for efficient carbon cycling

   6.   The rumen ecosystem—bacteria in the rumen anaerobically metabolize cellulose to smaller molecules that can be digested by the ruminant; microorganisms produce the majority of vitamins that are needed by the ruminant; methane also is produced in the process

C.   Cooperation

   1.   A mutually beneficial syntrophic relationship that is not obligatory (i.e., the organisms can survive independently); syntrophism is an association in which the growth of one organism either depends on or is improved by growth factors, nutrients, or substrates provided by another organism growing nearby; quorum sensing allows microorganisms to communicate as they form associations with plants and animals (some examples of cooperation are given below)

   2.   Linkage of the carbon cycle and the sulfur cycle by the relationship of sulfide-oxidizing autotrophic bacteria and heterotrophic organisms (e.g., other bacteria, crustaceans, nematodes, sponges, gastropods)

   3.   Linkage of the carbon cycle and the nitrogen cycle by the relationship of cellulolytic microbes and nitrogen-fixing bacteria

   4.   Degradation of 3-chlorobenzoate involves cooperation of three different microorganisms

   5.   Quorum sensing through specific autoinducers is a type of cooperation important for the establishment of biofilms

D.   Commensalism

   1.   A syntrophic relationship in which the microorganism (commensal) benefits, while the host is neither harmed nor helped

   2.   In many cases, the microorganism shares the same food source with the host

   3.   Can occur in situations in which waste products of one microorganism serve as the substrate for another

       a.   Nitrification requires the activity of two different species; one oxidizes ammonia to nitrite and the other oxidizes nitrite to nitrate

       b.   In anaerobic environments, interspecies hydrogen transfer occurs; this involves the activities of fermentative bacteria, which produce low-molecular-weight fatty acids; the fatty acids are degraded by anaerobes (e.g., *Syntrophobacter*), producing hydrogen gas; methanogens consume the hydrogen gas during methanogenesis, and this promotes further production of fatty acids and hydrogen gas

   4.   Can occur in situations where one microorganism modifies the environment, making it better suited for another microorganism

       a.   The common nonpathogenic strain of *Escherichia coli* lives in the human colon; this facultative anaerobe uses oxygen, creating an anaerobic environment in which obligate anaerobes (e.g., *Bacteroides*) can grow; *E. coli* derives no obvious benefit or harm

       b.   Succession of microorganisms in an environment—during milk spoilage, synthesis of acidic fermentation products by one population stimulates proliferation of acid-tolerant microorganisms; during biofilm formation, the first colonizer makes it possible for others to colonize

    c. Colonization of surfaces of plants and animals by normal flora—plant or animal produces organic substances, which are used by the normal flora of the host organism

E. Predation
1. Predator organism engulfs or attacks a prey organism; prey can be larger or smaller than predator; normally results in death of prey
2. Predatory microbes include bacteria (e.g., *Bdellovibrio, Vampirococcus,* and *Daptobacter*), ciliates (important microbial predators in aquatic environments and in wastewater treatment facilities) and fungi (e.g., fungi that trap nematodes)
3. Positive outcomes of predation
    a. Microbial loop—microbial predators mineralize the organic matter produced by autotrophs (primary producers) before it reaches the higher consumers; this returns nutrients to the primary producers and promotes their activity
    b. Ingestion of prey provides protective environment for the prey

F. Parasitism
1. One organism (parasite) benefits from another (host); there is a degree of coexistence between the host and parasite that can shift to a pathogenic relationship (a type of predation); once a parasite establishes the interaction with its host, over time it often loses unnecessary genetic information in a process called genomic reduction (some examples of parasitic interactions are given below)
2. Although once considered an example of a mutualistic interaction, lichens are now believed to be an example of a controlled parasitism between a fungus (ascomycete) and an alga (green alga) or cyanobacterium that only occurs when the two potential partners are nutritionally deprived
    a. Fungal partner (mycobiont) obtains nutrients from alga by hyphal projections (haustoria) that penetrate the algal cell wall, as well as oxygen for respiration
    b. Algal partner (phycobiont) is protected from excess light intensity and is provided with water, minerals, and a firm substratum in which it can grow protected from environmental stress

G. Ammensalism—one organism releases a specific compound that harms another organism (e.g., antibiotics, bacteriocins, antibacterial peptides, and acidic fermentation products)

H. Competition
1. Different organisms within a population or community try to acquire the same resources (e.g., nutrients, location)
2. If one of the two competing organisms can dominate, it will outgrow the other organism (competitive exclusion principle); in some cases the competing organisms share the limiting resource and co-exist at lower population levels

II. Human-Microbe Interactions
A. The normal microbial flora or microbiota are microorganisms normally associated with the human body; acquisition of normal flora is an adaptive process and the composition of the flora is dynamic
B. Pathogenicity—the ability to produce pathological changes (disease) as the result of a parasitic symbiosis between a microorganism and a host; a pathogen is any disease-producing microorganism
C. Gnotobiotic Animals
1. Gnotobiotic—an environment or animal in which all microbial species present are known or that is germ-free (e.g., mammalian fetuses *in utero* are free from microorganisms)
2. Gnotobiotic animals allow investigation of the interactions of animals with specific microorganisms that are deliberately introduced into the animal
3. Gnotobiotic animals are more susceptible to pathogens, but may be resistant to diseases caused by protozoa that use bacteria as a food source (e.g. *Entamoeba histolytica*) and dental caries

III. Normal Microbiota of the Human Body
A. Internal tissues are normally free of microorganisms; however, many other sites are colonized; normal microbiota are the microorganisms regularly found at any anatomical site
B. Reasons to acquire knowledge of normal human microbiota
1. It provides greater insight into possible infections resulting from injury to these sites

298

2. It increases understanding of the causes and consequences of growth by microorganisms normally absent from a specific body site
3. It aids awareness of the role these indigenous microorganisms play in stimulating the immune response

C. Skin
1. Resident microbiota multiply on or in the skin; transients usually are unable to multiply on skin
2. Skin is a mechanically strong barrier; few microbes can penetrate the thick outer layer of closely packed keratinocytes; shedding of outer epithelial cells removes many microbes that adhere to skin surface
3. Skin surface varies from one part of the body to another and generally is a hostile environment; skin surface undergoes periodic drying, is slightly acidic, is salty, and has antibacterial substances (e.g., lysozyme and cathelicidins)
4. Most skin bacteria are found on superficial cells, colonizing dead cells, or closely associated with oil and sweat glands
5. *Staphylococcus epidermidis* and corynebacteria (dry areas and sweat glands)
6. Gram-negative bacteria (moist areas)
7. Yeast (scalp)
8. Dermatophytic fungi (e.g., those causing ringworm and athletes' foot)
9. *Propionibacterium acnes* is prevalent in skin glands and is associated with acne vulgaris

D. Nose and nasopharynx
1. Nose—just inside the nares; *Staphylococcus epidermidis* and *S. aureus* are predominant; they are also found on skin of face
2. Nasopharynx—above the level of the soft palate; contains nonencapsulated strains of some of the same species that may cause clinical infection (e.g., streptococci and *Neisseria*); other species also are found

E. Oropharynx—between the soft palate and upper edge of the epiglottis; houses many different species, including staphylococci and streptococci

F. Respiratory tract—no normal microbiota; microbes are removed by three mechanisms: continuous stream of mucous and the activity of ciliated epithelial cells that move the microbes out, phagocytosis of microbes by alveolar macrophages, and the bactericidal effect of lysozyme

G. Eye—aerobic commensals are found on the conjunctiva

H. External ear—resembles microbiota of the skin; includes some fungi

I. Mouth
1. Contains those organisms that survive mechanical removal by adhering to surfaces such as the gums and teeth
2. Normal microbiota include streptococci, lactobacilli, and actinomycetes
3. Some contribute to the formation of dental plaque, dental caries, gingivitis, and periodontal disease

J. Stomach—most microorganisms are killed by acidic conditions unless they pass through very quickly or are resistant to gastric pH

K. Small intestine
1. Duodenum—few microorganisms present because of stomach acidity and inhibitory action of bile and pancreatic secretions; those that are found are gram-positive rods and cocci
2. Jejunum—*Enterococcus faecalis*, diphtheroids, lactobacilli, and *Candida albicans* are occasionally found
3. Ileum—microbiota resemble those of the colon (e.g., anaerobic gram-negative rods and *Enterobacteriaceae*)

L. Large intestine (colon)
1. Largest microbial population of the body
2. Over 300 different species have been isolated from human feces; most are anaerobes or facultative organisms growing anaerobically

3. Normal microbiota are excreted by peristalsis, segmentation, desquamation, and movement of mucus, but are replaced rapidly because of high reproductive rate; the microbial community is self-regulating and can be disturbed by stress, altitude, starvation, diet, parasite infection, diarrhea, and use of antibiotics or probiotics

M. Genitourinary tract
1. Kidneys, ureter, and bladder are normally free of microorganisms; though in both males and females a few microorganisms are found in distal portions of the urethra
2. Female genital tract hosts a complex microbiota in a state of flux due to menstrual cycle; *Lactobacillus acidophilus* predominates; it forms lactic acid and thereby maintains an acidic pH in the vagina

N. The relationship between normal microbiota and the host
1. Relationship with normal microbiota is usually mutually beneficial
2. Normal microbiota help repel invading pathogens by a number of mechanisms (e.g., competition, production of inhibitory chemicals)
3. Establishment of a stable microflora has been linked to the induction of immune competency in gnotobiotic organisms
4. Under some conditions, normal microbiota can become pathogenic; such microorganisms are referred to as opportunistic
5. Compromised host—host that is seriously debilitated and has lowered resistance; is often target of opportunistic microorganisms

TERMS AND DEFINITIONS

Place the letter of each term in the space next to the definition or description that best matches it.

_____ 1. A relationship in which one organism benefits while the other is neither harmed nor helped

_____ 2. An obligatory mutually beneficial association of two organisms

_____ 3. A physical relationship in which two organisms remain outside of each other

_____ 4. A physical relationship in which one organism lives inside another

_____ 5. A relationship in which one organism releases a specific compound that has a negative effect on another organism

_____ 6. A mutually beneficial relationship that is not obligatory

_____ 7. A relationship in which one organism engulfs another

_____ 8. An association in which the growth of an organism is dependent upon one or more growth factors, nutrients, or substrates provided by another organism in the same vicinity

_____ 9. Animals that are germfree or that associate with one or more known species of microorganisms

_____ 10. The microorganisms normally associated with a particular tissue or structure

_____ 11. The relationship in which one organism remains outside the other

_____ 12. The relationship in which one organism lives inside the other

_____ 13. Germfree

_____ 14. Microbial systems involving interactions between a host and more than one symbiont

_____ 15. Endosymbiotic dinoflagellates of invertebrates

_____ 16. A relationship in which one organism is harmed while the other benefits

_____ 17. The cyanobacterial partner in a lichen

_____ 18. The ability of an organism to cause disease

a. amensalism
b. axenic
c. commensalism
d. consortium
e. cooperation
f. ectosymbiosis
g. ectosymbiosis
h. endosymbiosis
i. endosymbiosis
j. gnotobiotic
k. mutualism
l. normal microbiota
m. parasitism
n. pathogenicity
o. phycobiont
p. predation
q. syntrophism
r. zooxanthellae

FILL IN THE BLANK

1. A relationship between any two organisms that spend all or part of their lives in association with each other is called _____. If the organisms remain outside of each other, it is called an _____; and if one organism is found within the other, it is called an _____. In many cases microorganisms live on both the inside and the outside of another organism; this is called an _____.

2. The discipline of _____ _____ is defined as the study of the interactions of microorganisms with their living and nonliving environment. The organism-organism interactions can take many forms. An example of a positive interaction is _____, an obligatory relationship that is mutually beneficial. An example of a negative interaction is _____, in which organisms compete for space or a limiting nutrient.

3. Herbivorous animals that have a stomach divided into four parts and chew a cud consisting of partially digested food are called _____. These animals have _____ associations with the microorganisms that inhabit the upper portion of the stomach (the _____). The microorganisms in this relationship are referred to as _____.

4. For many years, scientists thought _____ were single organisms, and they were even given genus and species names. It was later determined that they are actually associations of an ascomycete fungus and an alga or a cyanobacterium. The fungal partner in the association is called the _____, and the algal or cyanobacterial partner is called the _____. Once considered to be an example of a _____ relationship, they are now thought to be an example of a controlled _____.

5. Although _____ is usually thought of as a negative interaction, it has many beneficial effects on both the predator and its prey. For instance, it can lead to increased rates of nutrient cycling, which is critical for the functioning of the _____ _____. In this cyclic process, organic matter produced through photosynthetic and chemotrophic activity is mineralized before it reaches higher consumers, making inorganic nutrients available to the _____ _____ in the loop and promoting their activity.

6. If organisms compete for resources, two outcomes are often observed. In some cases one population dominates, excluding the other. This phenomenon is called the _____ _____ _____. However, in some cases the two populations may share the resource, each existing at a lower population level.

7. A _____ is a physical association of different organisms that are often similar in size. In many cases the relationship is mutually beneficial.

8. One interesting outcome of parasitism is the loss over time of some of the parasite's genes. This loss is referred to as _____ _____.

9. Organisms whose microbial inhabitants are completely known are said to be _____.

10. The mixture of microorganisms regularly found at any anatomical site is called the _____ _____. These microorganisms often offer some protection from transient microbes, but can become _____ and cause disease under certain circumstances. When they do so, they are termed

_____ _____. Individuals with lowered resistance are called _____ _____, and they are particularly vulnerable to such infections.

MULTIPLE CHOICE

For each of the questions below select the *one best* answer.

1. In lichens, the mycobiont sends projections of its hyphae across the algal cell wall in order to obtain nutrients from its photosynthetic partner. What are these projections called?
    a. mycelia
    b. hyphal extensions
    c. haustoria
    d. lichenthropes
2. Which of the following is NOT true about the coral reef-zooxanthellae mutualism?
    a. The zooxanthellae provide most of the energy for the corals.
    b. The pigments produced by the corals protect the algae from harmful ultraviolet radiation.
    c. The rate of calcification (reef-building) is at least ten times higher in the light than in the dark.
    d. All of the above are true about the coral reef-zooxanthellae mutualism.
3. In some symbiotic relationships, the microorganism benefits while the colonized plant or animal is neither helped nor harmed. What is the microorganism in such a relationship called?
    a. bacteriont
    b. commensal
    c. mutualist
    d. host
4. How are the endosymbionts of ruminant animals useful to their hosts?
    a. They digest the cellulose of plant cell walls and provide organic molecules that the ruminant can metabolize.
    b. They produce most of the vitamins needed by the ruminant.
    c. They are digested as nutrients.
    d. All of the above are ways in which the endosymbionts of ruminant animals are useful to their hosts.
5. Which of the following is a positive effect of predation?
    a. It contributes to the microbial loop.
    b. It may provide protection of prey from heat and damaging chemicals.
    c. It may aid pathogenicity.
    d. All of the above are positive effects of predation.
6. What are the microorganisms normally associated with a particular tissue or structure called?
    a. indigenous microbial populations
    b. normal microbiota
    c. normal flora
    d. All of the above are correct.
7. In the tube worm system, the bacterial partner obtains energy from what compound?
    a. hydrogen sulfide
    b. carbon dioxide
    c. carbohydrates
    d. methane
8. Which of the following can use methane as a sole carbon source?
    a. methanogens
    b. methanotrophs
    c. methylotrophs
    d. None of these can.
9. Which gram-positive anaerobe is the most abundant organism in oil glands?
    a. *Frankia*
    b. *Raffia*

302

    c.   *Pityrosporum*
    d.   *Propionibacterium*

10. Which of the following has the largest microbial community?
    a.   stomach
    b.   small intestine
    c.   large intestine
    d.   genitourinary tract

TRUE/FALSE

_____ 1. Mutual advantage is central to many organism-organism interactions.

_____ 2. The interactions of paired populations in ecosystems have no impact on the interactions of other pairs.

_____ 3. Most disease-causing bacteria from the intestinal tract of humans and other higher organisms do not survive long outside of their hosts.

_____ 4. The use of one organism to control the activity of another is called biocontrol.

_____ 5. Skin microbes can be characterized as either transients or residents.

_____ 6. Although gnotobiotic animals are generally more susceptible to infection by microbial and other parasitic organisms, they are almost completely resistant to dysentery caused by _Entamoeba histolytica_ because they lack the bacteria that this protozoan normally uses as a food source.

_____ 7. The male and female genitourinary tracts have nearly identical microbiota.

_____ 8. Gnotobiotic animals are usually germ-free.

_____ 9. Gnotobiotic animals are useful for studying the interactions of animals and specific species of microorganisms.

_____ 10. The lower respiratory tract has an active microbiota that helps provide protection.

_____ 11. Cathelicidins are antimicrobial peptides produced by sweat glands.

_____ 12. Normal microbiota of the skin are protected by the enzyme lysozyme.

CRITICAL THINKING

1. Explain the complex endosymbiont-tube worm relationship. Why has hydrogen sulfide been referred to as the "sunlight" of hydrothermal vent communities?

3. Compare and contrast mutualism, commensalism, and cooperation. Do the same for predation, competition, parasitism, and amensalism.

ANSWER KEY

Terms and Definitions

1. c, 2. k, 3. g, 4. i, 5. a, 6. e, 7. p, 8. q, 9. j, 10. l, 11. f, 12. h, 13. b, 14. d, 15. r, 16. m, 17. o, 18. n

Fill in the Blank

1. symbiosis; ectosymbiosis; endosymbiosis ecto/endosymbiosis 2. microbial ecology; mutualism; competition 3. ruminants; mutualistic; rumen; mutualists 4. lichens; mycobiont; phycobiont; mutualistic; parasitism 5. predation; microbial loop; primary producers 6. competitive exclusion principle 7. consortium 8. genomic reduction 9. gnotobiotic 10. normal microbiota; pathogens; opportunistic pathogens; compromised hosts

Multiple Choice

1. c, 2. d, 3. b, 4. d, 5. d, 6. d, 7. a, 8. b, 9. d, 10. c

True/False

1. T, 2. F, 3. T, 4. T, 5. T, 6. T, 7. F, 8. T, 9. T, 10. F, 11. T, 12. F

Part IX NONSPECIFIC RESISTANCE AND THE IMMUNE RESPONSE

31 Nonspecific (Innate) Host Resistance

CHAPTER OVERVIEW

Humans resist parasitic relationships by employing both nonspecific and specific mechanisms. The nonspecific resistance mechanisms are explored in this chapter. First there is a discussion of immunity and the cells and organs involved in human immune responses. The mechanisms of cellular immunity are presented and inflammation detailed. Physical barriers of innate resistance are discussed for several tissues and organs. The details of innate immunity are presented, including complement activation, cytokines, and acute phase proteins.

CHAPTER OBJECTIVES

After reading this chapter you should be able to:

- describe the cells, tissues, and organs of the immune system
- describe the physical and chemical barriers a pathogen must breach if a parasitic relationship is to be established
- describe inflammation and its role in providing nonspecific resistance
- describe the complement system and how it is activated
- describe how the complement system and phagocytosis function in inflammation and other nonspecific resistance mechanisms
- discuss cytokines and natural killer cells and their role in nonspecific resistance

CHAPTER OUTLINE

I. Overview of Host Resistance
   A. To establish infection, a pathogen must first overcome barrier defenses
   B. If a pathogen succeeds, the immune system offers protection
      1. The immune system is composed of widely distributed cells, tissues, and organs that recognize foreign materials and microorganisms
   C. Immunity—ability of a host to resist a particular disease; immunology—the science that deals with immune responses
      1. Two types of immune responses
         a. Nonspecific immune responses (also called innate or natural immunity)
            1) General resistance mechanisms inherited as part of the innate structure and function of each animal

306

        2)   Lack immunological memory

        3)   Nonspecific response occurs to same extent with each encounter

    b.   Specific immune response (also called acquired, adaptive, or specific immunity)

        1)   Resists a particular foreign agent by an immune response (e.g., production of antibodies) to specific antigens

        2)   Improves on repeated exposure

    c.   Multiple bridges occur between innate and adaptive immunity; a variety of white blood cells function in both systems

II.   Cells, Tissues, and Organs of the Immune System

    A.   Cells of the immune system

        1.   Leukocytes—white blood cells; arise from pluripotent stem cells in bone marrow and migrate to other body sites to mature and perform their functions; include all the cells described below

        2.   Monocytes and macrophages—highly phagocytic cells that constitute the monocyte-macrophage system

           a.   Monocytes—mononuclear phagocytic cells that circulate in blood for short time and can migrate to tissues where they mature into macrophages

           b.   Macrophages—larger than monocytes; have more organelles and possess receptors that allow them to discriminate self from nonself; surface molecules recognize common components of pathogens (e.g., lipopolysaccharide [LPS] from bacterial cell walls) respond to opsonization (chemical enhancement of phagocytosis)

        3.   Granulocytes—also called polymorphonuclear leukocytes (PMNs)

           a.   Basophils—nonphagocytic; upon stimulation, release chemicals (e.g., histamine, prostaglandins) that impact blood vessels (vasoactive); basophils play important roles in allergic responses

           b.   Eosinophils—mobile cells that migrate from bloodstream into tissue spaces; protect against protozoa and helminth parasites; also may have role in allergic responses

           c.   Neutrophils—highly phagocytic cells that rapidly migrate to sites of tissue damage and infection

        4.   Mast cells—found in connective tissue; contain granules with histamine and other chemicals that contribute to immune response; play important role in allergies and hypersensitivities

        5.   Dendritic cells—phagocytose microorganisms and kill viruses by secreting interferon-α; mature dendritic cells migrate to blood stream or lymphatic system where they interact with B cells and natural killer cells and present foreign antigens to T cells

        6.   Lymphocytes

           a.   Major cells of specific immune system; when activated can differentiate to stimulate the immune response, produce antibodies, or produce memory cells

           b.   Divided into three populations: T cells, B cells, and null cells (including natural killer cells)

           c.   B lymphocytes (or B cells) mature in bone marrow and disperse throughout lymphoid tissue; when activated, differentiate into plasma cells and produce antibodies

           d.   T lymphocytes (or T cells) mature in thymus gland and circulate in blood or lymphoid tissue; when activated, T cells do not produce antibodies, but stimulate the immune response by producing cytokine proteins

        7.   Natural Killer Cells

           a.   Natural killer (NK) cells are large, nonphagocytic, granular lymphocytes that destroy malignant cells and cells infected with microorganisms

           b.   Recognize and target in two ways:

               1)   Antibody-dependent cell-mediated cytotoxicity (ADCC)—receptors on NK cells link them to antibody-coated target cells

               2)   If NK cells bind class I major histocompatibility (MHC) molecule (a self antigen) on a cell's surface, killing is inhibited; if there is no class I MHC on the target cell (i.e., because cell is infected with virus or is malignant), then killing occurs through pore-forming proteins and cytotoxic enzymes (granzymes)

    B.   Organs and tissues of the immune system

1. Primary lymphoid organs and tissues
   a. Thymus—site of T cell maturation
   b. Bone marrow—site of B-cell maturation in mammals
   c. Bursa of Fabricus—site of B-cell maturation in birds
2. Secondary lymphoid organs and tissue
   a. Spleen—filters blood and traps blood-borne microorganisms and antigens; contains macrophages and dendritic cells that present antigens to T cells
   b. Lymph nodes—filter lymph and trap microorganisms and antigens; contain macrophages and dendritic cells that present antigens to T cells; T cells release cytokines that stimulate differentiation and proliferation of B cells in antibody-producing plasma cells and memory cells
   c. Some lymphoid tissue is closely associated with certain tissues such as skin-associated pymphoid tissue (SALT), mucous-associated lymphoid tissue (MALT), and bronchial-associated lymphoid tissue (BALT)

III. Phagocytosis
   A. Phagocytic cells (monocytes, tissue macrophages, dendritic cells, and neutrophils) phagocytose infecting organisms; phagocytosis is the process by which invaders are recognized, ingested, and killed
   B. Pathogen recognition can be opsonin-depedent or opsonin-independent
      1. Opsonin-independent recognition—uses nonspecific and specific receptors on the phagocytic cells to recognize and bind structures on the microorganism, or to signal the induction of host defense pathways
         a. Recognition based on the interaction of surface lectins on one cell and surface carbohydrates on the other cell
         b. Recognition based on protein-protein interactions between the Arg-Gly-Asp peptide sequence of microorganisms and receptors on the phagocytic cell
         c. Recognition based on hydrophobic interactions between bacteria and phagocytic cells
         d. Recognition based on the detection of conserved molecular structures that occur in patterns and are essential products of normal microbial physiology (pathogen-associated molecular patterns – PAMPs; e.g., LPS) by specific receptors on phagocytic cells (pattern recognition receptors – PRRs)
   C. Toll-like receptors (TLRs) —PRRs act exclusively as signaling receptors that bind PAMPs and communicate with the cell nucleus to elicit the appropriate response to different classes of pathogens
   D. Intracellular digestion—ingested microorganism is enclosed in phagosome, which then fuses with lysosome; digestion occurs in phagolysosome
      1. Lysosomal enzymes (e.g., lysozyme, phospholipase, proteases) hydrolyze microbial structural molecules
      2. Lysosomes of macrophages and neutrophils have enzymes that make toxic reactive oxygen intermediates (e.g., superoxide radical) during the respiratory burst that accompanies phagocytosis
      3. Macrophages, neutrophils, and mast cells form reactive nitrogen intermediates (e.g., nitric oxide) that are potent cytotoxic agents
      4. Neutrophil granules contain microbiocidal substances (e.g., defensins), which are delivered to the phagolysosome
   E. Exocytosis—the antigenic remains of invaders can be expelled from the cell (as neutrophils do) or further processed for antigen presentation on the lymphocyte cell surface (as macrophages and dendritic cells do)

IV. Inflammation
   A. Nonspecific response to tissue injury characterized by redness, heat, pain, swelling, and altered function of the tissue
   B. Inflammatory response
      1. Injured tissue cells release chemical signals that activate cells in capillaries

2. Interaction of selectins on vascular endothelial surface and integrins on neutrophil surface promotes neutrophil extravasation
3. Neutrophils attack pathogen
4. More neutrophils and other leukocytes are attracted to site of tissue damage to help destroy microorganisms

C. Numerous inflammatory mediators function in response
1. Kallikrein—an enzyme that catalyzes formation of bradykinin
2. Bradykinin
   a. Binds capillary walls, causing movement of fluid and leukocytes into tissue and production of prostaglandins (cause pain)
   b. Binds mast cells, causing release of histamine and other inflammation mediators
3. Histamine—promotes movement of more fluid, leukocytes, bradykinin, and kallikrein into tissue

D. During acute inflammation, pathogen is neutralized and eliminated by a series of events
1. Increase in blood flow and capillary dilation bring more antimicrobial factors and leukocytes into the area; these destroy the pathogen; dead cells also release antimicrobial factors
2. The rise in temperature stimulates the inflammatory response and may inhibit microbial growth
3. A fibrin clot often forms and may limit the spread of the invaders so that they remain localized
4. Phagocytes collect in the inflamed area and phagocytose the pathogen; chemicals stimulate release of neutrophils and increase the rate of granulocyte production

E. Chronic inflammation is characterized by its longer duration, dense infiltration of lymphocytes and macrophages, and formation of granulomas (in some cases)

V. Physical and Chemical Barriers in Nonspecific Resistance
A. Many factors influence host-microbe relationships (e.g., nutrition, age, genetic factors, hygiene)
B. Physical and mechanical barriers
1. Skin
   a. Provides a very effective mechanical barrier to microbial invasion, due to its thick, closely packed cells; frequent shedding and being acidic and salty
   b. If pathogen penetrates tissue under skin, it encounters skin-associated lymphoid tissue (SALT)
      1) Langerhans cells—specialized dendritic cells that phagocytose antigens, then migrate to lymph nodes and differentiate into interdigitating dendritic cells, a type of antigen-presenting cell; activate T cells, which interact with activated B cells to induce a humoral response
      2) Intraepidermal lymphocytes—function as T cells to destroy antigen
2. Mucous membranes
   a. Mucus secretions form a protective covering that contains antibacterial substances, such as lysozyme, lactoferrin, and lactoperoxidase
   b. Contain mucosal-associated lymphoid tissue (MALT)
      1) Several types, including gut-associated (GALT) and bronchial-associated (BALT)
      2) MALT operates by the action of M cells in the mucous membrane; M cells phagocytose antigen and transport it either to a pocket within the M cell containing B cells and macrophages or to lymphoid follicles containing B cells
3. Respiratory system
   a. Aerodynamic filtration deposits organisms onto mucosal surfaces, and microbes become entrapped in mucus (mucociliary blanket)
   b. Activity of ciliated epithelial cells transports microbes away from the lungs (mucociliary escalator); coughing, sneezing, and salivation also remove microorganisms
   c. Alveolar macrophages destroy those pathogens that get to the alveoli
4. Gastrointestinal tract
   a. Gastric acid kills most microorganisms
   b. In intestines, pancreatic enzymes, bile, intestinal enzymes, GALT, peristalsis, normal microbiota, lysozyme (produced by Paneth cells), and antibacterial peptides (cryptins) destroy or remove microorganisms

5. Genitourinary tract
   a. Kidneys, ureters, and urinary bladder are sterile due to multiple factors (e.g., pH and flushing action)
   b. Vagina produces glycogen, which is fermented by lactobacilli to lactic acid, thus lowering the pH
6. The eye-flushing action, lysozyme, and other antibacterial substances act as defenses

VI. Chemical Mediators in Nonspecific (Innate) Resistance
   A. Gastric juices, salivary glycoproteins, lysozyme, oleic acid on the skin, urea, and other chemicals have already been discussed
   B. Antimicrobial peptides
      1. Cationic peptides damage bacterial plasma membranes through electrostatic interactions
         a. Cathelicidin—one of a group of linear alpha-helical peptides (12 to 80 amino acids) produced by a variety of cells (e.g., neutrophils, respiratory epithelia)
         b. Defensins—a diverse group of disulfide-linked, open-ended peptides (29 to 42 amino acids) found in a variety of cells (e.g., neutrophils, Paneth cells)
         c. Histatins—larger peptides with regular structural repeats
   C. Bacteriocins—plasmid-encoded antibacterial peptides produced by normal bacterial flora; are lethal to related species through a variety of mechanisms
   D. Complement
      1. The complement system is a set of serum proteins that play a major role in the immune response; complement has three major physiological activities: defending against bacterial infections, bridging innate and adaptive immunity, and disposing of wastes
      2. During opsonization, the microorganism is coated with antibodies, mannan-binding protein, and/or complement proteins (together known as opsonins); this promotes recognition and phagocytosis
      3. Complement acts in a cascade fashion; the complement proteins are produced in an inactive form, and the activation of one (by cleavage of the protein) leads to the sequential activation of others
      4. There are three pathways of complement activation
         a. Alternative complement pathway—occurs in response to intravascular invasion by bacteria and some fungi; involves interaction of complement with the surface of the pathogen forming the membrane attack complex
         b. Lectin complement pathway (also called the mannan-binding lectin pathway)—occurs when macrophages stimulate liver cells to release acute phase proteins such as mannose-binding protein (a lectin), which then can activate complement via the alternative pathway or the classical pathway
         c. Classical pathway—results from antigen-antibody interactions that occur during specific immune responses
      5. Overview of complement activation and immune responses
         a. Bacteria at local tissue site interact with components of alternative pathway
         b. If bacteria persist or invade a second time, antibody responses activate the classical pathway
         c. Generation of complement fragments C3a and C5a leads to
            1) Activation of mast cells, which release their contents, causing hyperemia
            2) Release of neutrophils from bone marrow into circulation, and their chemotaxis to injury site
         d. Ultimately neutrophils and phagocytes ingest and destroy the bacteria

VII. Cytokines
   A. Cytokines are soluble proteins or glycoproteins that are released by one cell population and act as intercellular mediators
      1. Monokines—released from mononuclear phagocytes
      2. Lymphokines—released from T lymphocytes
      3. Interleukins—released from a leukocyte and act on another leukocyte

        4.   Colony-stimulating factors (CSFs) —stimulate growth and differentiation of immature
              leukocytes in the bone marrow
    B.   Cytokines can affect various cell populations
        1.   Autocrine function—affect the same cell responsible for its production
        2.   Paracrine function—affect nearby cells
        3.   Endocrine function—distributed by circulatory system to target cells
    C.   Exert their effects by binding to cell-surface receptors called cell-association differentiation antigens
          (CDs); possible effects include:
        1.   Stimulation of cell division
        2.   Stimulation of cell differentiation
        3.   Inhibition of cell division
        4.   Apoptosis—programmed cell death
        5.   Stimulation of chemotaxis and chemokinesis
    D.   Interferons (INFs) —Regulatory cytokines produced by certain eucaryotic cells in response to viral
          infection; in addition to protecting against viral infections, interferons also help regulate the immune
          response
    E.   Fever—results from disturbances in hypothalamic regulatory control, leading to increase of thermal
          "set point"
        1.   Most common cause of fever is viral or bacterial infection, usually due to action of an
              endogenous pyrogen (e.g., interleukin-1, interleukin-6, tissue necrosis factor), which induces
              secretion of prostaglandins; these reset the hypothalamic thermostat
        2.   Fever augments host's defenses three ways
            a.   Stimulates leukocytes so that they can destroy the microorganism
            b.   Enhances specific activity of the immune system
            c.   Enhances microbiostasis (growth inhibition) by decreasing available iron to the
                  microorganisms
    F.   Acute phase proteins—produced by liver in response to cytokines; act as opsonins and activate
          complement; collectins act as molecular scavengers that bind cellular debris

TERMS AND DEFINITIONS

Place the letter of each term in the space next to the definition or description that best matches it.

_____   1.   The oral administration of living microorganisms or substances to promote health and growth
_____   2.   The general ability of a host to resist a particular infection or disease
_____   3.   The study of immune responses
_____   4.   White blood cells
_____   5.   B cells that are actively secreting antibodies
_____   6.   Mucous secretions and ciliated epithelial cells that line the respiratory tract and trap invading
                organisms
_____   7.   The ciliary action that moves mucus-entrapped microbes up and away from the lungs
_____   8.   Peptides produced by Paneth cells that are toxic for some bacteria
_____   9.   Molecules on the vascular endothelium to which neutrophils can attach during the inflammatory
                response
_____   10.   Molecules on the surface of neutrophils that act as selectin receptors to mediate the attachment of
                neutrophils to the vascular endothelium during the inflammatory response
_____   11.   Serum proteins that act in a cascade fashion and play a major role in an animal's immune response
_____   12.   Coating of microorganisms with serum proteins, thereby preparing them for recognition and
                ingestion by phagocytic cells
_____   13.   A phagocytic vacuole
_____   14.   A vacuole formed by the fusion of a lysosome and a phagosome
_____   15.   Increased oxygen consumption needed to support the increased metabolic activity of phagocytosis
_____   16.   Broad-spectrum antimicrobial peptides released from neutrophils
_____   17.   Antimicrobial peptides produced by sweat glands

_____ 18. Soluble proteins or glycoproteins released by one cell population that act as intercellular signaling molecules

_____ 19. Soluble proteins that act as signaling molecules between populations of leukocytes

_____ 20. For human adults, an oral temperature above 98.6°F

_____ 21. Decreased availability of iron for microorganisms in a host animal

_____ 22. Increased availability of iron for microorganisms in a host animal

a.   cathelicidins
b.   complement system
c.   cryptins
d.   cytokines
e.   defensins
f.   fever
g.   hyperferremia
h.   hypoferremia
i.   immunity
j.   immunology
k.   integrins
l.   interleukins
m.   leukocytes
n.   mucociliary blanket
o.   mucociliary escalator
p.   opsonization
q.   phagosome
r.   phagolysosome
s.   plasma cells
t.   probiotics
u.   respiratory burst
v.   selectins

NONSPECIFIC DEFENSE MECHANISMS

Describe how the following help to provide resistance and indicate the body systems or structures where they function.

Defense Mechanism	Role Played in Providing Resistance	Where It Functions
alveolar macrophages		
bacteriocins		
BALT		
colicins		
cryptins		
GALT		
interdigitating dendritic cells		
intraepidermal lymphocytes		

lactoferrin		
Langerhans cells		
lysozyme		
M cells		
MALT		
mucociliary blanket		
Paneth cells		
peristalsis		
SALT		

FILL IN THE BLANK

1. Eucaryotic cells produce a group of related low-molecular-weight, regulatory cytokines in response to virus infection. These are called _____, and they are important in making host cells resistant to virus infections.

2. Acute _____ is the immediate response of the body to injury or cell death. The signs of this response are redness, warmth, pain, swelling, and altered function of the tissue. Chronic _____ differs not only in duration but also by the formation of new connective tissue and by a dense infiltration of lymphocytes and macrophages. It usually causes permanent tissue damage; under certain conditions, it can lead to the formation of a _____ in an attempt to isolate the site.

3. The system that protects vertebrates from pathogenic microorganisms is called the _____ system. This system produces two different types of responses to invading microbes and foreign material. One response offers resistance to any microorganism or foreign material and acts as a first line of defense. It has several names, including the _____ _____ response, _____ resistance, _____immunity, and _____ immunity. The second type of response resists a particular foreign agent and improves upon repeated exposure. It also has several names, including the _____ _____ response,_____ immunity, _____ immunity, and _____ immunity.

4. In mammals, _____ _____ mature in the _____ and _____ _____ mature in the bone marrow. In birds, the latter cells mature in the _____ _____ _____.

5. An important group of leukocytes are the _____ _____ _____, which are large, nonphagocytic lymphocytes that destroy malignant cells and cells infected with microorganisms. One way these cells recognize their targets involves receptors that bind antibodies, leading to killing by a process called _____ _____ _____. Another way they recognize their targets involves _____ receptors and _____ receptors.

6. Certain body cells recognize, ingest, and kill extracellular microorganisms by a process called _____. These cells use two mechanisms for recognition of microorganisms. The first involves the coating of microorganisms with serum components (_____) such as antibodies and/or C3b (one of the proteins of the _____ _____). This method of recognition is called the_____ recognition mechanism. The second recognition mechanism, called the _____ recognition mechanism, involves recognition of certain structures (e.g., lectins) on the surface of the microorganism. One example of this type of recognition involves the detection of conserved molecular structures that occur in patterns on a pathogen (e.g., LPS). These structures are called _____ _____ _____ (PAMPs). They are detected by receptors on phagocytic cells called

_____ _____ _____. Some PRRs function exclusively as signaling receptors (_____ _____), signaling the appropriate response to different classes of pathogens.

7. Soluble proteins or glycoproteins that are released by one population of cells and act as signaling molecules are called _____. If T cells release them, they are called _____. If they are released by monocytes, they are called _____. If they are produced by a leukocyte and act on another leukocyte, they are referred to as _____. Finally, if their effect is to stimulate the growth and differentiation of leukocytes, they are called _____ _____.

8. Activation of the complement system can proceed by three pathways. One, called the _____ complement pathway, plays an important role in the nonspecific defense against intravascular invasion by bacteria and some fungi. It begins with the cleavage of complement protein C3 into fragments C3a and C3b, and ultimately results in the formation of the _____ _____ _____, which creates a pore in the plasma membrane of the target cell. The second pathway is called the _____ complement pathway. In this pathway, macrophages ingest viruses, bacteria and other foreign material and then release chemicals that stimulate liver cells to secrete lectins, proteins that bind specific carbohydrates (e.g., mannose-binding protein). This ultimately generates an enzyme that cleaves complement protein C3. The third pathway is dependent on the reaction between antibodies and antigens and is called the classical pathway.

MULTIPLE CHOICE

For each of the questions below select the *one best* answer.

1. Why does the skin have an environment unfavorable for colonization by many microorganisms?
    a. It has a slightly acidic pH, which prevents growth of many microorganisms.
    b. It is subject to periodic drying, which inhibits the growth of many microorganisms.
    c. It is hyperosmotic because of salt produced by sweat glands.
    d. All of the above are reasons why the skin has an environment unfavorable for colonization by microorganisms.

2. Which of the following are the major cells of the specific immune system?
    a. monocytes
    b. macrophages
    c. lymphocytes
    d. mast cells

3. Which cells are phagocytic?
    a. macrophages
    b. neutrophils
    c. dendritic cells
    d. All of the above are phagocytic.

4. Basophils, eosinophils, and neutrophils share a common feature. What is it?
    a. All are phagocytic.
    b. All mediate allergic reactions.
    c. All have irregularly shaped nuclei with 2–5 lobes.

    d. All of the above are characteristic of these cells.

5. Which component of the monocyte-macrophage system enters the blood and circulates for about 8 hours before entering body tissues?
    a. monocytes
    b. macrophages
    c. PMNs
    d. mast cells

6. Which of the following is characteristic of mast cells?
    a. They have irregular shaped nuclei with 2–5 lobes.
    b. They are phagocytic.
    c. They produce cryptins.
    d. They contain granules with histamine and other substances that contribute to the inflammatory response.

7. Which of the following is a primary lymphoid organ or tissue?
    a. spleen
    b. lymph nodes
    c. thymus
    d. MALT

8. Where do lymphocytes encounter and bind antigen and as a result become activated to carry out their immune function?
    a. spleen
    b. lymph nodes
    c. MALT
    d. all of the above

314

9. Which of the following is NOT a way in which fever augments the host's defenses?
   a. It inhibits the microorganism's growth by raising the temperature above the optimum growth temperature.
   b. It inhibits the microorganism's growth by decreasing the availability of iron to the organism.
   c. It stimulates leukocytes into action so they can kill the microorganism.
   d. All of the above are ways in which fever augments the host's defenses.

10. Which of the following statements is true?
   a. The normal microbiota of the oral cavity is composed of microbes able to resist mechanical removal.
   b. The stomach contains many microbes because of its alkaline pH.
   c. The distal portion of the small intestine and the entire large intestine are largely free of microorganisms.
   d. All encounters between a microbe and a human host lead to disease.
   e. The normal microbiota of the body provide no defense against pathogenic microbes.

11. Which of the following mechanisms does NOT help protect the lungs from infection?
   a. Turbulent airflow deposits airborne pathogens on sticky mucosal surfaces.
   b. The mucociliary escalator moves organisms trapped in the mucociliary blanket away from the lungs.

   c. Coughing and sneezing forcefully expel organisms away from the lungs.
   d. All of the above help protect the lungs from infection.

12. What mechanisms are used to kill and destroy microorganisms within a phagolysosome?
   a. hydrolytic enzymes
   b. reactive oxygen intermediates
   c. reactive nitrogen intermediates
   d. All of the above are used.

13. Which of the following is a way the complement system aids in the defensive responses of the host organism?
   a. lysis of target cells
   b. attraction of phagocytic cells
   c. activation of phagocytic cells
   d. All of the above aid the defensive responses of the host.

14. What type of cytokine function occurs when the cytokine affects the same cell responsible for its production?
   a. autocrine function
   b. paracrine function
   c. endocrine function
   d. none of the above

15. Which of the following is NOT the result of the action of cytokines?
   a. stimulation of cell division
   b. apoptosis
   c. inhibition of cell division
   d. All of the above can result from the action of cytokines.

TRUE/FALSE

_____ 1. Important outcomes of inflammation are the walling off of the infected area by a fibrin clot and the elimination of the pathogen by phagocytosis.

_____ 2. One of the symptoms of inflammation is a generalized fever.

_____ 3. The human fetus in utero is normally free from bacteria and other microorganisms.

_____ 4. Age and nutrition do not impact host resistance.

_____ 5. Bacteriocins are antimicrobial peptides.

_____ 6. Opsonization prepares macrophges for tumor necrosis factor.

_____ 7. Interferons are cytokines that are important in responding to viral infections.

_____ 8. Macrophages release cytokines that stimulate the liver to produce acute phase proteins.

CRITICAL THINKING

1. The skin is constantly being exposed to pathogenic organisms. However, it is a very effective barrier against infection—it is not easily colonized and it is not readily penetrated. Discuss the various properties of the skin that make it such an effective barrier against colonization and penetration.

2. Inflammation is an important nonspecific defense in response to tissue injury. Describe the steps involved in the inflammation response. Contrast acute and chronic inflammation.

ANSWER KEY

Terms and Definitions

1. t, 2. i, 3. j, 4. m, 5. s, 6. n, 7. o, 8. c, 9. v, 10. k, 11. b, 12. p, 13. q, 14. r, 15. u, 16. e, 17. a, 18. d, 19. l, 20. f, 21. h, 22. g

Fill in the Blank

1. interferons 2. inflammation; inflammation; granuloma 3. immune; nonspecific immune; nonspecific; innate; natural; specific immune; specific; adaptive; acquired 4. T cells (T lymphocytes); thymus; B cells (B lymphocytes); bursa of Fabricus 5. natural killer cells; antibody-dependent cell-mediated cytotoxicity (ADCC); killer-activating; killer-inhibitory 6. phagocytosis; opsonization; complement system; opsonin-dependent; opsonin-independent; pathogen-associated molecular patterns; pattern recognition receptors (PRRs); toll-like receptors (TRRs) 7. cytokines; lymphokines; monokines; interleukins; colony-stimulating factors (CSFs) 8. alternative; membrane attack complex; lectin

Multiple Choice

1. d, 2. c, 3. d, 4. c, 5. a, 6. d, 7. c, 8. d, 9. a, 10. a, 11. d, 12. d, 13. d, 14. d, 15. d

True/False

1. T, 2. F, 3. T, 4. F, 5. T, 6. F, 7. T, 8. T

32 Specific (Adaptive) Immunity

CHAPTER OVERVIEW

This chapter focuses on specific (adaptive) immunity, a complex process involving interactions of the antigens of a pathogen with antigen-receptors and antibodies of a host. These interactions trigger a series of events that either destroy the pathogen or render it harmless. Most of the chapter is devoted to discussions of the functional cells and molecules of the specific immune system. During the discussion, the various connections between these cells and molecules are drawn and linked to other types of immune responses. The chapter continues with a discussion of the ways these responses protect higher animals against viral and bacterial pathogens. It concludes with a discussion of hypersensitivities (allergies), autoimmune diseases, and immunodeficiencies.

CHAPTER OBJECTIVES

After reading this chapter you should be able to:

- compare and contrast specific immunity and nonspecific immunity
- discuss antigens, haptens, superantigens, and CD antigens
- compare and contrast IgG, IgM, IgA, IgD, and IgE antibodies
- discuss the mechanisms by which antibody diversity is generated
- describe the clonal selection theory
- discuss the role of T-cell receptors and MHC molecules in the functioning of T cells
- describe the roles of cytotoxic T cells, T-helper cells, and T-suppressor cells in specific immunity
- describe B-cell activation
- describe the outcomes *in vivo* of antigen-antibody binding
- describe the activation of complement by the classical pathway
- describe the mechanisms used to establish immune tolerance
- list the ways antibodies, lymphocytes, and nonspecific defenses provide immunity to viral and bacterial pathogens
- discuss the four types of hypersensitivities (allergies) and the roles of various immune system components in mediating these hypersensitivities
- discuss autoimmune diseases and immune deficiencies
- describe the role of the immune system in transplant rejection

CHAPTER OUTLINE

I. Overview of Specific (Adaptive) Immunity
   A. Specific immune system
      1. Has three major functions
         a. Recognize anything that is nonself
         b. Respond to this foreign material (effector response)—involves the recruitment of various defense molecules and cells to either destroy foreign material or render it harmless
         c. Remember the foreign invader (anamnestic response)—a more rapid and intense response to foreign material that occurs upon later encounters with that material
      2. The characteristics of specificity and memory distinguish the specific immune response from nonspecific resistance

3. There are two arms of specific immunity
   a. Humoral (antibody-mediated) immunity—based on action of antibodies that bind bacteria, toxins, and extracellular viruses, tagging or marking them for destruction
   b. Cellular (cell-mediated) immunity—based on action of T cells that directly attack cells infected with viruses or parasites, transplanted cells or organs, and cancer cells
II. Antigens
   A. Prior to birth, the immune system removes most T cells specific for self-recognition determinants
   B. Antigens are substances, such as proteins, nucleoproteins, polysaccharides, and some glycolipids that elicit an immune response and react with products of that response
      1. Epitopes (antigenic determinant sites) are areas of an antigen that can stimulate production of specific antibodies and that can combine with them
      2. Valence—the number of epitopes on an antigen; determines number of antibody molecules an antigen can combine with at one time
      3. Antibodies have two antigen-binding sites and can cross-link multivalent antigens, causing agglutination (precipitation)
   C. Hapten—a small organic molecule that is not itself antigenic but that may become antigenic when bound to a larger carrier molecule
   D. Cluster of differentiation molecules (CDs)—functional cell surface proteins that are used to differentiate leukocyte subpopulations; concentration of these molecules in serum is usually low and elevated levels are associated with disease (e.g., various cancers, autoimmune diseases, HIV infection); levels in serum can be used in disease management
III. Types of Specific (Adaptive) Immunity
   A. Acquired immunity develops after exposure to antigen or after transfer of antibodies or lymphocytes from an immune donor
   B. Naturally acquired immunity
      1. Naturally acquired active immunity—an individual comes in contact with an antigen via a natural process (e.g., infection) and produces sensitized lymphocytes and/or antibodies that inactivate or destroy the antigen
      2. Naturally acquired passive immunity—transfer (e.g., transplacentally or in breast milk) of antibodies from one individual (where they were actively produced) to another (where they are passively received)
   C. Artificially acquired immunity
      1. Artificially acquired active immunity—deliberate exposure of an individual to a vaccine (a solution containing antigen) with subsequent development of an immune response
      2. Artificially acquired passive immunity—deliberate introduction of antibodies from an immune donor into an individual
IV. Recognition of Foreignness
   A. The immune system must be able to distinguish between resident (self) and foreign (nonself) cells
   B. Major histocompatability complex (MHC) is a group of genes that encode three classes of proteins; only class I and class II are involved in antigen presentation; called human leukocyte antigen (HLA) complex in humans and H-2 complex in mice
   C. Each person has two sets of MHC genes that are codominant; more closely related individuals have more closely related MHC genes
   D. Class I and II MHC
      1. Both class I and class II MHC molecules consist of two protein chains and are transmembrane proteins in the plasma membrane
      2. Both class I and class II MHC molecules fold into similar shapes, each having a deep groove into which a short peptide or other antigen fragment can bind
      3. The presence of a foreign peptide in this groove alerts the immune system and activates T cells or macrophages
      4. For class I molecules, the peptides are produced intracellularly (e.g., from replicating viruses) by antigen processing in the proteosome; proteins pumped from cytoplasm to endoplasmic reticulum, where they become associated with newly synthesized class I MHC molecules; the

peptide-class I MHC complex is then carried to and incorporated into the plasma membrane; detected by cytotoxic T cells

5. For class II molecules, exocytosis brings antigens into antigen-presenting cells (APCs) and produces fragments in phagolysosomes; these peptides combine with class II MHC and are delivered to cell surface; detected by T-helper cells

V. T Cell Biology

   A. T-cell receptors—bind to antigens only when an antigen-presenting cell presents antigen

   B. Types of T cells

      1. T cells originate from stem cells in the bone marrow that migrate to the thymus for further differentiation

      2. T-helper cells (T_H or $CD4^+$ cells)

         a. T_H0—undifferentiated precursors to T_H1 and T_H2 cells

         b. T_H1 and T_H2 cells—each produces and secretes a specific mixture of cytokines

         c. T_H1 cells promote cytotoxic T cell activity, activate macrophages, and mediate inflammation and type IV hypersensitivities

         d. T_H2 cells stimulate antibody responses in general and defend against helminths

      3. Cytotoxic T cells (CTLs or CD8+ T cells) attach by their T-cell receptor to virus-infected cells that display class I MHC proteins and viral antigens; are then stimulated by T-helper cells; activated cytotoxic T cells produce cytokines that limit viral reproduction and activate macrophages and other phagocytic cells; ultimately cytotoxic T cell destroys target cell; two mechanisms are:

         a. CD95 pathway—transmembrane signal transduction leads to initiation of apoptosis; CD95 (FAS) is encoded by the *fas* gene, a member of the tumor necrosis factor family of genes

         b. Perforin pathway—release of perforins that damage the target cell membrane, resulting in cytolysis of target cell

   C. T-cell activation—require two signals for activation: presentation of antigen by an antigen-presenting cell and binding of a T_H receptor (CD24) to a macrophage surface protein (B7)

   D. Superantigens—bacterial proteins that provoke a dramatic immune response by nonspecifically stimulating T cells to proliferate; T-cell proliferation occurs when superantigen interacts both with class II MHC molecules and T-cell receptors; this leads to release of massive quantities of cytokines, which can cause disease symptoms; superantigens are associated with various chronic diseases including rabies, staphylococcal food poisoning, and others

VI. B Cell Biology

   A. Have surface molecules important to their function

      1. Surface molecules include B-cell receptors (BCRs—IgM and IgD on surface of B cell), Fc receptors, and complement receptors

      2. Binding of receptors to target molecules is involved in activation of B cell and in phagocytosis, processing, and presentation of antigens

      3. Mature B cells are specific for a single antigen epitope with the population of B cells able to recognize 10^{13} different epitopes

      4. Upon activation, B cells differentiate into plasma cells and memory cells; they also can act as antigen-presenting cells

   B. B-cell activation

      1. T-dependent antigen-triggering-B cells specific for an epitope cannot develop into plasma cells without collaboration with helper T cells

         a. The APC presents antigen in its class II MHC to the helper T cell

         b. Cytokine production is induced by the B7-CD24 interaction between the APC and helper T cell

         c. Helper T cells directly associate with B cells that display the same antigen-MHC complex as presented on the APC

         d. The B cell also requires the antigen, recognized through its BCR, to help trigger the response

         e.   Additional cytokines are released and the B cells proliferate, differentiate into plasma cells, and produce antibodies

    2.   T-independent antigen triggering

         a.   Polymeric antigens have a large number of identical epitopes that can crosslink BCRs and cause cell activation; the antibodies produced have low affinity for antigen; no memory cells are produced

VII. Antibodies

    A.   Antibody (immunoglobulin, Ig)—glycoprotein that serves as the BCR on B cells; after B-cell activation, made by plasma cells in response to an antigen; recognizes and binds the antigen that caused its production; five different classes: IgG, IgA, IgM, IgD, and IgE

    B.   Immunoglobulin structure

        1.   Multiple antigen-combining sites (usually two; some can form multimeric antibodies with up to 10 combining sites)

        2.   Basic structure is composed of four polypeptide chains

             a.   There are two heavy chains and two light chains

             b.   Within each chain are a constant region (little amino acid sequence variation within the same class of Ig) and a variable region

        3.   The four polypeptides are arranged in the form of a flexible Y

             a.   Fc (crystallizable fragment) is stalk of the Y; contains site at which antibody can bind to a cell; composed only of constant region

             b.   Fab (antigen-binding fragments) are at the top of the Y; they bind compatible epitopes of an antigen; composed of both constant and variable regions

             c.   Domains—homologous units, each about 100 amino acids long, observed in heavy chains and in light chains

        4.   Light chain exists in two distinct forms kappa (κ) and lambda (λ)

        5.   There are five types of heavy chains: gamma (γ), alpha (α), mu (μ), delta (δ), and epsilon (ε); these determine, respectively, the five classes (isotypes) of immunoglobulins: IgG, IgA, IgM, IgD, and IgE

        6.   In IgG there are four subclasses, and in IgA there are two subclasses; these subclasses result from variations in the amino acid composition of the heavy chains; the variations are classified as:

             a.   Isotypes—variations normally present in all individuals

             b.   Allotypes—genetically controlled allelic forms of the immunoglobulin molecules; allelic forms that are not present in all individuals

             c.   Idiotypes—individual-specific immunoglobulin molecules that differ in the Fab segments

    C.   Immunoglobulin function

        1.   Fab region binds to antigen whereas Fc region mediates binding to host tissue, various cells of the immune system, some phagocytic cells, or the first component of the complement system

        2.   Binding of antibody to an antigen does not destroy the antigen, but marks (targets) the antigen for immunological attack and activates nonspecific immune responses that destroy the antigen

        3.   Antigen-antibody binding occurs within the pocket formed by folding the V_H and V_L regions of Fab; binding is due to weak, noncovalent bonds and, in most cases, shapes of epitope and binding site must be highly complementary (i.e., lock and key) for efficient binding; in at least one case, it is known that the antigen induces a shape change of the antigen-binding site (induced fit mechanism); high complementarity of epitope and binding site provides for the high specificity associated with antigen-antibody binding

        4.   Opsonization—coating a bacterium with antibodies or complement to stimulate phagocytosis

    D.   Immunoglobulin classes

        1.   IgG—major Ig in human serum; monomeric protein; 80% of Ig pool

             a.   Antibacterial and antiviral

             b.   Enhances opsonization; neutralizes toxins

             c.   Only IgG is able to cross placenta (naturally acquired passive immunity for newborn)

             d.   Activates the complement system by the classical pathway

             e.   Four subclasses with some differences in function

2. IgM—pentameric protein joined with J chain at F_c ends; 10% of Ig pool
   a. First antibody made during B-cell maturation and first antibody secreted during primary antibody response
   b. Never leaves the bloodstream
   c. Agglutinates bacteria and activates complement by classical pathway; enhances phagocytosis of target cells
   d. Some may be red blood cell agglutinins
   e. Up to 5% may be hexameric; hexameric form is better able to activate the complement system than pentameric IgM; bacterial cell wall antigens may directly stimulate B cells to produce hexameric form
3. IgA—15% of Ig pool
   a. Some monomeric forms in serum, but most is dimeric (using a J chain) and associates with a protein called the secretory component (secretory IgA or sIgA)
   b. sIgA is primary Ig of mucosal-associated lymphoid tissue; also found in saliva, tears, and breast milk (protects nursing newborns); helps rid the body of antigen-antibody complexes by excretion; functions in alternate complement pathway
4. IgD—monomeric protein; trace amounts in serum
   a. Does not activate the complement system and cannot cross the placenta
   b. Abundant on surface of B cells where it plays a role in signaling B cells to start antibody production
5. IgE—monomeric protein; less than 1% of Ig pool
   a. Skin-sensitizing and anaphylactic antibodies
   b. When an antigen cross-links two molecules of IgE on the surface of a mast cell or basophil, it triggers release of histamine; stimulates eosinophilia and gut hypermotility, which help to eliminate helminthic parasites

E. Antibody kinetics
   1. Primary antibody response—when exposed to antigen (by infection or vaccination), levels (titer) of antibody change over time
      a. Initial lag phase of several days
      b. Log phase—antibody titer rises logarithmically
      c. Plateau phase—antibody titer stabilizes
      d. Decline phase—antibody titer decreases because the antibodies are metabolized or cleared from the circulation
      e. IgM appears first, then IgG; relatively low antigen affinity
   2. Secondary antibody response (anamnestic response)—has shorter lag phase, higher antibody titer, and more IgG, that have high affinity for antigens

F. Diversity of antibodies—several mechanisms contribute to the generation of antibody diversity
   1. Combinatorial joining
      a. Ig genes are interrupted or split genes with many exons; in light-chain gene, there are three types of exons (C, V, and J); in heavy-chain gene, there are four types of exons (C, V, J, and D)
      b. During differentiation of B cells, one C exon, one V exon, and one J exon are joined together to make a functional light-chain gene; one C, one V, one J, and one D are joined together to make a functional heavy-chain gene; since there are numerous C, V, J, and D exons, many different combinations are possible (2×10^8)
   2. Somatic mutations—the V regions of germ-line DNA are susceptible to a high rate of somatic mutation during B-cell development
   3. Splice-site variability—the same exons can be joined at different nucleotides, thus generating different codons and the possible diversity
   4. The number of different antibodies possible is the product of the number of light chains possible and the number of heavy chains possible

G. Clonal selection
   1. Because of combinatorial joining and somatic mutation, there are a small number of B cells capable of responding to any given antigen; each group of cells is derived asexually from a

parent cell and is referred to as a clone; there is a large, diverse population of B-cell clones that collectively are capable of responding to many possible antigens

2. Identical antibody molecules, specific to each B cell and a single antigen, are integrated into the plasma membrane of B cell; when these bind the appropriate antigen, the B cell is stimulated to divide and differentiate into two populations of cells: plasma cells and memory B cells

    a. Plasma cells are protein factories that produce about 2,000 antibodies per second for their brief life span (5 to 7 days)

    b. Memory B cells can initiate antibody-mediated immune response if they are stimulated by being bound to the antigen; they circulate more actively from blood to lymph and have long life spans (years or decades); are responsible for rapid secondary response; are not produced unless B cell has been appropriately signaled by activated T-helper cell

VIII. Action of Antibodies

  A. Neutralization

    1. Toxin neutralization—antibody (antitoxin) binding to toxin renders the toxin incapable of attachment or entry into target cells

    2. Viral neutralization—binding prevents virus from binding to and entering target cells

    3. Adherence inhibition—sIgA prevents bacterial adherence to mucosal surfaces

    4. IgE and parasitic infection—in the presence of elevated IgE levels, eosinophils bind parasites and release lysosomal enzymes that lead to destruction of parasite

  B. Opsonization—enhancement of phagocytosis; results from coating of microorganisms or other material by antibodies or complement; forms bridge between the antigen and the phagocyte

  C. Immune complex formation—two or more antigen-binding sites per antibody molecule lead to cross-linking, forming molecular aggregates called immune complexes; these complexes are more easily phagocytosed

    1. Precipitation (precipitin) reaction—soluble particles are cross-linked, causing them to precipitate from solution; the antibody involved is called a precipitin antibody

    2. Agglutination reaction—particles or cells are cross-linked, forming an aggregate; the antibody involved is called an agglutinin.

    3. A variety of in vitro diagnostic assays rely on immune complex formation to detect the presence of antigen or antibody

IX. Summary: The Role of Antibodies and Lymphocytes in Resistance

  A. Response of a host to any particular pathogen may involve a complex interaction between host and pathogen, as well as the components of both nonspecific (innate) and specific (adaptive) immunity

  B. Immunity to viral infection

    1. Antibodies neutralize viruses

    2. Antibodies enhance phagocytosis

    3. Interferons shut down protein synthesis in virus-infected cells; interferons stimulate the activity of T cells and NK cells (natural killer cells are non-B, non-T lymphocytes with no antigen receptors)

    4. In cell-mediated immunity, activated macrophages and cytotoxic T lymphocytes (CTLs) destroy virus-infected cells by inducing apoptosis, and the production of granzymes and perforin

    5. Viruses released from cells can trigger humoral responses (neutralization and opsonization)

  C. Immunity to bacterial infections

    1. Inflammatory response destroys pathogens and recruits macrophages

    2. T_H2 cells activate B cells and trigger humoral response leading to opsonization, agglutination, neutralization, and complement activation (by classical pathway)

    3. Activated macrophages and cytotoxic T cells destroy cells infected with intracellular pathogenic bacteria and attack bacterial pathogens

X. Acquired Immune Tolerance

  A. Nonresponse to self; three mechanisms have been proposed: negative selection by clonal deletion, induction of anergy, and inhibition of immune response by T cells with suppressor/regulatory function

1. Negative selection by clonal deletion—T cells with ability to interact with self-antigens are destroyed in the thymus
2. Induction of anergy—an example of peripheral tolerance (tolerance that develops in areas other than thymus); lymphocytes that can interact with self-antigens are given incomplete activation signals, causing them to enter into an unresponsive state known as anergy

XI. Immune Disorders

A. Hypersensitivities—exaggerated or inappropriate immune responses that result in tissue damage to the individual

1. Type I hypersensitivity—includes allergic reactions
   a. Occurs immediately following second contact with responsible antigen (allergen); on first exposure, B cells form plasma cells that produce IgE (reagin), which binds to mast cells or basophils via Fc receptors and sensitizes them; upon subsequent exposure, the allergen binds to these IgE-bearing cells; physiological mediators released by this binding cause anaphylaxis (smooth muscle contraction, vasodilation, increased vascular permeability, and mucus secretion)
   b. Systemic anaphylaxis results from a massive release of these mediators, which cause respiratory impairment, lowered blood pressure, and serious circulatory shock; death can occur within a few minutes
   c. Localized anaphylaxis (atopy) includes hay fever (upper respiratory tract), bronchial asthma (lower respiratory tract), and hives (food allergy)
   d. Skin testing is used to identify allergens; small amounts of possible allergens are inoculated into skin; rapid inflammatory reaction indicates sensitivity
   e. Desensitization to allergens involves controlled exposure to the allergen in order to stimulate IgG production; IgG molecules serve as blocking antibodies that intercept and neutralize the allergen before it can bind to the IgE-bound mast cells

2. Type II hypersensitivity—generally cytolytic or cytotoxic reaction that destroys host cells
   a. IgG or IgM antibodies are directed against cell surface or tissue antigens; this stimulates complement pathway and a variety of immune effector cells
   b. Blood types (ABO blood groups) are determined by cell surface glycoproteins on red blood cells; types A, B, and AB have specific cell surface proteins while type O has none; ABO glycoproteins are self antigens; there is an intense immune response to transfusions of blood that does not match the host type
   c. Rh factor is another is another blood antigen that is either present or absent in an individual; incompatibility between Rh- mothers and Rh+ fetuses can induce anti-Rh antibodies that destroy fetal blood cells in a syndrome called erythroblastosis fetalis

3. Type III hypersensitivity
   a. Involves formation of immune complexes, which in the presence of excess antigen are not efficiently removed; their accumulation triggers complement-mediated inflammation
   b. Can cause inflammation and damage to blood vessels (vasculitis), kidney glomerular basement membranes (glomerulonephritis), joints (arthritis), and skin (systemic lupus erythematosus)

4. Type IV hypersensitivity—involves T_H1 lymphocytes or CTLs that migrate to and accumulate near the antigen; this is a time-delayed response
   a. Presentation of antigen to T_H1 or CTL cells causes the release of cytokines; these attract macrophages and basophils to the area, leading to inflammatory reactions that can cause extensive tissue damage
   b. Can be used diagnostically, as in the tuberculin skin test (for tuberculosis exposure)
   c. Examples of type IV hypersensitivities include allergic contact dermatitis (poison ivy, cosmetic allergies) and some chronic diseases (leprosy, tuberculosis, leishmaniasis, candidiasis, herpes simplex lesions)

B. Autoimmune diseases—autoimmunity is characterized by the presence of autoantibodies and is a natural consequence of aging; autoimmune disease results from activation of self-reactive T and B cells, which leads to tissue damage

1. Viruses, genetic factors, hormones, and psycho-neuro-immunological factors all influence the development of autoimmune disease
2. More common in older individuals; may involve viral or bacterial infections that cause tissue damage and the release of abnormally large quantities of antigen, or that cause some self-proteins to alter their form so that they are no longer recognized as self

C. Transplantation (tissue) rejection
1. Transplantation of tissue from one individual to another can be an allograft (donor and recipient are genetically different individuals of the same species) or xenograft (donor and recipient are different species)
2. Mechanisms of tissue rejection
   a. Foreign class II MHC antigens trigger T_H cells to help CTL cells destroy the graft; the CTL cells recognize the graft as foreign by detecting the class I MHC antigens of the graft
   b. T_H cells may react directly with the graft, releasing cytokines that stimulate macrophages to enter the graft and destroy it
3. Graft-versus-host reaction—immunocompetent cells in donor tissue (e.g., bone marrow) reject the immunosuppressed host

D. Immunodeficiencies—failure to recognize and/or respond properly to antigens
1. Primary (congenital) immunodeficiencies result from a genetic disorder
2. Secondary (acquired) immunodeficiencies result from infection by immunosuppressive microorganisms (e.g., AIDS, chronic mucocutaneous candidiasis)

TERMS AND DEFINITIONS

Place the letter of each term in the space next to the definition or description that best matches it.

_____ 1. The recruitment of defensive molecules and cells by the specific immune system in response to recognition of nonself
_____ 2. Immunity developed after exposure to antigen or after transfer of antibodies or lymphocytes from an immune donor
_____ 3. Immunity that occurs when an individual's immune system contacts an antigen during an infection
_____ 4. Immunity that occurs when antibodies from a pregnant woman pass across the placenta to her fetus
_____ 5. Immunity that results when an animal is given an antigen preparation (vaccine) to induce antibody formation and activation of lymphocytes
_____ 6. Immunity that results when antibodies produced in another animal or in vitro are introduced into a host
_____ 7. Substances that elicit an immune response and react with the products of that response
_____ 8. Antigenic determinant sites
_____ 9. Small molecules that can bind antibodies but cannot stimulate antibody formation
_____ 10. Functional cell surface proteins than are often used as a nomenclature system to differentiate leukocyte subpopulations.
_____ 11. A glycoprotein that is made in response to an antigen, and can recognize and bind to the antigen that caused its production.
_____ 12. Compact, self-folding, structurally independent regions of proteins
_____ 13. A substance that when mixed with an antigen enhances the rate and quantity of antibody production
_____ 14. The fluid that remains after blood clots
_____ 15. A serum obtained from an immunized host
_____ 16. Monoclonal antibodies that have been attached to a specific toxin or toxic agent
_____ 17. The general term for a collection of genes that encode proteins involved in immune function, including antigen presentation
_____ 18. A collection of genes in humans that encode proteins involved in immune function, including antigen presentation
_____ 19. A collection of genes in mice that encode proteins involved in immune function, including antigen presentation

_____ 20. A member of the tumor necrosis factor family of genes that encodes a protein important to the CD95 pathway used by cytotoxic T lymphocytes (CTLs)

_____ 21. Antigens that elicit B-cell responses but only with the aid of T-helper cells

_____ 22. Antigens that elicit B-cell responses without the aid of T-helper cells

_____ 23. The action of antibodies that makes a toxin unable to attach to its target cell and therefore unable to damage it

_____ 24. The action of antibodies that makes a virus unable to attach to its target cell and therefore unable to damage it

_____ 25. Precipitation of soluble antigens resulting from the cross-linking of antibody and antigen to form large aggregates

_____ 26. Cross-linking of antibody and antigen-bearing cells into large aggregates

_____ 27. The ability to produce antibodies against nonself-antigens, while not producing antibodies against self-antigens

_____ 28. An exaggerated or inappropriate immune response that results in tissue damage to the individual

_____ 29. An innocuous antigen that causes an allergic reaction

_____ 30. IgE molecules produced in response to an allergen

_____ 31. Transplants between genetically different individuals of the same species

_____ 32. Transplants between different species

_____ 33. Disease that develops when cells in transplanted tissue attack the host

    a.   acquired immune tolerance
    b.   acquired immunity
    c.   adjuvant
    d.   agglutination reaction
    e.   allergen
    f.   allografts
    g.   antibody (immunoglobulin)
    h.   antigens
    i.   antiserum
    j.   artificially acquired active immunity
    k.   artificially acquired passive immunity
    l.   cluster of differentiation molecules (CDs)
    m.   domains
    n.   effector response
    o.   epitopes
    p.   _fas_ gene
    q.   graft-versus-host disease
    r.   H-2 complex
    s.   haptens
    t.   human leukocyte antigen (HLA) complex
    u.   hypersensitivity
    v.   immunotoxins
    w.   major histocompatability complex (MHC)
    x.   naturally acquired active immunity
    y.   naturally acquired passive immunity
    z.   precipitation (precipitin) reaction
    aa.   reagin
    bb.   serum
    cc.   T-dependent antigens
    dd.   T-independent antigens
    ee.   toxin neutralization
    ff.   viral neutralization
    gg.   xenografts

FILL IN THE BLANK

1. Activation of the complement system can proceed by three different mechanisms. One is dependent on the reaction between antibodies and antigens and is called the _____ _____pathway. The others do not require the binding of antibodies to antigens. They are the alternative complement pathway and the lectin complement pathway.

2. One endpoint of complement activation is the formation of a _____ _____ complex, which creates a pore in the plasma membrane, leading to cytolysis.

3. When antibodies bind to certain types of antigens, those antigens lose the ability to bind to and enter their target cells. When the antigen is a toxin, this is called _____ _____, and the antibody or the serum containing the antibody is called an _____. In some cases, the antigen is a virus. In this case, the loss of the ability to bind to and enter a target cell is called _____ _____.

4. Two branches or arms of specific immunity are recognized. One is based on the action of soluble proteins called _____, which bind to bacteria, toxins, and extracellular viruses, thereby tagging them for destruction. This type of immunity is called either _____ immunity or _____ immunity. The other arm is based on the action of special kinds of T cells and is called _____ immunity or _____ immunity.

5. Since antibodies have at least two antigen-binding sites and most antigens have at least two determinants, cross-linking can occur, producing large aggregates called _____ _____. If this involves soluble antigens, the antibody is called a _____. If it involves cells or insoluble particles, the antibody is called an _____. If it specifically involves red blood cells, the antibody is called a _____.

6. The antibodies found in _____ are polyclonal antibodies; they are produced by several different B-cell clones and have different sensitivities. It is now possible to synthesize antibodies in vitro that have a single specificity. These are called _____ antibodies. They are produced by fusing plasma cells with _____ cells to form _____.

7. A widely accepted hypothesis to explain immunological specificity and memory is the _____ _____ theory. One tenet of this hypothesis is that each individual contains a B-cell _____ (a population of cells derived asexually from a single parent) that can respond to a specific antigen by producing the correct antibody. The reaction of the antibody with antigen causes the appropriate B cell to proliferate and differentiate to form two different populations: plasma cells and _____ _____ cells. The latter are responsible for the _____ _____ upon second exposure to the antigen.

8. _____ (also called _____ _____ sites) are the sites of an antigen that bind to an antibody or a T-cell receptor. The number of these sites on the surface of an antigen is called its _____, and this number determines the number of antibodies that can combine with the antigen at one time.

9. Antibodies, also called _____, have a basic structure composed of two heavy and two light chains. The heavy chains are distinct for each class of antibodies and five major classes are known (IgA, IgD, IgE, IgG, IgM). Both the light chains and the heavy chains contain two different regions. The _____ regions have amino acid sequences that do not vary significantly between antibodies of the same class. The _____ regions of different antibodies within the same class have significantly different amino acid sequences.

10. Variants of antibodies can be classified as: _____, variants that arise from variations in the heavy chain constant regions and are present in all individuals; _____, variants that represent allelic forms of antibodies and are not present in all individuals; and _____, variants that differ in the hypervariable region of the Fab.

11. $CD8^+$ cells are also called _____ _____ cells. They are activated by three signals, including exposure to interleukin secreted by _____ cells. Upon receiving the signals, the $CD8^+$ cells proliferate and differentiate to form active _____ _____ _____ (CTLs), which can attack virus-infected cells. These cells are also involved in the destruction of cancer cells—a process called _____ _____.

12. The primary immunoglobulin of mucosal-associated lymphoid tissue is _____ _____. This immunoglobulin consists of two monomers held together by a _____ _____, and it is further modified by a protein called the secretory component.

13. Even though B cells have receptors on their surfaces that are specific for one antigenic determinant, or _____, they usually cannot develop into plasma cells and begin secreting antibody without the help of _____ cells.

14. T cells that control the development and/or function of effector T cells are called _____ T cells. There are two subsets of these cells: _____ cells (CD4$^+$ cells) and _____ cells. The CD4$^+$ cells are further divided into three subsets: _____ cells produce cytokines that are involved in cellular immunity, _____ cells produce cytokines that promote humoral immunity and _____ cells are undifferentiated precursors of the other two subsets.

15. The ability to produce antibodies against nonself antigens while not producing antibodies against self-antigens is called_____ _____ _____. Three mechanisms have been proposed to explain this. One is that early in development, those lymphocytes capable of responding to self-antigens are destroyed; this is called negative selection by clonal deletion. A second mechanism involves induction of a nonresponsive state called _____. The third proposed mechanism involves inhibition of immune responses by _____cells.

16. When self-reactive T cells and B cells attack the body and cause tissue damage, the resulting disease is called an _____ disease.

17. Defects in one or more components of the immune system can result in its failure to recognize and respond properly to antigens; this is called an _____.

18. People with allergies sometimes undergo a process called _____ in which a series of doses of the _____ are injected beneath the skin; this results in the formation of IgG instead of IgE antibodies. The circulating IgGs act as blocking antibodies that intercept and neutralize the _____before it has time to interact with IgE-bound mast cells.

19. In _____ hypersensitivities, the initial exposure to a soluble _____ stimulates B cells to differentiate and begin to produce IgE, which is sometimes called _____. The IgE binds to mast cells and basophils and sensitizes them. When a second exposure to the _____ occurs, it attaches to the surface bound IgE and causes immediate degranulation. This results in smooth muscle contractions, vasodilation, increased vascular permeability, and mucous secretions, responses that are collectively called_____.

MULTIPLE CHOICE

For each of the questions below select the *one best* answer.

1. Which of the following is normally involved in inhibiting the adherence of bacteria to mucosal surfaces?
   a. IgA
   b. IgD
   c. IgE
   d. IgG
   e. IgM

2. What is a toxin-neutralizing antibody called?
   a. toxoid
   b. antitoxin
   c. antitoxoid
   d. neurotoxin

3. Which immunoglobulin is the major immunoglobulin in human serum?
   a. IgA
   b. IgD
   c. IgE
   d. IgG
   e. IgM

4. Which immunoglobulin has an extra peptide (J chain) that holds five monomeric units together in a pinwheel array?
   a. IgA
   b. IgD
   c. IgE
   d. IgG
   e. IgM
5. Which cells are responsible for the anamnestic response?
   a. plasma cells
   b. memory B cells
   c. T-helper cells
   d. cytotoxic T cells
6. Which body cells have class I MHC molecules?
   a. all cells of the body except red blood cells
   b. only macrophages
   c. only B cells
   d. only dendritic cells
7. Which of the following is NOT true of T-independent antigens?
   a. They usually present a large array of identical epitopes to a B cell specific for that determinant.
   b. The antibody produced is usually IgM with little switching to IgG.
   c. The antibody produced usually has a low binding affinity.
   d. All of the above are true of T-independent antigens.
8. Which of the following is true about class II MHC molecules?
   a. Their shape is significantly different than the shape of class I MHC.
   b. They are required for recognition of exogenous antigens by T-helper cells.
   c. All cells of the body have them, except red blood cells.
   d. All of the above are true about class II MHC molecules.
9. Which of the following types of cells can present antigens to T cells?
   a. macrophages
   b. dendritic cells
   c. B cells
   d. All of the above act as antigen-presenting cells.
10. Which immunoglobulin is important in providing resistance to protozoan and helminthic parasites?
    a. IgA
    b. IgD
    c. IgE
    d. IgG
    e. IgM
11. Which of the following is a pathway used by CTLs to destroy their target cells?
    a. CD95 pathway, which triggers apoptosis of the target cell
    b. perforin pathway, which can cause cytolysis or apoptosis of the target cell
    c. Both (a) and (b) are used by CTLs.
    d. Neither (a) nor (b) is used by CTLs.
12. Which of the following is a mechanism by which antibodies help provide immunity?
    a. opsonization
    b. formation of immune complexes, which are more rapidly phagocytosed
    c. activation of complement via the classical complement pathway
    d. All of the above are mechanisms by which antibodies help provide immunity.
13. Hives (skin eruptions associated with food allergies) are an example of which type of hypersensitivity?
    a. type I
    b. type II
    c. type III
    d. type IV

14. Allergic contact dermatitis, such as the rash associated with poison ivy, is an example of which type of hypersensitivity?
    a. type I
    b. type II
    c. type III
    d. type IV
15. Which of the following can influence the development of autoimmune disease?
    a. viral infection
    b. genetic makeup of the individual
    c. production of certain hormones (e.g., estrogen)
    d. stress
    e. All of the above influence the development of autoimmune diseases.
16. Which of the following is a type I hypersensitivity?
    a. hay fever
    b. allergic contact dermatitis
    c. systemic lupus erythematosus
    d. All are type I sensitivities.
17. Adverse reaction to a blood transfusion is which type of hypersensitivity?
    a. type I
    b. type II
    c. type III
    d. type IV

TRUE/FALSE

_____ 1. Once activated by antibodies binding to antigens, the actions of the other components of the immune system are not specific (i.e., those components will interact with anything in the vicinity).

_____ 2. Opsonizing antibodies must be against surface components, not internal components, if they are to effectively stimulate phagocytosis.

_____ 3. The titer of antibody in serum is the reciprocal of the highest dilution of serum showing a positive reaction in the test being used to detect antibody.

_____ 4. Combinatorial joining, somatic mutations, and variation in splicing generate antibody diversity.

_____ 5. A single pathogen will trigger only one type of immune response (cellular, humoral, or nonspecific).

_____ 6. A person's total B-cell population carries B-cell antigen receptors (BCRs) that are specific for a large number of antigens; however, each mature B cell possesses BCRs specific for only one epitope.

_____ 7. IgM (transmembrane form) is a component of BCRs.

_____ 8. IgD is a component for T-cell antigen receptors (TCRs).

_____ 9. TCRs can only recognize antigens on the surfaces of antigen-presenting cells; they cannot bind free antigen.

_____ 10. CTLs can kill their target cells either by cytolysis or by apoptosis (programmed cell death).

_____ 11. Superantigens stimulate T cells to proliferate nonspecifically, causing the release of massive quantities of cytokines, which, in time, can lead to tissue damage.

_____ 12. The CDs on all cell types are nearly identical; therefore the complement of CDs on a cell cannot be used to indicate any specific cell lineage.

_____ 13. The primary antibody response results in the secretion of high levels of IgG antibodies, which remain in the body for many years.

_____ 14. The presence of serum autoantibodies (autoimmunity) is a natural consequence of aging and does not necessarily lead to autoimmune disease.

_____ 15. Autoimmunity and autoimmune disease are different terms for the same phenomenon.

_____ 16. Autoimmune diseases result when self-reactive T and B cells attack the body and cause tissue damage.

CRITICAL THINKING

1. Draw the basic structure of all immunoglobulin molecules and label the following: Fc, Fab, constant regions, variable regions, antigen-binding sites, heavy chains, and light chains. Correlate the constant and variable regions of each chain to their coding regions in the germline DNA.

2. Macrophages and other phagocytic cells have relatively nonspecific targets. Explain how cytokines increase the apparent specificity of these effector cells. In your discussion indicate why this is an *apparent* and not an *actual* change in specificity, even though the result is indeed a more specific destruction of the invading organism.

3. CD4$^+$ cells are among the primary targets of human immunodeficiency virus (HIV). As HIV infection progresses, the number of CD4$^+$ cells declines, causing a weakening both the cellular and humoral immune responses. Explain why the targeting of this single type of lymphocyte has such broad impact on the functioning of the immune system.

ANSWER KEY

Terms and Definitions

1. n, 2. b, 3. x, 4. y, 5. j, 6. k, 7. h, 8. o, 9. s, 10. l, 11. g, 12. m, 13. c, 14. bb, 15. i, 16. v, 17. w, 18. t, 19. r, 20. p, 21. cc, 22. dd, 23. ee, 24. ff, 25. z, 26. d, 27. a, 28. u, 29. e, 30. aa, 31. f, 32. gg, 33. q

Fill in the Blank

1. classical complement 2. membrane attack 3. toxin neutralization; antitoxin; viral neutralization 4. antibodies; humoral; antibody-mediated; cellular; cell-mediated 5. immune complexes; precipitin; agglutinin; hemagglutinin 6. serum; monoclonal; myeloma; hybridomas 7. clonal selection; clone; memory B; anamnestic response 8. Epitopes; antigenic determinant; valence 9. immunoglobulins; constant; variable 10. isotypes; allotypes; idiotypes 11. cytotoxic T; T_H1; cytotoxic T lymphocytes; immune surveillance 12. secretory IgA; J chain 13. epitope; T_H2 14. regulator; T-helper; T-suppressor; T_H1; T_H2; T_H0; 15. acquired immune tolerance; anergy; T-suppressor 16. autoimmune 17. immunodeficiency 18. desensitization; allergen; allergen 19. type I; allergen; reagin; allergen; anaphylaxis

Multiple Choice

1. a, 2. b, 3. d, 4. e, 5. b, 6. a, 7. d, 8. b, 9. d, 10. c, 11. c, 12. d, 13. a, 14. d, 15. e, 16. a, 17. b

True/False

1. F, 2. T, 3. T, 4. T, 5. F, 6. T; 7. T, 8. F, 9. T, 10. T, 11. T, 12. F, 13. F, 14. T, 15. F, 16. T

33   Pathogenicity of Microorganisms

CHAPTER OVERVIEW

This chapter focuses on parasitism and pathogenicity. The development of a disease state is a dynamic process that is dependent on the virulence of the pathogen and the resistance of the host. This dynamic process is illustrated in the discussions of viral and bacterial pathogenesis. The chapter concludes with a discussion of mechanisms used by viruses and bacteria to evade host defenses.

CHAPTER OBJECTIVES

After reading this chapter you should be able to:

*   discuss the general characteristics of parasitic symbiosis
*   discuss the concepts of pathogens, disease, infection, and infectious disease
*   describe the stages that pathogens usually go through in order to cause disease
*   distinguish exotoxins from endotoxins
*   describe the modes of action of various toxins
*   describe the mechanisms used by viruses and bacteria to evade host defenses

CHAPTER OUTLINE

I.   Host-Parasite Relationships
  A.   A parasitic organism is one that lives at expense of its host; any organism that causes disease is a parasite
  B.   Host-parasite relationships
    1.   Ectoparasite—lives on the surface of the host
    2.   Endoparasite—lives within the host
    3.   Final host—the host on (or in) which the parasitic organism either gains sexual maturity or reproduces
    4.   Intermediate host—a host that serves as a temporary but essential environment for development of a parasitic organism
    5.   Transfer host—a host that is not necessary for development but that serves as a vehicle for reaching the final host
    6.   Reservoir host—an organism that is infected with a parasitic organism that can also infect humans
  C.   Infection—the state occurring when a parasite is growing and multiplying on or within a host
    1.   Infectious disease—a change from a state of health as a result of an infection by a parasitic organism
    2.   Pathogen—any parasitic organism that produces an infectious disease
    3.   Pathogenicity—the ability of a parasitic organism to cause a disease
    4.   Primary (frank) pathogen—organism that causes disease in a healthy host by direct interaction
    5.   Opportunistic pathogen—organism that is normally free-living or part of the host's normal microbiota, but adopts a pathogenic role under certain circumstances (host weakness)
  D.   Some infectious organisms can enter a latent state; this can be intermittent (e.g., cold sores) or quiescent (e.g., chickenpox/shingles)
  E.   The final outcome of most host-parasite relationships is dependent on three main factors:

1. The number of pathogenic organisms present
2. The virulence of the organism
3. The host's defenses or degree of resistance

F. Virulence—the degree or intensity of pathogenicity of an organism; it is determined by three characteristics of the pathogen (invasiveness, infectivity, and pathogenic potential)
   1. Virulence factors are individual characteristics that confer virulence (e.g., capsules, pili, toxins)
   2. Invasiveness—the ability of the organism to spread to adjacent tissues
   3. Infectivity—the ability of the organism to establish a focal point of infection
   4. Pathogenic potential—the degree to which the pathogen can cause morbid symptoms (e.g., toxigenicity)
   5. Virulence is often measured experimentally by lethal dose 50 (LD_{50}) or infectious dose 50 (ID_{50}), the dose needed to kill or infect 50% of a group of hosts, respectively

G. Disease also can result from exaggerated immunological responses to a pathogen (immunopathology)

II. Pathogenesis of Viral Diseases
   A. The steps for viral infectious processes often include the following:
      1. Maintain a reservoir
      2. Enter a host
      3. Contact and enter susceptible cells
      4. Replicate within the cells
      5. Release from host cells
      6. Spread to adjacent cells
      7. Engender a host immune response
      8. Be cleared from the body of the host, establish a persistent infection, or kill the host
      9. Be shed back into the environment
   B. Maintaining a reservoir—infectious agents must reside somewhere while seeking a susceptible host; the most common reservoirs for human pathogens are humans and animals
   C. Contact, entry, and primary replication
      1. Entrance can be gained through one of the body surfaces as the result of medical procedures (e.g., needle stick, blood transfusion) or by insect vectors (organisms that transmit pathogens from one host to another)
      2. Viral infection begins with adsorption of the virus to the host cell, followed by entrance into the cell to access the host's replicative machinery; viruses bind with ligands that complement receptors on the host cell surface and then:
         a. Viral genome enters directly
         b. Endocytosis of viral particle is followed by uncoating
         c. Viral envelope fuses with cell membrane, followed by uncoating
      3. Some viruses replicate at site of entry; some viruses spread to distant sites
   D. Release from host cells—host cells are either lysed or new virions bud off from the cell membrane
   E. Viral spread and cell tropism
      1. Mechanisms of viral spread vary, but most common routes are bloodstream and lymph system; presence of virus in blood is called viremia
      2. Tropisms—specificity of a virus to a type of cell, tissue, or organ; specificity usually results from presence of specific receptors or host cells that bind virus
   F. Virus-host interactions
      1. Cytopathic viruses—kill host cells through lysis causing necrosis or by initiating apoptosis, programmed cell death
      2. Noncytopathic viruses—do not immediately kill host cells, but produce latent or persistent infections; productive viruses release a few new virions during persistent infections; nonproductive viruses cause latent infections and do not release virions unless induced into a productive state
      3. The characteristics of clinical illness is determined by the extent of cell damage and the tissues involved

G. Host immune response—both humoral and cellular components of immune response control viral infection

H. Recovery from infection—host either succumbs or recovers; recovery mechanisms involve numerous components of immune system and the importance of any individual component varies with the virus

I. Virus shedding—infected host is infectious (contagious); necessary to maintain a source of viruses in a population of hosts; often occur at same body surface used for entry; for some viral infections humans and other animals are dead-end hosts (no shedding of virus)

III. Overview of Bacterial Pathogenesis
    A. The steps for bacterial infectious processes often include the following (Note: the first five influence infectivity and invasiveness; toxigenicity plays a major role in the sixth step):
        1. Maintain a reservoir (a place to live before and after causing an infection)
        2. Initially be transported to the host
        3. Adhere to, colonize, or invade the host
        4. Initially evade host defense mechanisms
        5. Multiply or complete its life cycle on or in the host
        6. Damage the host
        7. Leave the host and return to the reservoir or enter a new host

    B. Maintaining a reservoir of the bacterial pathogen—most common reservoirs for human pathogens are humans, animals, and the environment

    C. Transport of the bacterial pathogen to the host
        1. Direct contact (e.g., coughing, sneezing, body contact)
        2. Indirect transmission
            a. Vehicles include soil, water, and food
            b. Vectors include living organisms that transmit a pathogen (e.g., insects)
            c. Fomites—inanimate objects contaminated with a pathogen and able to spread it

    D. Attachment and colonization by the bacterial pathogen
        1. Bacterium must be able to adhere to and colonize (but not necessarily invade) host cells and tissue
        2. Depends on ability of bacterium to successfully adhere to host and compete with normal microbiota for essential nutrients
        3. Adherence structures such as pili, fimbriae, and specialized adhesion molecules that facilitate attachment to host cell receptors are important virulence factors

    E. Invasion of the host tissues
        1. Penetration of the host's epithelial cells or tissues
            a. Pathogen-associated mechanisms involve the production of lytic substances that:
                1) Attack the ground substance and basement membranes of integuments and intestinal linings
                2) Degrade carbohydrate-protein complexes between cells or on cell surfaces
                3) Disrupt cell surfaces
            b. Passive mechanisms of entry involve:
                1) Breaks, lesions, or ulcers in the mucous membranes
                2) Wounds, abrasions, or burns on the skin surface
                3) Arthropod vectors that penetrate when feeding
                4) Tissue damage caused by other organisms
                5) Endocytosis by host cells
        2. Invasion of deeper tissues can be accomplished by production of specific products or enzymes that promote spreading (these are one type of virulence factor) or by entry into the circulatory system

    F. Growth and multiplication of the bacterial pathogen
        1. Pathogen must find an appropriate environment
        2. Presence of bacteria in bloodstream is called bacteremia
        3. Release of toxins by bacteria into bloodstream can cause septicemia
        4. Facultative intracellular pathogens can reside within host cells or in the environment, while obligate intracellular pathogens are incapable of living outside of the host

G. Leaving the host—must be able to leave host or disease cycle will be interrupted and the bacterium will not be perpetuated; most bacteria leave host by passive mechanisms (e.g., in feces, urine, or saliva)

H. Regulation of bacterial virulence factor expression

    1. Many virulence genes can be transferred by horizontal gene transfer; some transfer processes result in insertion of virulence genes into chromosome; this leads to the formation of different clonal types, some that cause disease and some that don't

    2. Some bacteria are adapted to both free-living state and parasitic state; these bacteria have complex signal transduction pathways that regulate virulence genes; expression of virulence genes may be under control of phages or under control of environmental factors

    3. Pathogenicity islands

        a. Large segments of DNA that carry virulence genes acquired during evolution by horizontal gene transfer; are not present in nonpathogenic members of same genus or species

        b. Have unique sequence characteristics including several open reading frames with putative genes, a different G + C content than the host, and insertion-like elements at their ends

        c. Type III secretion system enables gram-negative bacteria to secrete and inject virulence proteins into the cytoplasm of hosts; an injectosome anchored by a basal body includes a helical protein needle, the tip of which forms a pore in the host cell membrane; the type III secretion system is induced by contact with the host cell

IV. Toxigenicity

A. The capacity of an organism to produce a toxin

B. Intoxications—diseases that result from the entry of a specific toxin into the host

    1. Toxin—a specific substance, often a metabolic product of the organism, that damages the host in some specified manner

    2. Toxemia—symptoms caused by toxins in the blood of the host

C. Exotoxins

    1. Soluble, heat-labile proteins produced by and released from a pathogen; generally associated with gram-positive bacteria; may damage the host at some remote site

    2. Toxins can be inactivated by neutralizing antibodies (antitoxins) or by chemical means that create immunogenic toxoids (used for vaccines)

    3. Can be grouped into four types based on structure and physiological activities

        a. AB toxins can be separated into two distinct portions: one that binds the host cell and one that causes toxicity (e.g., diphtheria toxin—binds host cell surface receptor by the B portion and is taken into the cell by the formation of clathrin-coated vesicles; toxin is then cleaved, releasing A fragment, which enters cytosol; the A fragment inhibits protein synthesis)

        b. Specific host site exotoxins: neurotoxins damage nervous tissue (e.g., botulinum toxin and tetanus toxin), enterotoxins damage the small intestine (e.g., cholera toxin), and cytotoxins do general tissue damage (e.g., shiga toxin); some host-site-specific exotoxins are also AB toxins (e.g., cholera toxin)

        c. Membrane-disrupting exotoxins—two subtypes, those that bind cholesterol in the host cell membrane and then form a pore (e.g., leukocidins and hemolysins) and those that are phospholipases (e.g., gas gangrene-associated toxin)

        d. Superantigens—pathogen proteins (e.g., staphylococcal enterotoxin) that provoke massive cytokine releases, causing endothelial cell damage, circulatory shock, and multiorgan failure

    4. Roles of exotoxins in disease—can cause disease when they are ingested as preformed exotoxins (e.g., staphylococcal food poisoning), when produced after colonization of host (e.g., cholera), and when produced at a wound site (e.g., gas gangrene)

D. Endotoxins—LPS of many gram-negative bacteria

    1. Released only when the microorganism lyses or divides

    2. Usually capable of producing fever, shock, blood coagulation, weakness, diarrhea, inflammation, intestinal hemorrhage, and/or fibrinolysis; many of these effects are indirect

and are mediated by host molecules and cells (e.g., macrophages, endogenous pyrogens, host cytokines)

V. Host Defense Against Microbial Invasion
    A. Primary defenses—host strategies to prevent pathogen entry including physical and chemical barriers, although these are not 100% effective
    B. Secondary defenses—a variety of innate defenses including complement activation, inflammation, fever, and phagocytosis
    C. Factors influencing host defenses include age, stress, nutritional deficiencies, and genetic background

VI. Microbial Mechanisms for Escaping Host Defenses
    A. Evasion of host defenses by viruses
        1. Antigenic drift—mutations cause change in antigenic sites on the virion (e.g., influenza virus)
        2. Infection of T cells (e.g., HIV)
        3. Fusion of host cells—allows spread from cell to cell without exposure to antibody-containing fluids (e.g., HIV, measles virus, cytomegalovirus)
        4. Infection of neurons having little or no MHC molecules (e.g., herpesvirus)
        5. Production and release of antigens that bind neutralizing antibodies—ties up neutralizing antibodies so there is insufficient antibody to bind complete viral particle (e.g., hepatitis B virus)
    B. Evasion of host defenses by bacteria
        1. Evading the complement system
            a. Capsules prevent complement activation
            b. Lengthened O chains in LPS prevent complement activation
            c. Serum resistance—features on surface of bacterium prevent formation of membrane-attack complex (e.g., *Neisseria gonorrhoeae*)
        2. Resisting phagocytosis
            a. Capsules
            b. Specialized proteins (e.g., M protein of *Streptococcus pyogenes*)
            c. Prevention of phagolysosome formation (e.g., *Chlamydia*)
            d. Production of leukocidins (e.g., staphylococci)
            e. Production of enzymes that destroy complement-derived chemoattractants for phagocytes
        3. Survival inside phagocytic cells—escape from phagosome before it merges with lysosome
        4. Evading the specific immune response
            a. Capsules that are not antigenic (e.g., *Streptococcus pyogenes*)
            b. Phase variation—alteration in antigens (e.g., *N. gonorrhoeae*)
            c. Production of IgA proteases (e.g., *N. gonorrhoeae*)
            d. Production of proteins that interfere with antibody-mediated opsonization (e.g., staphylococcal protein A)

TERMS AND DEFINITIONS

Place the letter of each term in the space next to the definition or description that best matches it.

_____ 1. An organism that supports the growth of a parasitic organism

_____ 2. A host on or in which a parasitic organism either attains sexual maturity or reproduces

_____ 3. A host that serves as a temporary but essential environment for certain stages of the development of a parasitic organism

_____ 4. A host that is used by a parasitic organism as a vehicle for reaching a final host

_____ 5. A host infected by a parasitic organism that can also infect humans

_____ 6. The colonization of a host by a parasitic organism

a. adhesins
b. antitoxin
c. bacteremia
d. endogenous pyrogen
e. endotoxin unit
f. endotoxins
g. exotoxins
h. final host
i. fomites
j. hemolysins
k. host
l. immunopathology
m. infection
n. infectious disease
o. infectivity
p. intermediate host
q. intoxications
r. invasiveness

336

_____ 7. A change from a state of health due to the presence of an organism or its products in the host

_____ 8. An organism that causes infectious disease

_____ 9. An organism that causes disease in a healthy host by a direct interaction

_____ 10. An organism that causes disease only when the host's resistance is impaired

_____ 11. The degree or intensity of pathogenicity

_____ 12. The ability of a pathogen to spread to adjacent or other tissues

_____ 13. The ability of a pathogen to establish a focal point of infection

_____ 14. The degree that a pathogen causes damage

_____ 15. A pathogen's ability to produce toxins

_____ 16. Damage to the host resulting from exaggerated immune responses to second or chronic exposures to a pathogen

_____ 17. Organisms (e.g., insects) that transfer pathogens from one host to another

_____ 18. A place where a pathogen lives before and after causing an infection

_____ 19. Inanimate objects that harbor and transmit pathogens

_____ 20. Molecules on the surface of a pathogen that mediate attachment to host cells or tissues

_____ 21. Products or structural components of a pathogen that contribute to its virulence or pathogenicity

_____ 22. The presence of viable bacteria in the bloodstream

_____ 23. Disease caused by the presence of pathogens or bacterial toxins in the blood

_____ 24. Diseases that result from the entry of preformed toxins into the body of the host

_____ 25. A substance, often a metabolic product of the organism, that alters the metabolism of host cells with deleterious effects on the host

_____ 26. The conditions caused by toxins that have entered the blood of the host

_____ 27. Soluble proteins that are released by a pathogen into the host; they may damage the host at some site remote from the site of infection

_____ 28. Antibodies capable of neutralizing toxins

s. leukocidins
t. LPS-binding proteins
u. opportunistic pathogen
v. pathogen
w pathogenic potential
x. primary (frank) pathogen
y. reservoir
z. reservoir host
aa. septicemia
bb. toxemia
cc. toxigenicity
dd. toxin
ee. toxoids
ff. transfer host
gg. vector
hh. virulence
ii. virulence factors

_____ 29. Chemically altered forms of toxins that are immunogenic but are no longer toxic

_____ 30. Toxins that kill leukocytes

_____ 31. Toxins that kill erythrocytes

_____ 32. Substances that are part of the cell wall of gram-negative bacteria and cause toxic reactions

_____ 33. A substance released by macrophages during infection that triggers fever

_____ 34. Proteins in the plasma membranes of macrophages and monocytes that bind endotoxin

_____ 35. A unit of measure used to indicate the level of endotoxin contamination in drugs, media, or other products.

FILL IN THE BLANK

1. By convention, an organism that causes harm or that lives at the expense of another organism is called a _____ organism; when just the term _____ is used, it refers to harmful protozoa and helminths.

2 The relationship between a harmful organism and its host is called _____. If the harmful organism lives on the surface of its host, it is called an _____; while if it lives within its host, it is called an _____.

3. Some pathogens release substances that kill leukocytes. These are called _____.

4. Many pathogens possess mechanical, chemical, or molecular abilities to damage the host and cause disease. Based on the way damage occurs, two distinct categories of disease can be recognized. If the disease is due to the pathogen's growth, metabolism, reproduction, or tissue alterations, it is called an _____ disease. However, if the disease results from the entry of a preformed toxin into the host, it is called an _____. The term _____ is used if specific symptoms are caused by toxins in the blood of the host.

5. Soluble proteins produced and released by an organism that can act at sites remote from the site of infection are called _____. They are divided into four types based on their structure and physiological activities. One type is the _____ toxin, which has two parts, one that binds the target cell and another that has toxic activity. A second type may also have this structure, but is distinguished by its specificity for certain host sites. Those toxins that target the nervous system are called _____; those that cause cell death are called _____; and those that damage the intestinal tract are referred to as _____. A third type does not have two different portions. These toxins target cell membranes and are called _____ _____. The fourth type is the superantigen that stimulates T cells to release cytokines.

6. Many bacteria have large segments of DNA, called _____ _____, that carry genes responsible for virulence. An example of virulence genes carried this way is a set of genes that code for proteins of the _____ _____ _____ _____. These proteins enable gram-negative bacteria to secrete and inject virulence proteins into the cytoplasm of host cells.

7. Viruses are usually spread throughout the body by the bloodstream and lymphatic system. The presence of viruses in blood is called _____. As viruses are spread, they specifically infect certain cells, tissues, and organs. These specificities are called _____.

8. The term _____ refers to the ability of an organism to cause disease. The term _____ refers to the intensity of this ability and it is measured experimentally by the _____ (the dose of pathogens that kills 50% of an experimental group of hosts) or by the _____ (the dose of pathogens that infects 50% of an experimental group of hosts.

9. A number of bacterial pathogens produce _____, which lyse red blood cells. This can be observed by culturing the bacteria on blood agar plates. A complete zone of clearing around a bacterial colony is called _____ _____, and a partial clearing is called _____ _____.

10. Some gram-negative bacteria secrete and inject _____ _____ into the cytoplasm of hosts using a protein complex called an _____ that is part of a _____ _____ secretion system.

MULTIPLE CHOICE

For each of the questions below select the *one best* answer.

1. Which of the following has no effect on the outcome of the host-parasite relationship?
   a. the number of parasites on or in the host
   b. the virulence of the parasite
   c. the defenses of the host
   d. All of the above affect the outcome of the host-parasite relationship.

2. What is an organism that transmits a parasitic organism from one host to another called?
   a. fomite
   b. vector
   c. transmitter
   d. carrier

3. Which of the following is a membrane-disrupting exotoxin?
   a. leukocidin
   b. streptolysin-O
   c. streptolysin-S
   d. phospholipase
   e. All of the above are membrane-disrupting exotoxins.

4. Which of the following is a factor that helps determine the virulence of a pathogen?
   a. invasiveness
   b. infectivity
   c. pathogenic potential
   d. All of the above are correct.

5. Which of the following is NOT a mode of action associated with endotoxins?
   a. shock
   b. paralysis
   c. diarrhea
   d. All of the above are modes of action associated with endotoxins.
6. When is endotoxin released from a gram-negative bacterium?
   a. when the bacterium lyses
   b. when the bacterium reproduces
   c. Both (a) and (b) are correct.
   d. Neither (a) nor (b) is correct.
7. *Neisseria gonorrhoeae* uses a mechanism called serum resistance to evade host defenses. Which host defense is evaded by this mechanism?
   a. formation of the membrane attack complex
   b. phagocytosis
   c. destruction within a phagosome
   d. the action of secretory IgA

8. Which of the following is a large segment of DNA that carries virulence genes acquired through gene transfer?
   a. fomite
   b. pathogenicity island
   c. peripheral transducer
   d. immune complex

TRUE/FALSE

_____ 1. Parasitic organisms are metabolically dependent on their hosts.
_____ 2. Some pathogens have mechanisms that allow them to penetrate into host epithelial cells, while others can enter only if the epithelial cells have been damaged by other means (wounds, insect bites, etc).
_____ 3. Generally, exotoxins tend to be more heat stable than endotoxins.
_____ 4. A pathogen is generally found in the area of the host's body that provides the most favorable conditions for its growth and reproduction.
_____ 5. Endotoxin effects can be mediated by cytokines released from monocytes and macrophages to which the endotoxin has bound.
_____ 6. Colonization specifically refers to the multiplication of a pathogen on or within a host.
_____ 7. For most pathogenic bacteria, many clonal types exist in the environment.
_____ 8. Genes for virulence are expressed constitutively.
_____ 9. Exotoxins can cause disease only after the bacterium colonizes its host.

TYPES OF INFECTIONS ASSOCIATED WITH PARASITIC ORGANISMS

Using your own words, complete the following table by describing each type of infection.

Type of Infection	Description
Abscess	
Acute	
Bacteremia	
Chronic	
Covert	
Cross	
Focal	
Fulminating	
Iatrogenic	
Latent	
Localized	
Mixed	
Nosocomial	
Opportunistic	
Overt	
Phytogenic	
Primary	
Pyogenic	
Secondary	
Sepsis	
Septicemia	
Septic shock	
Sever sepsis	
Sporadic	
Subclinical (unapparent or covert)	
Systemic	
Toxemia	
Zoonosis	

CRITICAL THINKING

1. Compare and contrast exotoxins and endotoxins. Discuss the chemical and physiological characteristics of the molecules, as well as their mechanisms of pathogenesis.

2. Compare and contrast viral and bacterial pathogenesis. For each, give specific examples of mechanisms by which they colonize, invade, do harm, and evade host defenses.

ANSWER KEY

Terms and Definitions

1. k, 2. h, 3. p, 4. ff, 5. z, 6. m, 7. n, 8. v, 9. x, 10. u, 11. hh, 12. r, 13. o, 14. w, 15. cc, 16. l, 17. gg, 18. y, 19. i, 20. a, 21. ii, 22. c, 23. aa, 24. q, 25. dd, 26. bb, 27. g, 28. b, 29. ee, 30. s, 31. j, 32. f, 33. d, 34. t, 35. e

Fill in the Blank

1. parasitic; parasite 2. parasitism; ectoparasite; endoparasite 3. leukocidins 4. infectious; intoxication; toxemia
5. exotoxins; AB; neurotoxins; cytotoxins; enterotoxins; membrane-disrupting exotoxins
6. pathogenicity islands; type III secretion system 7. viremia; tropisms 8. pathogenicity; virulence; LD_{50}; ID_{50}
9. hemolysins; beta hemolysis; alpha hemolysis 10. virulence proteins; injectosome; type III

Multiple Choice

1. d, 2. b, 3. e, 4. d, 5. b, 6. c, 7. a, 8. b

True/False
1. T, 2. T, 3. F, 4. T, 5. T, 6. T, 7. F, 8. F, 9. F

34 Antimicrobial Chemotherapy

CHAPTER OVERVIEW

The control or the destruction of microorganisms that reside within the bodies of humans and other animals is of tremendous importance. This chapter introduces the principles of chemotherapy and discusses the ideal characteristics for successful chemotherapeutic agents (including the concept of selectively damaging the target microorganism while minimizing damage to the host). The chapter also presents characteristics of some commonly used antibacterial, antifungal, antiviral, and antiprotozoan drugs.

CHAPTER OBJECTIVES

After reading this chapter you should be able to:

- discuss the various ways in which antimicrobial agents can damage pathogens while causing minimal damage to the host
- discuss the various factors that influence the effectiveness of a chemotherapeutic agent
- discuss the increasingly serious problem of drug-resistant pathogens
- discuss the modes of action and selectivity of antifungal, antiviral, and antiprotozoan agents

CHAPTER OUTLINE

I.   Introduction
    A.   Chemotherapeutic agents are chemical agents used to treat disease
    B.   Antibiotics are microbial products or their derivatives that kill or inhibit susceptible microorganisms
    C.   Synthetics—drugs that are not microbially synthesized
II. The Development of Chemotherapy
    A.   Paul Ehrlich (1904–1909) —aniline dyes and arsenic compounds
    B.   Gerhard Domagk, and Jacques and Therese Trefouel (1939) —sulfanilamide
    C.   Ernest Duchesne (1896) discovered penicillin; however, this discovery was not followed up and was lost for 50 years
    D.   Alexander Fleming (1928) accidentally discovered the antimicrobial activity of penicillin on a contaminated plate; however, follow-up studies convinced him that penicillin would not remain active in the body long enough to be effective
    E.   Howard Florey and Ernst Chain (1939) aided by the biochemist, Norman Heatley, worked from Fleming's published observations, obtained a culture from him, and demonstrated the effectiveness of penicillin
    F.   Selman Waksman (1944) —streptomycin; this success led to a worldwide search for additional antibiotics, and the field has progressed rapidly since then
III. General Characteristics of Antimicrobial Drugs
    A.   Selective toxicity—ability to kill or inhibit microbial pathogen with minimal side effects in the host
        1.   Therapeutic dose—the drug level required for clinical treatment of a particular infection
        2.   Toxic dose—the drug level at which the agent becomes too toxic for the host (produces undesirable side effects)
        3.   Therapeutic index—the ratio of toxic dose to therapeutic dose: the larger the better
    B.   Chemotherapeutic agents can occur naturally, be synthetic, or semisynthetic (chemical modifications of naturally occurring antibiotics)
    C.   Drugs with narrow spectrum activity are effective against a limited variety of pathogens; drugs with broad-spectrum activity are effective against a wide variety of pathogens
    D.   Drug can be cidal (able to kill) or static (able to reversibly inhibit growth)

E. Minimal inhibitory concentration (MIC) is the lowest concentration of the drug that prevents growth of a pathogen; minimal lethal concentration (MLC) is the lowest drug concentration that kills the pathogen

IV. Determining the Level of Antimicrobial Activity
   A. Dilution susceptibility tests—a set of broth-containing tubes are prepared; each tube in the set has a specific antibiotic concentration; to each is added a standard number of test organisms
      1. The lowest concentration of the antibiotic resulting in no microbial growth is the MIC
      2. Tubes showing no growth are subcultured into tubes of fresh medium lacking the antibiotic to determine the lowest concentration of the drug from which the organism does not recover; this is the MLC
   B. Disk diffusion tests
      1. Disks impregnated with specific drugs are placed on agar plates inoculated with the test organism; clear zones (no growth) will be observed if the organism is sensitive to the drug; the size of the clear zone is used to determine the relative sensitivity; zone width also is a function of initial concentration, solubility, and diffusion rate of the antibiotic
      2. Kirby-Bauer method is most commonly used disk diffusion test; test results are determined using tables that relate zone diameter to the degree of sensitivity
   C. The Etest
      1. Especially useful for testing anaerobic microorganisms
      2. Makes use of special plastic strips that contain a concentration gradient of an antibiotic; each strip is labeled with a scale of MIC values; after incubation an elliptical zone of inhibition is observed and its intersection with the strip is used to determine the MIC
   D. Measurement of drug concentrations in the blood can be done using microbiological, chemical, immunological, enzymatic, and/or chromatographic assays

V. Antibacterial Drugs
   A. Inhibitors of cell wall synthesis are effective because bacterial cell walls have unique structures not found in eucaryotic cells
      1. Penicillins—inhibit cell wall synthesis; many types have been identified or synthesized including ampicillin, carbenicillin, and methicillin; they differ in spectrum of activity and administration route but all have a β-lactam ring that is crucial for activity; resistance is an increasing problem, often due to penicillinase; some patients are allergic to these antibiotics
      2. Cephalosporins—inhibit cell wall synthesis; broad spectrum of activity; they contain a β-lactam ring, but are not subject to degradation by penicillinase; they can be given to some patients with penicillin allergies; they include cephalothin, cefoxitin, and ceftriazone
      3. Vancomycin and teicoplanin—glycopeptide antibiotics that block peptidoglycan synthesis; vancomycin is particularly important as the last line of defense against antibiotic-resistant staphylococcal and enterococcal infections
   B. Protein synthesis inhibitors exploit the differences between procaryotic and eucaryotic ribosomes
      1. Aminoglycosides—contain cyclohexane ring and amino sugars; includes kanamycin, streptomycin, neomycin, and gentamycin; can be quite toxic to patients
      2. Tetracyclines—contain a four-ring structure with side chains; very broad spectrum that includes intracellular parasites and mycoplasmas
      3. Macrolides—12- to 22-carbon lactone rings linked to sugars; broad spectrum similar to that of penicillin; includes erythromycin
      4. Chloramphenicol—has a broad spectrum but is quite toxic
   C. Metabolic antagonists are structural analogs of metabolic intermediates that act as antimetabolites, inhibiting metabolic pathways
      1. Sulfonamides or sulfa drugs—inhibit folic acid synthesis in bacteria (humans don't synthesize folic acid, so are not affected); resistance is increasing and many patients are allergic to these drugs; includes p-aminobenzoic acid (PABA)
      2. Trimethoprim-synthetic antibiotic that blocks folic acid production; broad spectrum often combined with sulfa drugs
   D. Nucleic acid synthesis inhibitors block enzymes of transcription and translation; generally not as selectively toxic

343

1. Quinolones—synthetic drugs that inhibit bacterial DNA gyrase or topoisomerase II, thereby disrupting replication, repair, and other processes involving DNA; broad spectrum; includes nalidixic acid and ciprofloxacin (Cipro)

VI. Factors Influencing the Effectiveness of Antimicrobial Drugs
    A. Drug's ability to reach the site of infection—this is greatly influenced by the mode of administration (e.g., oral, topical, parenteral), but can also be influenced by exclusion from the site of infections (e.g., blood clots or necrotic tissue protects bacterium)
    B. Susceptibility of pathogen—influenced by growth rate and by inherent properties (e.g., whether or not pathogen has target of the drug)
    C. Factors influencing drug concentration in the body—must exceed the pathogen's MIC for the drug to be effective; this will depend on the amount of drug administered, the route of administration, the speed of uptake, and the rate of clearance (elimination) from the body
    D. Drug resistance has become an increasing problem

VII. Drug Resistance
    A. Mechanisms of drug resistance
       1. Prevent entrance of drug (e.g., alter drug transport into cell)
       2. Pump the drug out of the cell once it has entered
       3. Enzymatic inactivation of the drug—chemical modification of the drug by cellular enzymes can render it inactive before it has a chance to damage the cell
       4. Alteration of target enzyme or organelle—modification of the target so that it is no longer susceptible to the action of the drug
       5. Use of alternative pathways and increased production of the target metabolite have been used by some organisms to minimize the effects of the drug
    B. The origin and transmission of drug resistance
       1. Spontaneous mutations in chromosomal genes; these are then inherited by progeny of the resistant mutant
       2. Transfer of R plasmids
       3. Other genetic elements can carry one or more resistance genes
          a. Composite transposons (e.g., Tn5, Tn9, Tn10, Tn21, Tn551, and Tn4001)
          b. Integrons—genetic elements that contain a site into which genes can be inserted and an integrase gene that allows for incorporation of the integron into the bacterial chromosome; gene cassettes (a unit of genetic material that contains a set of resistance genes; gene cassettes are usually a linear part of a transposon, plasmid, or bacterial chromosome, and they are able to move resistance genes from one recombination site to another) can be captured by integrons and spread to other sites and organisms
          c. Conjugative transposons
       4. Several strategies can be used to discourage emergence of drug resistance (e.g., administration of high doses, simultaneous treatment with two different drugs, limited use of broad-spectrum antibiotics)
       5. Drug resistance has become an increasing problem; new drugs are constantly being developed and new treatment methods (e.g., phage treatment of bacterial infections) are being explored

VIII. Antifungal Drugs
    A. Fungal infections are more difficult to treat than bacterial infections because the greater similarity of fungi and host limits the ability of a drug to have a selective point of attack; furthermore, many fungi have detoxification systems that inactivate drugs
    B. Superficial mycoses are infections of superficial tissues and can often be treated by topical application of antifungal drugs such as miconazole, nystatin, and griseofulvin, thereby minimizing systemic side effects
    C. Systemic mycoses are more difficult to treat and can be fatal; amphotericin B and flucytosine have been used with limited success; amphotericin B is highly toxic and must be used with care; flucytosine must be converted by the fungus to an active form, and animal cells are incapable of this; some selectivity is possible, but severe side effects have been observed with both drugs

IX. Antiviral Drugs
    A. Selectivity is a problem because viruses use the metabolic machinery of the host

B. Antiviral drugs target specific steps of life cycle, including viral uncoating and DNA replication (e.g., amantadine, vidarabine, acyclovir, cidofovir, and azidothymidine)
C. Anti-HIV drugs target reverse transcriptase (e.g., AZT) and viral polypeptide processing (the protease inhibitors)
D. Tamiflu is a neuraminidase inhibitor that is used to treat influenza

X. Antiprotozoan drugs
A. Mechanisms of action for antiprotozoan drugs are largely unknown; as protozoans and humans are both eucaryotes, selective toxicity is difficult to achieve
B. Chloroquine and mefloquine—used to treat malaria; variety of mechanisms proposed
C. Metronidazole—used to treat *Entamoeba* infections; appears to interact with DNA
D. Atovaquone—an analog of ubiquinone that interferes with electron transport chain

TERMS AND DEFINITIONS

Place the letter of each term in the space next to the definition or description that best matches it.

_____ 1. The ratio of therapeutic dose to toxic dose
_____ 2. Compounds used in the treatment of disease that kill or prevent the growth of microorganisms at concentrations low enough to avoid undesirable damage to the host
_____ 3. Chemotherapeutic agents that are natural products of microorganisms
_____ 4. Activities of a chemotherapeutic agent that damage the host either by inhibiting the same process in the host as in the target organism or by damaging other processes
_____ 5. Describes an antibiotic that attacks many different pathogens
_____ 6. Describes an antibiotic that is effective against only a limited variety of pathogens
_____ 7. Susceptibility tests that involve the inoculation of a set of dilutions of an antimicrobial agent with a test microorganism
_____ 8. Term used to describe chemotherapeutic agents that reversibly inhibit the growth of microorganisms
_____ 9. Term used to describe chemotherapeutic agents that kill microorganisms
_____ 10. Lowest concentration of a drug necessary to prevent the growth of a particular microorganism
_____ 11. Lowest concentration of a drug necessary to kill a particular microorganism
_____ 12. An immune system chemical that inhibits virus replication
_____ 13. Drugs that block the function of metabolic pathways
_____ 14. Small circular DNA molecules that can exist separately from the chromosome or be integrated into it
_____ 15. Plasmids that bear one or more resistance genes
_____ 16. A set of resistance genes that can be captured by an integron and transferred as a unit
_____ 17. A genetic element that has a site at which genes can be inserted and a gene for integrase, which allows the element to insert into other DNA molecules

a. antibiotics
b. antimetabolites
c. broad-spectrum drug
d. chemotherapeutic agents
e. cidal
f. dilution susceptibility tests
g. gene cassette
h. integron
i. interferon
j. minimal inhibitory concentration
k. minimal lethal concentration
l. narrow-spectrum drug
m. plasmids
n. R plasmids
o. side effects
p. static
q. therapeutic index

DRUGS AND THEIR DISCOVERERS

Match the following scientists with their discoveries.

_____ 1. Use of the arsenic compound Salvarsan as a treatment for syphilis

_____ 2. Use of Prontosil Red (sulfanilamide) as a treatment for streptococcal and staphylococcal infections

_____ 3. First discovered penicillin, but discovery was lost

_____ 4. Rediscovered penicillin, but did not pursue the significance of it

_____ 5. Co-discoverer of the therapeutic value of penicillin

_____ 6. Co-discoverer of the therapeutic value of penicillin

_____ 7. Discovered streptomycin and stimulated intense search for other antibiotics

_____ 8. Biochemist who helped with work demonstrating the therapeutic value of penicillin

a. Chain
b. Domagk
c. Duchesne
d. Ehrlich
e. Fleming
f. Florey
g. Heatley
h. Waksman

IMPORTANT CHEMOTHERAPEUTIC AGENTS

For each of the drugs below, provide the requested information.

Chemotherapeutic Agent	Structural Features	Mechanism of Action	Cidal/Static	Targeted Organism(s)
Acyclovir				
Adenine arabinoside (vidarabine)				
Amantadine				
Amphotericin B				
Aminoglycosides (e.g., streptomycin)				
AZT (zidovudine)				
Cephalosporins				
Chloramphenicol				
Chloroquine				
Griseofulvin				
Macrolides (e.g., erythromycin)				
Nystatin				
Penicillin				
Quinolones				
Sulfonamides				
Tetracyclines				
Vancomycin				

FILL IN THE BLANK

1. The _____ _____ is the ratio of the therapeutic dose to the toxic dose. The larger this ratio, the greater the _____ _____ and the less likely it is that the chemotherapeutic agent will cause _____ _____.

2. A number of useful drugs act as _____; they block the functioning of metabolic pathways by inhibiting key enzymes.

3. Antibiotics that are taken by a route other than by mouth are said to have a _____ route of administration.

4. The discovery of HIV and AIDS has heightened the urgency for discovering antiviral chemotherapeutic agents. One of the first anti-HIV agents was _____, which is an inhibitor of HIV reverse transcriptase (RT). This drug is a _____ _____ of thymine nucleotides and therefore blocks activity of RT and replication of HIV. Another HIV protein, HIV protease, also is a good target for therapy. Today several _____ _____ _____ have been developed, including saquinvir, indinavir, and ritonavir.

MULTIPLE CHOICE

For each of the questions below select the *one best* answer.

1. The most selective antibiotics are those that interfere with bacterial cell wall synthesis. Why is this?
   a. because bacterial cell walls have a unique structure not found in eucaryotic cells
   b. because bacterial cell wall synthesis is easy to inhibit, while animal cell wall synthesis is more resistant to the actions of the drugs
   c. because animal cells do not take up the drugs
   d. because animal cells inactivate the drugs before they can do any damage

2. Which of the following will NOT have an effect on the concentration achieved in the blood by a particular antibiotic?
   a. the route of administration
   b. the speed of uptake from the site of administration
   c. the rate at which the drug is eliminated from the body
   d. All of the above will affect the concentration of the drug in the blood.

3. What is penicillinase?
   a. an enzyme that modifies penicillins, making them more potent
   b. an enzyme that cleaves the beta-lactam ring of penicillin, rendering it inactive
   c. a semi-synthetic form of penicillin
   d. none of the above

4. Which of the following is NOT a common mechanism by which microorganisms develop drug resistance?
   a. enzymatic inactivation of the drug
   b. exclusion of the drug from the cell
   c. use of an alternative pathway to bypass the drug-sensitive pathway
   d. All of the above are common mechanisms by which microorganisms develop drug resistance.

5. Which of the following is used to discourage the development of drug resistance?
   a. sufficiently high drug doses to destroy any resistant mutants that may have arisen spontaneously
   b. use of two drugs simultaneously with the hope that each will prevent the emergence of resistance to the other
   c. avoidance of indiscriminate use of drugs
   d. All of the above are used to discourage development of drug resistance.

6. What is the drug level required for the clinical treatment of a particular infection called?
   a. therapeutic dose
   b. toxic dose
   c. therapeutic index
   d. None of the above is correct.

7. Which of the following is NOT a reason that treatment of fungal infections generally has been less successful than treatment of bacterial infections?
   a. Fungi use the metabolic machinery of the host and therefore cannot be selectively attacked.
   b. Fungi are more similar to human cells than are bacteria, and many drugs that inhibit or kill fungi are toxic for humans.
   c. Fungi have detoxifying systems that rapidly inactivate many drugs.
   d. All of the above are reasons that treatment of fungal infections has been less successful than treatment of bacterial infections.

8. Which of the following affects the size of the clear zone in a Kirby-Bauer test?
   a. the initial concentration of the drug
   b. the solubility of the drug
   c. the diffusion rate of the drug
   d. All of the above are correct.

9. For which organisms is antibiotic resistance becoming a major problem?
   a. bacterial diseases
   b. fungal diseases
   c. both (a) and (b)
   d. neither (a) nor (b)

10. Which of the following is not an antiprotozoan drug?
   a. chloroquine
   b. metronidazole
   c. pentamidine
   d. amantadine

TRUE/FALSE

_____ 1. A drug that disrupts a microbial function not found in animal cells usually has a lower therapeutic index.

_____ 2. Static agents do not kill infectious organisms and therefore are not useful as chemotherapeutic agents.

_____ 3. Protein synthesis inhibitors have a high therapeutic index because they can usually discriminate between procaryotic and eucaryotic ribosomes; however, their therapeutic index is not as high as that of cell wall synthesis inhibitors.

_____ 4. Sulfonamides and other drugs that inhibit folic acid synthesis have a high therapeutic index because humans must obtain folic acid in their diets while microorganisms synthesize their own.

_____ 5. The fungus *Candida albicans* is normally present in various parts of the body and can cause problems (superinfection) when bacterial competition is eliminated by antibiotic treatment.

_____ 6. Isoniazid is a narrow-spectrum antibiotic. However, it is considered useful because it is one of the few drugs that are effective against tuberculosis.

_____ 7. Drugs with highly toxic side effects are usually used only in life-threatening situations where suitable alternatives are not available.

_____ 8. There are few effective antiviral drugs because viruses use the metabolic machinery of their hosts, making it difficult to identify a selective point of attack.

_____ 9. One way in which organisms may exhibit resistance to a drug is to pump the drug out of the cell immediately after it has entered.

_____ 10. One approach to limiting the development of drug resistance is to use multiple drugs simultaneously at high doses.

_____ 11. One of the most serious threats to the successful treatment of disease is the spread of drug-resistant pathogens.

CRITICAL THINKING

1. Antibiotics are natural products of certain microorganisms. What advantages might these antibiotics provide for the organisms that produce them?

2. Viruses generally use the metabolic machinery of their hosts. Therefore, they should present no selective point of attack for potential antiviral drugs. Yet, recently there have been several antiviral drugs developed that have a reasonably high therapeutic index. Explain.

ANSWER KEY

Terms and Definitions

1. q, 2. d, 3. a, 4. o, 5. c, 6. l, 7. f, 8. p, 9. e, 10. j, 11. k, 12. i, 13. b, 14. m, 15. n, 16. g, 17. h

Drugs and Their Discoverers

1. d, 2. b, 3. c, 4. e, 5. f, 6. a, 7. h, 8. g

Fill in the Blank

1. therapeutic index; selective toxicity; side effects 2. antimetabolites 3. parenteral 4. AZT; structural analog; HIV protease inhibitors

Multiple Choice

1. a, 2. d, 3. b, 4. d, 5. d, 6. a, 7. a, 8. d, 9. c, 10. d

True/False

1. F, 2. F, 3. T, 4. T, 5. T, 6. T, 7. T, 8. T, 9. T, 10. T, 11.T

35 Clinical Microbiology and Immunology

CHAPTER OVERVIEW

This chapter describes the field of clinical microbiology, which is concerned with the detection and identification of pathogens that are the etiological agents of infectious disease. Identification may be based on the results of some combination of morphological, physiological, biochemical, and immunological procedures. Time may be critical in life-threatening situations. Therefore, rapid identification systems and computers can be used to greatly speed up the process. The chapter closes with a discussion of *in vitro* antigen-antibody interactions. These are particularly useful in diagnostic procedures.

CHAPTER OBJECTIVES

After reading this chapter you should be able to:

- describe the functions and/or services performed by clinical microbiology laboratories
- discuss the need for proper specimen selection, collection, handling, and processing
- discuss the types of procedures used to identify microorganisms in specimens
- discuss the use and advantages of computers in clinical microbiology laboratories
- describe the various *in vitro* antigen-antibody interactions, and give examples of diagnostic tests based on them

CHAPTER OUTLINE

I. Specimens
    A. Clinical microbiologists are microbiologists whose main function is to isolate and identify microorganisms from clinical specimens, and to do so as rapidly as possible
    B. Specimen—human material that is tested, examined, or studied to determine the presence or absence of specific microorganisms
        1. Because of safety concerns, specimens must be handled carefully; universal safety precautions have been recommended by the CDC to address safety issues in specimen handling
        2. Specimens should be:
            a. Representative of the diseased area
            b. Adequate in quantity for a variety of diagnostic tests
            c. Devoid of contamination, particularly by microorganisms indigenous to the skin and mucous membranes
            d. Forwarded promptly to the clinical laboratory
            e. Obtained prior to the administration of any antimicrobials
    C. Collection
        1. Sterile swabs—used to collect specimens from skin and mucous membranes; associated with greater risk of contamination and have limited volume capacity, so their use is generally discouraged
        2. Needle aspiration—used to collect blood and cerebrospinal fluid; skin surface microorganisms must be excluded by the use of stringent antiseptic techniques; anticoagulants are used to prevent blood clotting
        3. Intubation—used to collect specimens from stomach
        4. Catheterization—used to collect urine

     5.   Clean-catch midstream urine—first urine voided is not collected because it is likely to be contaminated with surface organisms

     6.   Sputum—mucous secretions expectorated from the lungs, bronchi, and/or trachea

  D.   Handling—includes any special additives (e.g., anticoagulants) and proper labeling

  E.   Transport—should be timely; temperature control may be needed; supplementation may be needed to support microbial survival or to inhibit normal microbial flora; special treatment may be needed for anaerobes

II.   Identification of Microorganisms from Specimens

  A.   Microscopy

     1.   Direct examination of specimen, or examination of specimen after various staining procedures that are specific for types of bacteria, fungi, or protists

     2.   Hybridomas produce monoclonal antibodies specific for a single epitope; these can be fluorescently labeled and used for rapid, accurate culture confirmation

     3.   Immunofluorescence satins specimens with fluorescent dyes that emit visible colored light when excited

        a.   Direct immunofluorescence—after fixation to a slide, the specimen is stained with a fluorescently labeled antibody directed at cell surface antigens

        b.   Indirect immunofluorescence—after a known antigen is attached to a slide, a patient's serum (antiserum) is applied; if antibodies specific for the antigen are present, these can be detected with a fluorescently labeled secondary antibody

  B.   Growth and biochemical characteristics

     1.   Viruses—identified by isolation in cell (tissue) culture, by immunodiagnosis, and by molecular detection

        a.   Viral cultivation

            1)   Cell cultures—viruses are detected by cytopathic effects (observable morphological changes in host cells) or by hemadsorption (binding of red blood cells to infected cells)

            2)   Embryonated eggs—virus can be inoculated into allantoic cavity, amniotic cavity, or the chorioallantoic cavity; virus is detected by development of pocks on the chorioallantoic membrane, by development of hemagglutinins in the allantoic and amniotic fluid, and by death of the embryo

            3)   Laboratory animals (e.g., suckling mice)—observed for signs of disease or death

        b.   Serological tests (e.g., monoclonal antibody-based immunofluorescence) can be used to detect virus in tissue-vial cultures

     2.   Fungi

        a.   Direct microscopic examination with fluorescent dyes

        b.   Examination of cultures

        c.   Serological tests for antifungal antibodies

        d.   Yeast can be identified by the use of rapid ID methods

     3.   Parasites—identified by directly examining specimens for eggs, cysts, larvae, or vegetative cells; some serological tests are available

     4.   Bacteria (other than rickettsias, chlamydiae, and mycoplasmas)

        a.   Isolation and growth of bacteria are required before many diagnostic tests can be used

        b.   Initial identity may be suggested by source of specimen; microscopic appearance and Gram reaction; pattern of growth on selective, differential, and other media; and by hemolytic, metabolic, and fermentative properties

        c.   After pure cultures are obtained, specific biochemical tests can be done and a dichotomous key used for identification

     5.   Rickettsias—identified by immunoassays or by isolation (the latter can be hazardous)

     6.   Chlamydiae—identified by Giemsa staining, immunofluorescent staining of tissues with anti-chlamydia monoclonal antibodies, DNA sequencing, and PCR

     7.   Mycoplasmas—identified immunologically or by the use of DNA probes

  C.   Rapid methods of identification

     1.   Manual biochemical systems such as the API 20E system for enterobacteria

              a.   Consists of 20 microtube inoculation tests

              b.   Results are converted to a seven- or nine-digit profile number

              c.   The number is compared to the *API Profile Index* to determine the name of the bacterium

       2.   Biosensors can be based on microfluidic antigen sensors, rapid PCR, sensitive spectroscopy systems, and liquid crystal amplification of immune complexes

       3.   Monoclonal antibodies are available for the detection of many microorganisms; the antibodies can be specific to the species or strain level

    D.   Bacteriophage typing—based on the fact that host range specificities of bacteriophages are dependent upon surface receptors on the bacteria

    E.   Molecular methods and analysis of metabolic products

       1.   Some of these procedures have been discussed (e.g., protein comparisons, enzyme characterizations, nucleic acid-base composition, nucleic acid hybridization, and nucleic acid sequencing)

       2.   Nucleic acid-based detection methods—ssDNA molecules that have been cloned from organism or prepared by PCR technology can be used in hybridization procedures; rRNA genes can be sequenced to identify bacterial strains (ribotyping)

       3.   Gas-liquid chromatography (GLC)used to identify specific microbial metabolites, cellular fatty acids, and products of pyrolysis; usually for nonpolar substances that are extractable in ether

       4.   Plasmid fingerprinting—separation and detection of the number, molecular weight, and restriction patterns of different plasmids, that are often consistently present in a strain of bacteria

III. Clinical Immunology

    A.   Detection of antigens and antibodies may be valuable diagnostically; interpretation of immunologic test results can be difficult

    B.   Serotyping—antigen-antibody specificity is used to differentiate among various strains (serovars) of an organism found in serum samples

    C.   Agglutination—visible clumps or aggregates of cells or of coated latex microspheres; if red blood cells are agglutinated, the reaction is called hemagglutination

       1.   Widal Test—direct agglutination test for diagnosing typhoid fever

       2.   Latex agglutination tests are used in pregnancy tests and to diagnose mycotic, helminthic, and bacterial infections

       3.   Viral hemagglutination inhibition tests are used to diagnose influenza and other viral infections

       4.   Agglutination tests can be used to measure antibody titer (the reciprocal of the greatest dilution showing agglutination reaction)

    D.   Complement fixation—used to detect the presence of serum antibodies to a pathogen; currently used to diagnose certain viral, fungal, rickettsial, chlamydial, and protozoan diseases

    E.   Enzyme-linked immunosorbent assay (ELISA)—involves linking enzymes to an antibody

       1.   Double antibody sandwich assay—detects antigens in a sample

           a.   Wells of a microtiter plate are coated with antibody specific to the antigen of interest

           b.   Test sample is placed in well; if it contains the antigen of interest, the antigen will be retained in the well after washing

           c.   Second antibody is added; it is conjugated to an enzyme and is specific to the antigen; the second antibody will be retained in the well after washing if the antigen was retained in the previous step

           d.   Substrate of enzyme is added; reaction only occurs if conjugated enzyme (and therefore antigen) is present in the well; produces a colored product that can be detected

       2.   Indirect immunosorbent assay—detects serum antibody

           a.   Well of a microtiter plate is coated with antigen specific to the antibody of interest

           b.   Test serum is added; if antibodies are present, they will bind antigen and will be retained after washing

           c.   An antibody against the test immunoglobulin is added; the second antibody is conjugated to an enzyme and will only be retained in the well after washing if the test antibody is present in the well

        d.   Substrate of the enzyme is added; reaction only occurs if conjugated antibody (and
             therefore test antibody) are present in the well; the colored product of the reaction can be
             detected spectrophotometrically
    F.   Immunoblotting (Western Blot)—proteins are separated by electrophoresis, blotted to nitrocellulose
         sheets, then treated with solution containing enzyme-tagged antibodies
    G.   Immunoprecipitation—soluble antigens form insoluble immune complexes that can be detected
    H.   Immunodiffusion—involves the precipitation of immune complexes in an agar gel
        1.   Single radial immunodiffusion (RID) assay is quantitative
        2.   Double diffusion assay (Ouchterlony technique)—lines of precipitation form where antibodies
             and antigens have diffused and met; determines whether antigens share identical determinants
    I.   Immunoelectrophoresis—antigens are first separated by electrophoresis according to charge, and
         are then visualized by the precipitation reaction; greater resolution than diffusion assay
    J.   Flow cytometry and fluorescence
        1.   Detects single or multiple microorganisms on the basis of a cytometric parameter or by means
             of fluorochromes
        2.   Flow cytometer forces cells through a laser beam and measures light scatter or fluorescence as
             the cells pass through the beam; cells can be tagged with fluorescent antibody directed against
             specific surface antigen
    K.   Radioimmunoassay (RIA)—purified antigen labeled with a radioisotope competes with unlabeled
         antigen sample for antibody binding
IV. Susceptibility Testing
    A.   Thought by many to be the most important testing done
    B.   Used to help physician decide which drug(s) and which dosage(s) to use
V. Computers in Clinical Microbiology
    A.   Test ordering—specific requests, patient data, and accession number
    B.   Result entry, often using PDAs
    C.   Report printing—flexible format to meet the needs of physician
    D.   Laboratory management
    E.   Interfaced with automated instruments

TERMS AND DEFINITIONS

Place the letter of each term in the space next to the definition or description that best matches it.

_____ 1.   A dacron-tipped polystyrene applicator used for obtaining specimens from skin or mucous
           membranes
_____ 2.   The insertion of a tube into a body canal or hollow organ
_____ 3.   A tubular instrument used for withdrawing or introducing fluids from or into a body cavity
_____ 4.   Mucous secretions expectorated from lungs, bronchi, and trachea
_____ 5.   An observable change that occurs in cells as a result of viral infection
_____ 6.   Phenomenon that occurs because of an alteration of the membrane of a virus-infected cell so that red
           blood cells will adhere to it
_____ 7.   A method of identifying bacteria based on the bacteriophage that infect them
_____ 8.   A method of strain typing that involves preparing a Southern Blot of chromosomal DNA cleaved
           with restriction endonucleases, and then probing the blot with specific rRNA probes
_____ 9.   Visible aggregates or clumps formed by agglutination reactions
_____ 10. An agglutination reaction-based test that is used to diagnose typhoid fever
_____ 11. The agglutination of red blood cells by viruses
_____ 12. A colorless substrate that is acted on by an enzyme to produce a colored end product
_____ 13. A process in which a suspension of cells is forced through a laser beam; it can be used to detect,
           count, separate, and characterize cells in the suspension

_____ 14. A process wherein a fluorescent dye is attached to an antibody and used to detect the antigen specific for that antibody

_____ 15. The use of in vitro antibody-antigen reactions to differentiate strains of microorganisms

     a.   agglutinates
     b.   bacteriophage typing
     c.   catheter
     d.   chromogen
     e.   cytopathic effect
     f.   flow cytometry
     g.   hemadsorption
     h.   immunofluorescence
     i.   intubation
     j.   ribotyping
     k.   serotyping
     l.   sputum
     m.   swab
     n.   viral hemagglutination
     o.   Widal test

FILL IN THE BLANK

1. The major concern of the _____ _____ is to rapidly isolate and identify microorganisms from clinical specimens.

2. Viral replication in cell cultures is detected in two ways: by the observation of _____ _____ (observable changes in cell morphology) and by _____ (alterations in the plasma membrane that enable red blood cells to adhere firmly to virus-infected cells).

3. One of the most widely used serological tests is the _____. It involves the linkage of an antibody molecule to an enzyme whose activity can be detected by the formation of a colored product.

4. When a precipitation reaction occurs in an agar gel, it is called _____. One assay based on this phenomenon is the _____ _____ _____ assay. This assay is based on the diffusion of antigen out of a well into agar containing an antibody. In another assay, the _____ _____ _____ assay, also known as the _____ _____, antigens are placed in a set of wells in agar and antibodies are placed in another well. Both antibodies and antigens diffuse into the agar.

5. Serological procedures used to differentiate different strains of microorganisms are called _____. An important example of this is the _____ _____, which is used to classify streptococci based on the antigenic nature their cell walls. One way of detecting these differences is the _____ _____, in which mixing antiserum with a solution of streptococci causes capsular swelling if the antiserum is specific for that particular serovar.

6. Some mixtures of antigens are very complex, making it difficult to detect a particular antigen. In such cases, _____ is useful. In this process, antigens are first separated based on their electrical charge and then are visualized by precipitation reaction.

MULTIPLE CHOICE

For each of the questions below select the *one best* answer.

1. Which of the following is used to collect a urine sample only under very limited conditions?
   a. clean-catch midstream
   b. needle aspiration
   c. catheterization
   d. intubation

2. Which of the following should be written or imprinted on the culture request form?
   a. patient information
   b. type or source of sample
   c. physician's choice of tests
   d. All of the above should be on the culture request form.

3. Which of the following is NOT normally used in the identification of microorganisms?
   a. microscopic examination
   b. growth or biochemical characteristics
   c. immunological techniques that detect antibodies or microbial antigens
   d. All of the above are used in the identification of microorganisms.

4. Which of the following is NOT normally used to culture viruses?
   a. growth on artificial media
   b. growth in cell cultures
   c. growth in embryonated hen's eggs
   d. growth in whole animals

5. Which of the following is normally used to detect spirochetes in skin lesions in early syphilis?
   a. bright-field microscopy
   b. phase-contrast microscopy
   c. dark-field microscopy
   d. immunofluorescence microscopy

6. Which of the following is NOT likely to be useful in the identification of bacteria?
   a. source of the culture specimen
   b. growth patterns on selective and differential media
   c. hemolytic, metabolic, and fermentative properties
   d. All of the above are useful in the identification of bacteria.

7. What is the basic principle underlying plasmid fingerprinting?
   a. Microbial isolates of the same strain contain the same number of plasmids with the same molecular weights.
   b. Microbial isolates of different strains have different plasmids (either in number, molecular weight, or both).
   c. Both (a) and (b) are correct.
   d. Neither (a) nor (b) is correct.

8. Which of the following is NOT a reason that immunological tests for antibodies against a particular infectious organism might yield negative results?
   a. The person might not be infected with the particular organism.
   b. The organism might be poorly immunogenic and may not stimulate sufficient antibody production to be detectable.
   c. There might not have been sufficient time since the onset of infection for an antibody response to develop.
   d. All of the above might yield negative results.

9. Which type of catheter is the best choice when multiple samples are required over a prolonged period?
   a. hard catheter
   b. French catheter
   c. Foley catheter
   d. soft catheter

10. Which of the following is NOT true about the use of DNA:rRNA hybrids for identification as compared to the use of DNA:DNA hybrids?
    a. DNA:rRNA hybrids are more sensitive, and therefore, fewer microorganisms are required.
    b. DNA:rRNA hybrids are more specific, and they show less cross-hybridization to other species.
    c. DNA:rRNA hybrids are formed more rapidly; test requires two hours or less for results.
    d. All of the above are true about the use of DNA:rRNA hybrids as compared to the use of DNA:DNA hybrids.

11. Which type of cell culture for cultivating viruses makes use of transformed cells, generally epithelial in origin?
    a. primary cultures
    b. secondary cultures
    c. semicontinuous cell cultures
    d. continuous cell cultures
12. When complement binds to an antibody-antigen complex, it becomes used up and is no longer available to lyse sensitized red blood cells. This can be used diagnostically in an assay. What is this assay called?
    a. complement utilization assay
    b. complement titration assay
    c. complement fixation assay
    d. complement complexation assay
13. In one immunological assay, antigens are separated by electrophoresis through a polyacrylamide gel and then are transferred to a sheet of nitrocellulose. This is then probed with enzyme-tagged antibodies. What is this called?
    a. immunoprecipitation
    b. immunodiffusion
    c. ELISA
    d. immunoblotting
14. Immunological assays can be used to differentiate strains of microorganisms. What are these assays called?
    a. serotyping
    b. immunotyping
    c. isotyping
    d. antigen typing

TRUE/FALSE

_____ 1. For best results, specimens should be obtained prior to the administration of any antimicrobial agents.

_____ 2. Needle aspirations of blood samples usually employ an anticoagulant in order to prevent entrapment of microorganisms in a clot, which would then make isolation difficult.

_____ 3. The Gram stain is used for bacteria that have cell walls, while the acid-fast stain is used primarily for wall-less bacteria.

_____ 4. Rickettsias are routinely isolated and identified by culture methods because these are relatively inexpensive and safe to use.

_____ 5. Bacteria can usually be identified by morphological examination, and biochemical tests are only needed for confirmation of identification.

_____ 6. Detection of an elevated antibody titer is used to indicate an active, ongoing infection.

_____ 7. Strains of bacteria that are infected by different phage isolates are referred to as phagovars.

_____ 8. The combination of gas-liquid chromatography (for separation of molecules) with mass spectrometry, nuclear magnetic resonance spectroscopy, and other analytical techniques for identification of separated components makes the identification of possible etiological agents of infectious diseases directly from body fluid specimens without first isolating the infectious organism.

_____ 9. Rapid ID systems such as the API 20E system identify bacteria based on substrate utilization characteristics.

_____ 10. Ribotyping involves probing Southern Blots of endonuclease digested chromosomal DNA with rRNA genes probes.

_____ 11. Immunoprecipitation reactions will occur as long as the antibody and antigen are present in nearly equal amounts.

CRITICAL THINKING

1. Nucleic acid-based detection methods have a great deal of power; however, microscopy and biochemical tests are still widely used. Why? Contrast these two approaches and compare them to current immunoassays for detection of pathogens.

2. Describe the various ways that computers can be used in the clinical microbiology laboratory. How does each use contribute to the more efficient running of the laboratory? Are there any disadvantages to the use of computers? If so, what are they?

3. Many clinical microbiologists feel that susceptibility tests are among the most important tests performed. Why do you think they feel this way?

4. A patient presents with a productive cough, fever, and complaints of a sore throat. What specimens might be useful for diagnosing the illness? How should the specimens be collected and handled? What organisms might be responsible for the illness?

ANSWER KEY

Terms and Definitions

1. m, 2. i, 3. c, 4. l, 5. e, 6. g, 7. b, 8. j, 9. a, 10. o, 11. n, 12. d, 13. f, 14. h, 15. k

True/False

1. T, 2. T, 3. F, 4. F, 5. F, 6. F, 7. T, 8. T, 9. T, 10. T, 11. T

Fill in the Blank

1. clinical microbiologist 2. cytopathic effect; hemadsorption 3. ELISA 4. immunodiffusion; single radial immunodiffusion (RID); double diffusion agar; Ouchterlony technique; 5. serotyping; Lancefield system; Quellung reaction 6. immunoelectrophoresis

Multiple Choice

1. b, 2. d, 3. d, 4. a, 5. c, 6. d, 7. c, 8. d, 9. c, 10. b, 11. d, 12. c, 13. d, 14. a

36 The Epidemiology of Infectious Disease

CHAPTER OVERVIEW

This chapter discusses the epidemiological parameters used to institute effective control, prevention, and eradication measures within an affected or potentially affected population, including a discussion of vaccines and immunization procedures. It also discusses the epidemiology of hospital-acquired (nosocomial) infections and bioterrorism, which have been of increasing concern in recent years.

CHAPTER OBJECTIVES

After reading this chapter you should be able to:

- define epidemiology and explain how it relates to infectious diseases
- discuss the statistical parameters used to define and describe various infectious diseases
- discuss the need to identify the etiologic agent in order to trace the origin and manner of spread of an infectious disease
- describe the five epidemiological links in the infectious disease cycle
- describe the various types of vaccines and give specific examples of each
- discuss bioterrorism agents and preparedness efforts
- discuss the increase in nosocomial infections in recent years and the consequences of these infections

CHAPTER OUTLINE

I.   Introduction
    A.   Epidemiology—the science that evaluates the occurrence, determinants, distribution, and control of health and disease in a defined human population
    B.   Health—an organism and all its parts functioning normally; a state of physical and mental well-being, not merely the absence of disease
    C.   Disease—an impairment of the normal state of an organism, or any of its components, that hinders the performance of vital functions
    D.   Epidemiologist—one who practices epidemiology (a disease detective)
II.   Epidemiological Terminology
    A.   Epidemiologists use a variety of terms to describe different types of disease occurrence
       1.   Sporadic disease—occurs occasionally at irregular intervals in a human population
       2.   Endemic disease—maintains a relatively steady, low-level frequency at a moderately regular interval
       3.   Hyperendemic disease—a gradual increase in frequency above the endemic level, but not to the epidemic level
       4.   Outbreak—an epidemic-like increase in frequency, but in a very limited (focal) segment of the population
       5.   Epidemic—sudden increase in frequency above the endemic level
       6.   Index case—the first case in an epidemic
       7.   Pandemic—a long-term increase in frequency in a large (usually worldwide) population
III. Measuring Frequency: The Epidemiologist's Tools

A. Measures of frequency are usually expressed as fractions; the numerator equals the number of individuals experiencing the event; the denominator is the number of individuals in whom event could have occurred

B. Statistics—the mathematics of collection, organization, and interpretation of numerical data

C. Morbidity rate—the number of new cases in a specific time period per unit of population

D. Prevalence rate—number of individuals infected at any one time per unit of population

E. Mortality rate—number of deaths from a disease per number of cases of the disease

IV. Recognition of an Infectious Disease in a Population

A. Infectious disease is a disease resulting from infection by microbial agents; a communicable disease is an infectious disease that can be transmitted from person to person

B. Epidemiologists study the natural history of an infectious disease

1. What organism caused the disease?

2. What is the source and/or reservoir of the disease?

3. How is the disease transmitted?

4. What host and environmental factors facilitated development of the disease within a defined population?

5. How can the disease best be controlled or eliminated?

C. Recognition involves various surveillance methods to monitor the population for disease occurrence and for demographic analysis

D. Surveillance involves identification of the signs and/or symptoms of a disease

1. Signs—objective changes in the body (e.g., fever, rash) that can be directly observed

2. Symptoms—subjective changes (e.g., pain, appetite loss) experienced by the patient

3. Disease syndrome—a set of signs and symptoms that is characteristic of a disease

E. Characteristic patterns of an infectious disease

1. Incubation period—the period after pathogen entry but before signs and symptoms appear

2. Prodromal stage—onset of signs and symptoms but not yet clear enough for diagnosis

3. Period of illness—disease is most severe and has characteristic signs and symptoms

4. Period of decline (convalescence)—signs and symptoms begin to disappear

F. Remote sensing and geographic information systems: charting infectious disease—map-based tools that can be used to study the distribution, dynamics, and environmental correlates of microbial disease

1. Remote sensing—gathering digital images of Earth's surfaces from satellites and transforming data to maps

2. Geographic information system (GIS)—data management system

3. This approach works best with diseases clearly associated with mapped environmental variables (e.g., vegetation types, elevation, precipitation)

G. Correlation with a single causative organism—uses methods described in previous chapters for the organism's isolation and identification

V. Recognition of an Epidemic

A. Common-source epidemic—characterized by a sharp rise to a peak and then a rapid but not as pronounced decline in the number of cases; usually results from exposure of all infected individuals to a single, common contaminated source, such as food or water

B. Propagated epidemic—characterized by a gradual increase and then a gradual decline in the number of cases; usually results from the introduction of one infected individual into a population, who then infects others; these in turn infect more, until an unusually large number of individuals within the population are infected

C. Herd immunity—the resistance of a population to infection and to the spread of an infectious organism due to the immunity of a large percentage of the population; this limits the effective contact between infective and susceptible individuals

D. Antigenic shift—genetically determined changes in the antigenic character of a pathogen so that it is no longer recognized by the host's immune system (e.g., new flu strains); smaller changes are called antigenic drift; can lead to increases in disease frequency because the population of susceptible hosts increases

VI. The Infectious Disease Cycle: Story of a Disease
   A.   What pathogen caused the disease?—Epidemiologists must determine the etiology (cause) of a disease; Koch's postulates (or modifications of them) are used if possible; the clinical microbiology laboratory plays an important role in the isolation and identification of the pathogen
   B.   What was the source and/or reservoir of the pathogen?
      1.   Source—location from which organisms are immediately transmitted to the host
      2.   Period of infectivity—the time during which the source is infectious or is disseminating the organism
      3.   Reservoir—site or natural environmental location where organism is normally found
      4.   Carrier—an infected individual who is a potential source of a pathogen
         a.   Active carrier—a carrier with an overt clinical case of the disease
         b.   Convalescent carrier—an individual who has recovered from the disease but continues to harbor large numbers of the pathogen
         c.   Healthy carrier—an individual who harbors the pathogen but is not ill
         d.   Incubatory carrier—an individual who harbors the pathogen but is not *yet* ill
         e.   Casual (acute, transient) carriers—any of the above carriers who harbor the pathogen for a brief period (hours, days, or weeks)
         f.   Chronic carriers—any of the above carriers who harbor the pathogen for long periods (months, years, or life)
      5.   Zoonoses—infectious diseases that occur in animals and can be transported to humans; the animals serve as a reservoir for the disease; transmission from animal to host can be direct or indirect
   C.   How was the pathogen transmitted?
      1.   Airborne transmission—suspended in air; travels a meter or more
         a.   Droplet nuclei—may come from sneezing, coughing, or vocalization
         b.   Dust particles—may be important in airborne transmission because microorganisms adhere readily to dust
      2.   Contact transmission—touching between source and host
         a.   Direct (person to person)—physical interaction between infected person and host
         b.   Indirect—involves an intermediate
         c.   Droplets—large particles that travel less than one meter through the air
      3.   Vehicle transmission—inanimate materials or objects are involved in transmission
         a.   Common vehicle transmission—a single vehicle serves to spread the pathogen to multiple hosts, but does not support its reproduction
         b.   Fomites—common vehicles such as surgical instruments, bedding, eating utensils
         c.   Food and water also are common vehicles
      4.   Vector-borne transmission—living transmitters, such as arthropods or vertebrates
         a.   External (mechanical) transmission—passive carriage of the pathogen on the body of the vector with no growth of the pathogen during transmission
         b.   Internal transmission—carried within the vector
            1)   Harborage—pathogen does not undergo morphological or physiological changes within the vector
            2)   Biologic—pathogen undergoes morphological or physiological changes within the vector
   D.   Why was the host susceptible to the pathogen? —depends on defense mechanisms of the host and the pathogenicity of the pathogen
   E.   How did the pathogen leave the host?
      1.   Active escape—movement of pathogen to portal of exit (e.g., many helminths)
      2.   Passive escape—excretion in feces, urine, droplets, saliva, or desquamated cells
VII. Virulence and the Mode of Transmission
   A.   There is evidence of a correlation between mode of transmission and degree of virulence
      1.   A pathogen that is spread by direct contact (e.g., rhinoviruses) cannot afford to make the host so ill it cannot be spread effectively

2. If mode of transmission does not depend on health and mobility of host, then greater degree of virulence is possible

3. A pathogen that is vector-borne needs to be able to replicate extensively within host; therefore, host is likely to remain healthy long enough for transmission to occur

B. Evidence also suggests that the pathogen's ability to survive outside the host is correlated with virulence

    1. Pathogens that do not survive well outside the host and that do not use a vector are likely to be less virulent

    2. Pathogens that can survive for long periods of time outside the host tend to be more virulent

C. Human cultural patterns and behavior also are correlated with virulence

VIII. Emerging and Reemerging Infectious Diseases and Pathogens

A. New diseases have emerged in the past few decades (e.g., AIDS, hantavirus pulmonary syndrome, and many "old" diseases that have increased in frequency, such as tuberculosis)

B. The Centers for Disease Control (CDC) has defined such diseases as "new, reemerging, or drug-resistant infections whose incidence has increased in the last two decades or whose incidence threatens to increase in the near future"; these diseases are the focus of systematic epidemiology, which is concerned with the ecological and social factors that influence the development and emergence of disease

C. Reasons for increases in emerging and reemerging infectious diseases

    1. Economic and military forces can cause population shifts and disruptions of normal public health measures

    2. Changes in sexual behavior, use of IV drugs, and changes in food preferences affect disease transmission

    3. Increasing population density

        a. Overcrowding increases possibility of exposure

        b. Sanitary measures and health-care systems can become overburdened

        c. Encroachment and destruction of natural habitats exposes humans to new pathogens

        d. Introduction of pathogens into new environments can alter transmission and exposure patterns

    4. Drug resistance has increased dramatically in nosocomial pathogens due to excessive or inappropriate use of antimicrobial therapy

    5. Rapid transportation systems aid in the spread of disease out of areas where they are endemic

IX. Control of Epidemics

A. Reduce or eliminate the source or reservoir of infection through:

    1. Quarantine and isolation of cases and carriers

    2. Destruction of an animal reservoir, if one exists

    3. Treatment of sewage to reduce contamination of water

    4. Therapy that reduces or eliminates infectivity of individuals

B. Break the connection between the source and susceptible individuals through sanitization, disinfection, vector control, and other measures; examples include:

    1. Chlorination of water supplies

    2. Pasteurization of milk

    3. Supervision and inspection of food and food handlers

    4. Destruction of insect vectors with pesticides

C. Reduce the number of susceptible individuals—increase herd immunity

    1. Passive immunization to give temporary immunity following exposure

    2. Active immunization to protect individuals and host population

D. Vaccines and immunization

    1. Vaccine—a preparation containing one or more antigens of a pathogen; immunization results from the immunity stimulated by the delivery of vaccines; vaccinomics is the application of genomics and bioinformatics to vaccine development

    2. More efficient immune responses are obtained by mixing antigens with adjuvants, nontoxic materials that help to stimulate a strong immune response

3. Vaccines and vaccination have a long history starting with Jenner's use of cowpox as a vaccine against smallpox; today there are many vaccines, and vaccination is one of the most cost-effective methods for preventing microbial disease
4. Immunization practices depend on the age of the individual and the risk group to which the individual belongs
    a. Children begin a vaccination series at 2 months of age; the series protects against numerous childhood disease (e.g., measles, mumps, rubella)
    b. Adults living in close quarters, having reduced immunity, traveling in other countries, and working in certain professions (e.g., health care provider) may receive additional immunizations

E. Types of vaccines and their characteristics
   1. Whole-cell vaccines
       a. Consist of whole organisms that have been inactivated (killed) or attenuated (live but avirulent); in general, attenuated whole-organism vaccines are most effective and easy to use, and they provide more complete immunity
       b. Though considered the "gold-standard," numerous problems are associated with whole-cell vaccines
           1) Fail to shield against some diseases
           2) Attenuated vaccines can cause disease in immunocompromised individuals
           3) Attenuated viruses can revert to virulence
           4) Molecules unimportant to establishing immunity can trigger allergic reactions; contaminants in preparation also can cause allergic reactions
   2. Acellular or subunit vaccines—vaccines containing specific, purified macromolecules derived from pathogen (capsular polysaccharides, recombinant surface antigens, and inactivated exotoxins called toxoids); avoid many of common risks associated with whole-cell vaccines
   3. Recombinant-vector vaccines—vaccines containing genetically engineered viruses or bacteria, having genes that encode major antigens from a pathogen; elicit both humoral and cellular immunity
   4. DNA vaccines—vaccines containing recombinant DNA molecules (usually a plasmid); the DNA is taken up by muscle cells after injection and enters host nuclei; the antigen gene is then expressed, producing antigenic proteins that elicit both humoral and cellular immunity; currently several human trials of DNA vaccines are underway

F. Role of the public health system: epidemiological guardian—a network of health professionals involved in surveillance, diagnosis, and control of epidemics (e.g., The Centers for Disease Control and the World Health Organization)

X. Bioterrorism Preparedness
   A. Bioterrorism is the intentional or threatened use of microorganisms or toxins from living organisms to produce death or disease in humans, animals, and plants
   B. The list of biological agents that could pose the greatest public health risk is short, and includes viruses, bacteria, parasites, and toxins; these are mostly invisible, odorless, tasteless, and difficult to detect, allowing the biocriminals time to escape
   C. Indicators of a bioterrorism attack include a sudden increase in cases of an unusual disease in humans or animals
   D. Biological weapons are more destructive than chemical weapons; the United States government recently launched an initiative to create a biological weapons defense; global organizations also are working to detect and respond to bioterrorist attacks

XI. Global Travel and Health Considerations
   A. In developed countries, effective public health systems are in place; because of this, travel in developed countries does not pose a health risk
   B. In underdeveloped countries, such public health systems do not exist, and people traveling in those countries are at greater risk
   C. Travelers can take several kinds of precautions in underdeveloped countries; of these, vaccinations are very important

XII. Nosocomial Infections
  A. Produced by infectious agents that develop within a hospital or other clinical care facility and that are acquired by patients while they are in the facility; infections that are incubating within the patient at the time of admission are not considered nosocomial
  B. Source
      1. Endogenous—patient's own microbiota
      2. Exogenous—microbiota other than the patient's
      3. Autogenous—cannot be determined to be endogenous or exogenous
  C. Control, prevention, and surveillance—should include proper handling of the patient and the materials provided to the patient, as well as monitoring of the patient for signs of infection
  D. The hospital epidemiologist—an individual (usually a registered nurse) responsible for developing and implementing policies to monitor and control infections and communicable disease; usually reports to an infection control committee or other similar group

TERMS AND DEFINITIONS

Place the letter of each term in the space next to the definition or description that best matches it.

_____ 1. The science that evaluates the determinants, occurrence, distribution, and control of health and disease in a defined population
_____ 2. A disease that occurs occasionally at irregular intervals
_____ 3. A disease that maintains a relatively steady, low-level frequency
_____ 4. A disease that gradually increases in frequency
_____ 5. Term that describes a disease that rapidly increases in frequency
_____ 6. Term that describes a disease that shows a long-term increase in a large (usually worldwide) population
_____ 7. The first case in an epidemic
_____ 8. A disease of animals that can be transmitted to humans
_____ 9. The mathematics of the collection, organization, and interpretation of numerical data
_____ 10. The number of new cases of a disease during a specific time period per the number of individuals in the population
_____ 11. The total number of cases of a disease at any one time per the number of individuals in the population
_____ 12. The number of deaths from a given disease per number of infected individuals
_____ 13. A disease capable of being transmitted from one individual or reservoir to another
_____ 14. The location from which the causative organism is immediately transmitted to a host
_____ 15. The location at which the causative organism is normally found
_____ 16. An infected individual who is a potential source of infection for others
_____ 17. Inanimate objects involved in the transmission of an infectious organism
_____ 18. Living transmitters of an infectious organism
_____ 19. Infections that are acquired in a hospital or other health-care facility
_____ 20. An impairment of the normal state of an organism or any of its components that hinders the performance of vital functions
_____ 21. A major genetically determined change in the antigenic character of a pathogen
_____ 22. Small genetically determined changes in the antigenic character of a pathogen
_____ 23. Objective changes in the body
_____ 24. The study of ecological and social factors that influence the development and emergence of disease
_____ 25. Subjective changes experienced by the patient
_____ 26. A nosocomial infection for which it is not known if the pathogen was endogenous or exogenous
_____ 27. The period after pathogen entry but before the onset of signs and symptoms
_____ 28. A set of signs and symptoms that are characteristic of a particular disease
_____ 29. The period in which signs and symptoms begin but are not yet clear enough for a diagnosis
_____ 30. An acute febrile condition that can follow childbirth

_____ 31. The condition in which an organism performs its vital functions normally or properly; a state of physical and mental well-being

_____ 32. The application of genomics and bioinformatics to vaccine development

_____ 33. A solution used to induce active immunity; it contains whole killed or live attenuated organisms

_____ 34. A solution used to induce active immunity; it contains a macromolecule purified from a pathogen

_____ 35. Inactivated exotoxins

_____ 36. A solution used to induce active immunity; it contains a genetically engineered nonvirulent virus or bacterium carrying the gene for an antigen from a pathogen

a. antigenic drift
b. antigenic shift
c. autogenous infection
d. carrier
e. communicable disease
f. disease
g. disease syndrome
h. endemic disease
i. epidemic
j. epidemiology
k. health
l. hyperendemic disease
m. incubation period
n. index case
o. macromolecule (subunit) vaccine
p. morbidity rate
q. mortality rate
r. nosocomial infections
s. pandemic
t. prevalence rate
u. prodromal stage
v. puerperal fever
w. recombinant-vector vaccine
x. reservoir
y. signs
z. source
aa. sporadic disease
bb. statistics
cc. symptoms
dd. systematic epidemiology
ee. toxoids
ff. vaccinomics
gg. vectors
hh. vehicles
ii. whole-cell vaccine
jj. zoonosis

FILL IN THE BLANK

1. Epidemics that are characterized by a sharp rise to a peak and then a rapid but not as pronounced decline in the number of individuals infected, and that usually result from exposure of all infected individuals to a single contaminated source such as food or water are called a _____ epidemic. An epidemic that is characterized by a slow and prolonged rise and then a gradual decline in the

number of infected individuals, and results from the introduction of a single infected individual into a population is called a _____ epidemic.

2. A _____ _____ is an illness caused by a pathogen that can be transmitted from an infected host or _____ (site where the pathogen is normally found). The immediate location from which the pathogen is transmitted to a host (either directly or through an intermediate) is called the _____.

3. A _____ is an infected individual who can transmit a pathogen to other hosts. If the infected individual has an overt clinical case of the disease, the individual is referred to as an _____ carrier. If the person has recovered from the infectious disease but can still transmit the pathogen, the person is referred to as a _____ carrier. If the individual can transmit the organism but is not ill, the individual is a _____ carrier. If the individual is not yet ill but will become ill, the individual is referred to as an _____ carrier.

4. The spread of a pathogen in the air for a meter or more from the source is called _____ _____.

5. The transmission of pathogens by a coming together or a touching of the source and the host is called _____ transmission.

6. Living transmitters of pathogens are called _____, and this method of transmission is called _____ transmission. If the pathogen is carried within the transmitter's body and does not undergo morphological or physiological changes, this is called _____ transmission, and if the pathogen does undergo morphological or physiological changes, this is called _____ transmission.

7. The resistance of a population to infection and spread of a pathogen because of the immunity of a large percentage of the population is referred to as _____ immunity.

8. Emerging and reemerging diseases are the focus of a new field called _____ _____, which focuses on the ecological and social factors that influence the development of these diseases. One way to obtain ecological data is _____ _____, a process of gathering digital images of the Earth's surface from satellites and transforming the data into maps. The large quantity of data generated is managed by a _____ _____ _____, which facilitates analysis and interpretation of the data.

9. Types of vaccines called _____ vaccines can contain either inactivated (killed) organisms or living but avirulent organisms. The latter are said to be _____.

MULTIPLE CHOICE

For each of the questions below select the *one best* answer.

1. Which of the following is the major concern of epidemiologists?
   a. the discovery of factors essential to disease occurrence
   b. the development of methods for disease prevention
   c. Both (a) and (b) are major concerns.
   d. Neither (a) nor (b) is a major concern.

2. Which term refers to a disease that has gradually risen above the endemic level but is not at an epidemic level?
   a. outbreak
   b. hyperendemic disease
   c. sporadic disease
   d. pandemic

3. Which term refers to a sudden but focal occurrence of a disease?
   a. outbreak
   b. hyperendemic disease
   c. sporadic disease
   d. pandemic

4. Which term refers to a disease that is of epidemic proportions and encompasses a broad (usually worldwide) population?
   a. outbreak
   b. hyperendemic disease
   c. sporadic disease
   d. pandemic

5. Which term refers to a disease that occurs occasionally at irregular intervals?
   a. outbreak
   b. hyperendemic disease
   c. sporadic disease
   d. pandemic

6. In a propagated epidemic, as the number of infected individuals increases, the number of susceptible individuals decreases. When the number of susceptible individuals decreases below a critical level, the probability of spread from an infected individual to a susceptible host becomes so small that the disease can no longer sustain propagation. What is this phenomenon called?
   a. population resistance
   b. susceptibility diminution
   c. herd immunity
   d. propagation termination

7. Which of the following is NOT a mechanism by which the number of susceptible individuals in a population increases?
   a. birth of new individuals
   b. migration of susceptible individuals into the population
   c. evolution of disease-causing organisms to forms that are no longer recognized by host immune mechanisms
   d. All of the above are ways that number of susceptible individuals in a population increases.

8. Which term refers to carriers that harbor a pathogen for only a brief time (hours, days, or weeks)?
   a. casual carriers
   b. acute carriers
   c. transient carriers
   d. All of the above are this type of carrier.

9. Which of the following is NOT normally a way that humans or animals introduce airborne organisms into the air?
   a. sneezing
   b. evaporation of sweat
   c. coughing
   d. vocalization (talking)

10. Which of the following does NOT refer to inanimate objects involved in the transmission of disease-causing organisms?
    a. vectors
    b. vehicles
    c. fomites
    d. All of the above refer to inanimate objects involved in the transmission of disease-causing organisms.

11. Which of the following is NOT used to reduce or eliminate the source of infection?
    a. treatment of sewage to reduce water contamination
    b. destruction of vectors by spraying insecticides
    c. destruction of an animal reservoir
    d. All of the above eliminate the source of an infection.

12. Which of the following contributes to the emergence of new diseases?
    a. rapid transportation systems and the mobility of the population
    b. ecological disruption such as loss of predators and/or rainforest destruction
    c. increased drug usage and sexual promiscuity
    d. All of the above contribute to the emergence of new diseases.

13. How does the CDC define new and reemerging diseases?
    a. They are defined as drug-resistant infections that have increased in incidence within the last 20 years.
    b. They are defined as new diseases that threaten to increase in incidence in the near future.
    c. They are defined as old diseases, thought to be conquered, that threaten to increase in incidence in the near future.
    d. The CDC uses all of the above to define new and reemerging diseases.

TRUE/FALSE

_____ 1. A measured increase in the morbidity rate of a disease may forecast the need for health-care professionals to prepare for an increase in mortality.

_____ 2. Epidemiologists monitor usage of specific antibiotics, antitoxins, vaccines, and other prophylactic measures in order to recognize the existence of a particular disease in the population.

_____ 3. Demographic data, such as movements of specific populations during various times of the year, is not a particularly useful predictor of disease occurrence.

_____ 4. Public health systems consist of organizations such as the CDC and WHO, as well as individuals working at local levels.

_____ 5. Generally the goal of public health officials is to make sure that at least 70% of the population is immunized against a particular disease-causing organism. This provides enough herd immunity to offer at least partial protection to those not immunized.

_____ 6. Most zoonoses are caused by wild animal populations because there are so many of them.

_____ 7. Nosocomial infections are those that are incubating in a patient at the time of admission and then become apparent while the patient is hospitalized.

_____ 8. Pathogens that are spread by direct contact tend to be more virulent than those that are vector-borne.

_____ 9. Direct contact spread of a pathogen implies that there is actual physical contact between the current host and the new host. Therefore, transfer of a respiratory pathogen by a sneeze or a cough is not considered direct transmission.

_____ 10. Because public health systems in other countries are not always as well developed as those in the USA, it is important to take precautions such as updating or getting additional immunizations when traveling elsewhere in the world.

_____ 11. Space travel decreases cell-mediated immunity; therefore, quarantines and other preventative measures are used to help protect astronauts from an illness while in space.

_____ 12. Although nosocomial infections are of increasing concern, hospitals are not currently required to have an infection control system in place.

_____ 13. In order to maintain its life cycle, a pathogen must escape from the host either passively (i.e., secreted out of the host in feces, urine, droplets, salvia, or desquamated cells) or actively (pathogen moves on its own to a portal of exit and then escapes).

_____ 14. The manifestations of an infectious disease are the same in all hosts belonging to the same species.

_____ 15. Bioterrorism is largely an imaginary threat.

_____ 16. A large number of pathogens or their products are suitable for development as bioweapons.

_____ 17. In general, vaccination is not a cost-effective method for providing immunity.

_____ 18. Vaccines composed of specific purified macromolecules avoid some of the risks associated with whole-organisms vaccines.

TYPES OF VACCINES

Type of Vaccine		Description	Example
Whole-organism vaccine	inactivated (killed)		
	live, attenuated		
Macromolecule vaccine	capsular polysaccharide		
	recombinant surface antigens		
	toxoid		
Recombinant-vector vaccine			
DNA vaccine			

CRITICAL THINKING

1. Explain the concept of herd immunity. How can it develop naturally? How can it be stimulated artificially? Cite examples of both methods.

2. Explain antigenic drift and how it overcomes herd immunity and leads to new outbreaks of the same disease.

3. The authors present a number of surveillance methods used by epidemiologists. Discuss these methods in the context of a particular disease of past or current importance (e.g., smallpox, AIDS) to show how the data collected by these methods increase our knowledge of the disease process.

ANSWER KEY

Terms and Definitions

1. j, 2. aa, 3. h, 4. l, 5. i, 6. s, 7. n, 8. jj, 9. bb, 10. p, 11. t, 12. q, 13. e, 14. z, 15. x, 16. d, 17. hh, 18. gg, 19. r, 20. f, 21. b, 22. a, 23. y, 24. dd, 25. cc, 26. c, 27. m, 28. g, 29. u, 30. v, 31. k, 32. ff, 33. ii, 34. o, 35. ee, 36. w

Fill in the Blank

1. common-source; propagated 2. communicable disease; reservoir; source 3. carrier; active; convalescent; healthy; incubatory 4. airborne transmission 5. contact 6. vectors; vector-borne; harborage; biologic 7. herd 8. systematic epidemiology; remote sensing; geographic information system (GIS) 9. whole-cell; attenuated

Multiple Choice

1. c, 2. b, 3. a, 4. d, 5. c, 6. c, 7. d, 8. d, 9. b, 10. a, 11. d, 12. d, 13. d

True/False

1. T, 2. T, 3. F, 4. T, 5. T, 6. F, 7. F, 8. F, 9. F, 10. T, 11. T, 12. F, 13. T, 14. F, 15. F, 16. F, 17. F, 18. T

37 Human Diseases Caused by Viruses and Prions

CHAPTER OVERVIEW

This chapter discusses viruses and prions that are pathogenic to humans, with emphasis on those diseases occurring in the United States.

CHAPTER OBJECTIVES

After reading this chapter you should be able to:

- describe those viral diseases that are transmitted through the air and that directly or indirectly involve the respiratory system
- discuss viral diseases transmitted by arthropod vectors
- discuss viruses requiring direct contact for transmission
- discuss viral diseases that are food- or waterborne
- discuss prion diseases

CHAPTER OUTLINE

I. Airborne Diseases
   A. Chickenpox (varicella) and shingles (Herpes zoster)
      1. Chickenpox
         a. Caused by the enveloped DNA varicella-zoster virus (VZV), a member of *Herpesviridae*; is acquired by inhaling virus-laden droplets into the respiratory system
         b. Incubation period is 10 to 23 days after which small vesicles appear on face and upper trunk; vesicles fill with pus, rupture, and crust with scabs; intensely itchy
         c. Can be prevented or infection shortened with attenuated vaccine or the drug acyclovir
         d. Infection confers permanent immunity from chickenpox, but does not rid individual of virus; instead, virus enters a latent stage in the nuclei of neurons in the dorsal root ganglion
      2. Shingles
         a. If an adult who harbors the virus becomes compromised due to stress, age, or illness, etc., the virus can emerge and cause sensory nerve damage and painful vesicle formation (postherpetic neuralgia), a condition known as shingles (Herpes zoster)
         b. Treated with acyclovir, famciclovir, or other antiviral agents in immunocompromised patients (e.g., AIDS patients)
   B. Influenza (flu)
      1. Caused by orthomyxoviruses; four groups are known, influenza A, B, or C, and Thogoto viruses
      2. Have a negative-strand RNA genome with 7 to 8 linear segments that can undergo frequent antigenic variation, particularly in surface hemagglutinin (HA) and neuraminidase (NA).
         a. Antigenic drift—small variations due to the accumulation of point mutations of HA and NA; lead to local increases in cases every 2–3 years
         b. Antigenic shift—large variations due to reassortment of RNA genome segments between two strains (typically an animal and human strain) into progeny virions; the greater change leads to epidemics or pandemics

3. Animal reservoirs are important (e.g., chickens and pigs) and contribute to antigenic shifts; designate strains by the 16 HA and 9 NA antigenic forms known (e.g., H5N1)
4. Virus is acquired by inhalation or ingestion of virus-contaminated respiratory secretions; it enters host cells by receptor-mediated endocytosis
5. Influenza is characterized by chills, fever, headache, malaise, and general muscular aches and pains; diagnosis can be confirmed by rapid serological tests
6. Treatment is focused on alleviating symptoms, but some antiviral drugs (Relenza, Tamiflu) have been shown to decrease duration and symptoms of type A influenza

C. Measles (rubeola)
   1. A skin disease with respiratory spread caused by a negative-strand enveloped RNA *Morbillivirus*, a member of family *Paramyxoviridae*
   2. After 10–21 day incubation, cold-like symptoms develop, followed by a rash of small, raised spots; useful diagnostically are Koplick's spots, which form in the mouth; on rare occasions can develop into subacute sclerosing panencephalitis
   3. MMR (measles, mumps, and rubella) vaccine is used for prevention

D. Mumps
   1. Caused by mumps virus, a helical, enveloped negative-strand RNA *Rubulavirus* in the family *Paramyxoviridae*
   2. Spread in saliva and respiratory droplets; portal of entry is the respiratory tract
   3. Causes swelling of salivary (parotid) glands; meningitis and inflammation of testes are complications, especially in postpubescent male
   4. Therapy is supportive and the MMR vaccine is used for prevention

E. Respiratory syndromes and viral pneumonia
   1. Acute respiratory syndromes
      a. Caused by a variety of viruses collectively referred to as acute respiratory viruses
      b. Associated with rhinitis, tonsillitis, laryngitis, and bronchitis; immunity resulting from infection is incomplete and reinfection is common
   2. Viral pneumonia is clinically nonspecific, and symptoms may be mild or severe (death is possible)
   3. Respiratory syncytial virus (RSV) is the most dangerous cause of respiratory infection in young children; it is a member of the RNA virus family *Paramyxoviridae*

F. Rubella (German measles)
   1. Caused by rubella virus, a ssRNA virus of family *Togoviridae*
   2. Virus is spread by respiratory droplets, and the resulting infection is mild in children (a rash), but disastrous for pregnant women in first trimester; in pregnant women it causes congenital rubella syndrome, which leads to fetal death, premature delivery, or congenital defects
   3. No treatment is indicated; a vaccine (MMR) is available

G. Severe acute respiratory syndrome (SARS)
   1. Highly contagious, helical enveloped, positive-strand RNA coronavirus
   2. Causes sudden severe illness with high fever, flu-like discomfort, and a dry cough that will most likely develop into pneumonia; transmitted by respiratory droplets
   3. Relatively recent health threat; no treatment available; diligent monitoring and rapid detection seen as preventive measures

H. Smallpox (variola)
   1. Highly contagious disease caused by variola (major or minor) virus, a dsDNA orthopoxvirus of the *Poxviridae*; a brick-shaped virus that is slightly larger than the smallest bacterium; replicates in host cell's cytoplasm
   2. Virus is transmitted by aerosol or contact; it rapidly moves from respiratory tract membranes to regional lymph nodes, where is replicates in the monocyte-macrophage system during the incubation period; after 12–14 days the prodromal period begins, followed in 2–3 days by clinical manifestations of severe fever, prostration, vesicular rash, toxemia, and septic shock; historically 20 to 50% fatal
   3. Long history of vaccination using live, related vaccinia virus; an important prevention method, but discontinued when virus declared eradicated

4. Suspected cases of smallpox should be managed in a negative-pressure room and the patient should be vaccinated, particularly if the disease is in the early stages
5. Virus was eradicated as the result of a vigorous worldwide vaccination program; eradication was made possible for several reasons
   a. Disease has easily identifiable clinical features
   b. There are virtually no asymptomatic carriers
   c. It infects only humans (there are no animal or environmental reservoirs)
   d. It has a short period of infectivity
6. There is concern that the virus could be used as a bioweapon by terrorists

II. Arthropod-Borne Diseases
   A. General features of arthropod-borne diseases
      1. Viruses multiply in tissues of insect vectors without producing disease, and vector acquires a lifelong infection
      2. Three clinical syndromes are common
         a. Undifferentiated fevers, with or without a rash
         b. Encephalitis—often with a high case fatality rate
         c. Hemorrhagic fevers—frequently severe and fatal
      3. Infection provides permanent immunity; for many of the diseases, no vaccines are available; treatment is usually supportive
   B. Equine encephalitis
      1. Caused by a positive-strand enveloped RNA *Alphavirus* in the family *Togaviridae*
      2. Transmitted by mosquitoes from animal reservoirs; presents as fever, headache, meningitis, and encephalitis, progressing to seizures, paralysis, coma, and death
      3. Treatment is supportive; no vaccine available; prevention relies on mosquito avoidance
   C. Tick-borne encephalitis
      1. Caused by positive-strand RNA *Flavivirus* in the family *Togaviridae*
      2. Transmitted by *Ixodes* ticks or by consuming unpasteurized dairy products
      3. After an incubation period of 7–14 days, fever develops for 2–4 days accompanied by flu-like symptoms; about 20–30% of patients develop central nervous system problems (meningitis and encephalitis) after an 8-day remission; can produce long-lasting deficits
      4. Vaccine is available but not routinely administered; prevention relies on tick avoidance
   D. Rift Valley fever
      1. Caused by a negative-strand RNA *Phlebovirus* in the family *Bunyaviridae*
      2. Transmitted by mosquitoes, it is widespread in livestock
   E. West Nile fever (encephalitis)
      1. Caused by a positive-strand ssRNA flavivirus indigenous to the Middle East, Africa, and southwest Asia; first appeared in USA in 1999
      2. Transmitted to humans by *Culex* mosquitoes that have fed on infected birds; can also be transmitted by blood transfusions and organ transplants
      3. Produces mild febrile disease in 20% of infected individuals; can produce meningitis or encephalitis in 0.7% of infected individuals
      4. Diagnosis is by a serological test that detects a rise in neutralizing antibodies in the patient's serum; PCR methods also are available
      5. No vaccine is available; prevention relies on mosquito abatement
   F. Yellow Fever
      1. Mosquito-borne (*Aedes*); there are two patterns of transmission
         a. Urban cycle—human-to-human transmission
         b. Sylvatic cycle—monkey-to-monkey and monkey-to-human transmission
      2. Abrupt onset after 3–6 days of incubation; fever, pain in back, stomach, and extremities, vomiting among symptoms; can progress to liver and renal failure with jaundice and hemorrhaging
      3. Diagnosis by culture, immunohistochemistry, ELISA, and PCR; prevention and control are by vaccination and vector control

III. Direct Contact Diseases

A. Acquired immune deficiency syndrome (AIDS)
   1. Caused by human immunodeficiency virus (HIV), a positive-strand enveloped lentivirus within the family *Retroviridae*; believed to have evolved in Africa from viruses that infect other primates
   2. Disease occurs worldwide, but certain groups are more at risk; these include homosexual/bisexual men, intravenous drug users, transfusion patients and hemophiliacs, prostitutes, and newborn children of infected mothers
   3. Virus is acquired by direct exposure of the person's bloodstream to body fluids containing the virus; also can be transmitted via breast milk
   4. Virus targets $CD4^+$ cells such as T-helper cells, macrophages, dendritic cells, and monocytes
   5. After penetration and uncoating, the RNA is copied into DNA by reverse transcriptase; the DNA provirus then is integrated into the host cell chromosome
   6. CDC classification scheme for the stages of HIV-related conditions
      a. Acute infection stage is a brief illness 2 to 8 weeks after infection with some fever, malaise, and a macular rash; large burst of new virions in blood
      b. Asymptomatic stage may last 6 months to 10 years; lower levels of virus in blood; attack of lymphoid tissue continues and some immune system dysfunction
      c. Chronic symptomatic stage can last for months to years; virus proliferation decreases $CD4^+$ populations; significant immune dysfunction ensues with development of opportunistic infections such as candidiasis and Kaposi's sarcoma
      d. AIDS follows with severe immune dysfunction and increased susceptibility to opportunistic infections such as *Pneumocystis* pneumonia, toxoplasmosis, cryptococcal meningitis, and histoplasmosis; nervous system effects can include dementia and ataxia; cancer rates increase with Kaposi's sarcoma (caused by human herpesvirus 8; HHV-8), carcinoma of the mouth and rectum, and B-cell lymphomas
   7. A harmless virus (GB virus C) discovered in 1995 appears to prolong the lives of those who are infected with HIV; the mechanism by which it does so is not known
   8. Diagnosis is by viral antigen detection or more commonly by viral antibody detection (ELISA); PCR and Western blots can be used
   9. Four types of antiviral agents are used to treat HIV disease, often in combination
      a. Nucleoside analogues that inhibit HIV reverse transcriptase (e.g. AZT, ddC, and 3TC)
      b. Nonnucleoside inhibitors of HIV reverse transcriptase (e.g., delaviridine)
      c. Inhibitors of HIV protease (e.g., ritonavir and indinavir)
      d. Fusion inhibitors prevent HIV entry into cells (e.g., enfuvirtide)
   10. Vaccines are currently under investigation
   11. Prevention and control involve screening of blood and blood products, education, and protected sexual practices (use of condoms)
B. Cold sores—fever blisters
   1. Caused by herpes simplex type 1 (HSV-1), an enveloped dsDNA virus
   2. Blister at site of infection is due to viral- and host-mediated tissue destruction
   3. Lifetime latency is established when virus migrates to trigeminal nerve ganglion; is periodically reactivated in times of physical or emotional stress
   4. Herpetic keratitis—recurring infections of the cornea; can result in blindness
   5. Drugs are available that are effective against cold sores (e.g., acyclovir and vidarabine); diagnosed by ELISA or PCR
C. Common cold
   1. Caused by many different rhinoviruses as well as other viruses; many do not confer durable immunity
   2. Understanding rhinovirus structure has suggested approaches to developing vaccines and drugs
   3. At one time, common cold was thought to be spread by explosive sneezing, but now it is believed to be primarily spread by hand-to-hand contact; treatment is supportive
D. Cytomegalovirus inclusion disease
   1. Caused by an enveloped dsDNA virus in the family *Herpesviridae*

2. Most infections are asymptomatic but infection can be serious in immunologically compromised individuals; virus persists in the body and is shed for several years in saliva, urine, semen, and cervical secretions
3. Infected cells have intranuclear inclusion bodies as well as cytoplasmic inclusions
4. Diagnosis is by viral isolation and serological tests
5. Some antiviral agents are available for treatment; these are only used in high-risk patients; prevention is by avoiding close personal contact with infected individual and by using blood or organs from seronegative donors

E. Genital herpes
1. Caused by herpes simplex type 2 (HSV-2), an enveloped dsDNA virus that is a member of *Herpesviridae*; virus is most frequently transmitted by sexual contact
2. Disease has active and latent phases
   a. Active phase—the virus rapidly reproduces; can be symptom free or painful blisters in the infected area may occur; other symptoms (fever, burning sensation, genital soreness) also may occur; blisters heal spontaneously
   b. Latent phase—after resolution of active phase virus retreats to nerve cells; the viral genome resides in the nuclei of host cells and can be periodically reactivated
3. Congenital (neonatal) herpes is spread to an infant during vaginal delivery; therefore, children of infected females should be delivered by caesarean section
4. There is no cure, but acyclovir decreases healing time, duration of viral shedding, and duration of pain

F. Human herpesvirus 6 infection
1. Etiologic agent of exanthem subitum (rash) in infants, a short-lived disease characterized by a high fever of 3 to 4 days' duration, followed by a macular rash; $CD4^+$ cells are the main sites of viral replication and the tropism of the virus is wide and includes $CD8^+$ T cells, natural killer cells, and probably epithelial cells; transmission is probably by way of saliva
2. Virus produces latent and chronic infections and can be reactivated in immunocompromised individuals, leading to pneumonitis; virus has been implicated in a variety of other diseases, including chronic fatigue syndrome and lymphadenitis; diagnosis is by immunofluorescence or enzyme immunoassay; there is neither treatment nor prevention currently available

G. Human parvovirus B19 infections
1. Mild symptoms (fever, headaches, chills, malaise) in most normal adults; erythema infectiosum in children; joint disease in some adults; serious aplastic crisis in immunocompromised individuals or those with sickle-cell disease or autoimmune hemolytic anemia; anemia and fetal hydrops (the accumulation of fluid in the tissues) in infected fetuses
2. Spread by a respiratory route
3. Antiviral antibodies are the principal means of defense, and treatment is by means of commercial anti-B19 immunoglobulins; infection is usually followed by lifelong immunity

H. Leukemia—certain leukemias (adult T-cell leukemia and hairy-cell leukemia) are caused by retroviruses (HTLV-1 and HTLV-2, respectively) and are spread similarly to HIV; they are often fatal and there is no effective treatment; IFN-α n3 (Alferon N) has shown some promise in treating disease

I. Mononucleosis (infectious)
1. Caused by the Epstein-Barr virus (EBV), a herpesvirus (enveloped dsDNA virus), that is spread by mouth-to-mouth contact ("kissing disease") or by shared bottles and glasses; virus replicates briefly in epithelial cells of throat and eventually infects memory B cells; causes enlargement of lymph nodes and spleen, sore throat, headache, nausea, general weakness and tiredness, and a mild fever; disease is self-limited
2. Treatment is largely supportive and requires plenty of rest; diagnosis is made by serological tests
3. EBV is also associated with Burkitt's lymphoma and nasopharyngeal carcinoma in certain parts of the world

J. Viral hepatitides

1. Hepatitis is any inflammation of the liver; currently 11 viruses are recognized as causing hepatitis; some have not been well characterized
2. Hepatitis B (serum hepatitis)
   a. Caused by hepatitis B virus (HBV), an enveloped, double-stranded, circular DNA virus in the family *Hepadnaviridae*
   b. Virus is transmitted by blood transfusions, contaminated equipment, unsterile needles, or any body secretion; transplacental transmission to fetus also occurs
   c. Most cases are asymptomatic; sometimes fever, appetite loss, abdominal discomfort, nausea, and fatigue develop; death can result from liver cirrhosis or HBV-related liver cancer
   d. Control measures involve excluding contact with contaminated materials, passive immunotherapy within seven days of exposure, and vaccination of high-risk groups
3. Hepatitis C
   a. Caused by hepatitis C virus (HCV), an enveloped ssRNA virus within the family *Flaviviridae*
   b. Virus is spread by intimate contact with virus-contaminated blood, in utero from mother to fetus, by the fecal-oral route, sexually, or through organ transplants
   c. Diagnosis is by serological tests
   d. Treated with interferon
4. Hepatitis D
   a. Is caused by hepatitis D virus (HDV) (formerly called the Delta agent), which only causes disease if the individual is coinfected with hepatitis B virus (which provides a coat protein)
   b. Diagnosis is by serological tests; treatment is difficult and often involves administration of alpha interferon; prevention and control are by the use of the hepatitis B vaccine
5. Recently, hepatitis F and hepatitis G (HGV) have been identified and are currently being investigated; HGV is a member of the *Flaviviridae* and can be transmitted parenterally or sexually

K. Warts
   1. Caused by papillomaviruses
   2. Treatment involves removal of warts, physical destruction, or injection of interferon
   3. Some papillomaviruses play a major role in the pathogenesis of epithelial cancers of the male and female genital tracts; a vaccine was recently developed

IV. Food-Borne and Waterborne Diseases
   A. Gastroenteritis (viral) —acute viral gastroenteritis
      1. Inflammation of the stomach or intestines caused by Norwalk viruses and Noroviruses, adenoviruses, rotaviruses, caliciviruses, and astroviruses
      2. Main transmission route is fecal-oral route; after an incubation period of 1 to 2 days, typical symptoms are mild (diarrhea with headache and fever) to severe (watery diarrhea with cramps)
      3. Seen most frequently in infants with occasional fatal dehydration; disease is leading cause of childhood death in developing countries
      4. Viral gastroenteritis is usually self-limited; treatment is supportive and includes fluid replacement and antiperistaltic medication
   B. Hepatitis A
      1. Spread by fecal contamination of food or drink, or by infected shellfish that live in contaminated water
      2. Caused by the hepatitis A virus (HAV), a naked, positive-strand RNA *Hepatovirus* in the family *Picornaviridae*
      3. Mild intestinal infections sometimes progress to liver involvement; most cases resolve in four to six weeks and produce strong immunity
      4. Control is by hygienic measures and sanitary disposal of excreta; a killed vaccine (Havrix) is available, greatly reducing incidence of the disease
   C. Hepatitis E

1. Implicated in many epidemics in developing countries in Asia, Africa, and Central and South America
2. Caused by hepatitis E virus (HEV), a naked, positive-strand RNA virus
3. Infection is associated with fecal-contaminated drinking water; HEV enters the blood from the gastrointestinal tract, replicates in the liver, is released from hepatocytes into the bile, and is subsequently excreted in the feces
4. HEV, like HAV, usually runs a benign course and is self-limiting; can be fatal (15 to 25%) in pregnant women in their last trimester
5. Diagnosis is by ELISA or reverse transcriptase PCR; prevention is aimed at improving the level of health and sanitation in affected areas

D. Poliomyelitis (infantile paralysis)
1. Polio is caused by poliovirus, a naked, positive-strand *Picornaviridae*; it is stable in the environment and remains infectious in food and water
2. Once ingested, virus multiplies in throat and intestinal mucosa; subsequently enters bloodstream and causes viremia (99% of viremia cases are transient and asymptomatic or with mild clinical disease including fever and vomiting); with extended viremia, can enter central nervous system (less than 1% of cases), leading to paralysis
3. Vaccines have been extremely effective in preventing and controlling the disease; global eradication may be possible in the near future

V. Zoonotic Diseases
A. Ebola and Marburg hemorrhagic fevers
1. Viral hemorrhagic fevers include a group of severe, multisystem diseases; overall host vascular system is damaged leading to hemorrhaging and dysfunction
2. Ebola hemorrhagic fever—caused by Ebola virus, a negative-strand RNA *Filoviridae*; after an incubation period of 2 to 21 days abrupt and severe illness is observed with fever, joint pain, and weakness, leading to diarrhea and vomiting, rash, red eyes, and bleeding; 80% of cases result in death; transmission can be through direct contact with bodily fluids; only supportive treatment is available
3. Marburg hemorrhagic fever—caused by an RNA *Filoviridae*; rare and severe disease with symptoms similar to Ebola hemorrhagic fever; no treatment currently exists

B. Hantavirus pulmonary syndrome
1. Caused by an enveloped, negative-strand RNA *Bunyaviridae*
2. Transmitted by inhalation of virions in wild rodent urine, feces, or saliva; potentially fatal disease in humans
3. Prevention relies on rodent control and avoidance

C. Lassa fever—caused by negative-strand RNA *Arenaviridae*; most cases (80%) mild fever, but in others causes severe multisystem disease with high mortality; transmitted in rodent feces and urine and person-to-person

D. Lymphocytic choriomeningitis—caused by a negative-strand RNA *Arenaviridae*; typically asymptomatic or mild febrile disease; in utero effects can be devastating; transmitted by rodent excretions or through organ transplant

E. Nipah virus—negative-strand RNA *Paramyxoviridae* similar to measles and mumps viruses; begins with fever and headache and in some cases progresses to neurological disease, coma, and death

F. Rabies
1. Caused by a number of different strains of neurotropic, negative-strand RNA *Lyssavirus* of the family *Rhabdoviridae*
2. Transmitted by bites of infected animals; aerosols in caves where bats roost; or by scratches, abrasions, open wounds, or mucous membranes contaminated with saliva of infected animals
3. Virus multiplies in skeletal muscle and connective tissue, then migrates to central nervous system, causing a rapidly progressing encephalitis
4. In the past, diagnosis depended on the observation of characteristic Negri bodies (masses of virus particles or unassembled viral subunits); today diagnosis is based on immunological tests, virus isolation, as well as the detection of Negri bodies

5. Symptoms progress and death results from destruction of the part of the brain that regulates breathing
6. Vaccines conferring short-term immunity are available and must be given soon after exposure (postexposure vaccination is effective because of the long incubation period of the virus); prevention and control involve annual preexposure vaccination of dogs and cats, postexposure vaccination of humans, and frequent preexposure vaccination of humans at special risk

VI. Prion Diseases (Transmissible Spongiform Encephalopathies)
   A. Caused by a proteinaceous infectious particle (a prion) that is a misfolded brain protein
   B. Primary symptom is dementia and motor dysfunction; characteristic spongiform degeneration of the brain and formation of amyloid plaques; symptoms progress for months or years, leading to death
   C. Group of diseases that includes kuru, Creutzfeldt-Jakob disease, Gertsmann-Straussler disease, fatal familial insomnia, and variants of bovine spongiform encephalopathies (mad cow disease)

TERMS AND DEFINITIONS

Place the letter of each term in the space next to the definition or description that best matches it.

_____ 1. An occasional complication of influenza and chickenpox in children under 14 years of age
_____ 2. A condition that involves the central nervous system and is associated with influenza infections
_____ 3. Lesions in the oral cavity that are red with a bluish-white center and that are diagnostic of measles
_____ 4. Inflammation of the epididymis and testes
_____ 5. Characteristic feature of brain tissue degenerated by prion diseases
_____ 6. Programmed cell death
_____ 7. Cold sores of the lips, mouth, and gums
_____ 8. Discharge from the nostrils; runny nose
_____ 9. Disease characterized by uncontrolled growth of white blood cells
_____ 10. Masses of viruses or unassembled viral subunits found in neurons infected with rabies virus
_____ 11. The only disease to be eradicated worldwide
_____ 12. Neurodegenerative disorders caused by prions
_____ 13. Warts on feet
_____ 14. Common warts on hands and fingers
_____ 15. Flat growths caused by papillomaviruses
_____ 16. Sexually transmitted warts caused by HPV 6, HPV 11, and HPV 42
_____ 17. Small changes in the antigenic nature of the HA and NA proteins of the influenza virus
_____ 18. Sudden dramatic changes in the antigenic nature of the HA and NA proteins of the influenza virus

a. amyloid plaques
b. anogenital condylomata (venereal warts)
c. antigenic drift
d. antigenic shift
e. apoptosis
f. coryza
g. flat (plane) warts
h. gingivostomatis
i. Guillain-Barré syndrome (French polio)
j. Koplik's spots
k. leukemia
l. Negri bodies
m. orchitis
n. plantar warts
o. Reyes syndrome
p. smallpox
q. spongiform encephalopathies
r. verrucae vulgaris

FILL IN THE BLANK

1. Individuals who recover from _____ are immune to this disease, but are not free of the virus. Instead, the virus is latent in certain neurons. In adults, the virus can be reactivated leading to a disease syndrome called _____ _____, which includes the production of painful vesicles. This reactivated disease is called _____.

2. The arthropod-borne viruses can cause three different clinical syndromes: undifferentiated fevers, with or without a rash; encephalitis , and _____ fevers.

3. The once common childhood disease _____ is characterized by _____ _____ (bright red spots with a bluish-white speck in the center that are observed in the mouth). On rare occasions, this disease leads to a progressive degeneration of the central nervous system called _____ _____ _____.

4. Human immunodeficiency virus (HIV) is the causative agent of _____.

5. Herpes simplex type 2 virus is the causative agent of _____ _____. It is most frequently spread by sexual contact, but can be passed to a newborn during vaginal delivery, resulting in _____ herpes.

6. German measles, also called _____, is generally a mild disease of children. However, if a pregnant woman contracts the disease, it can lead to a wide array of congenital defects collectively called _____ _____ _____.

7. A group of viral diseases called _____ _____ are characterized by massive hemorrhaging either locally or throughout the body. They include _____ _____ _____ _____, caused by an African virus that was first recognized in Germany; the deadly _____ _____ _____ _____, first discovered in Zaire and Sudan in the 1970s; _____ _____ _____, imported to the U.S. from Korea; and _____ _____ _____, discovered in the four-corners area of the U.S. and caused by a virus called a variety of names, including Sin Nombre virus and _____ _____ _____.

8. Any infection that results in inflammation of the liver is called _____. HBV causes _____. Serum from individuals infected with this virus contains three distinct antigenic particles. One, called the _____ particle is the infective form of the virus and the other two are unassembled components of it. In 1977 a cytopathic hepatitis agent termed the _____ _____ was discovered. It has been renamed _____ _____ virus, and the disease it causes is now called _____ _____.

9. The human cytomegalovirus (HCMV) causes _____ _____ disease. The disease gets its name from the observation that infected cells contain _____ _____ _____ as well as cytoplasmic inclusions.

10. Viral _____ _____ such as Ebola and _____ are severe multisystem diseases that damage the _____ system leading to dysfunction and _____.

11. Severe _____ _____ _____ causes sudden severe illness and is highly _____, making it a true public health threat. It most likely leads to _____ and has a high rate of mortality.

VIRAL and PRION DISEASES OF HUMANS

Disease or Virus	Mode of Transmission	Type of Treatment	Effective Vaccine? (yes or no)
AIDS (HIV)			
Chickenpox (*Varicella*)/shingles (*zoster*)			
Cold sore (fever blister, *Herpes labialis*)			
Colorado tick fever			
Common cold			
Creutzfeldt-Jakob disease			
Cytomegalovirus inclusion disease			
Ebola hemorrhagic fever			
Equine encephalitis			
Erythema infectiosum (fifth disease, human parvovirus B19)			
Exanthem subitum (roseola infantum, sixth disease, human herpesvirus 6)			
Fatal familial insomnia			
Genital herpes			
Hantavirus pulmonary syndrome			
Hemorrhagic fevers (Ebola, Korean, and Marburg)			
Hepatitis A			
Hepatitis B			
Hepatitis C			
Hepatitis D			
Hepatitis E			
Hepatitis G			
Infectious mononucleosis			
Influenza (flu)			
Lassa fever			
Measles (rubeola)			
Mumps			
Poliomyelitis (polio, infantile paralysis)			
Rabies			
Respiratory syncytial virus			
Rift Valley fever			

Rubella (German measles)			
Severe acute respiratory syndrome			
Smallpox			
Tick-borne encephalitis			
Warts			
West Nile virus			
Yellow fever			

MULTIPLE CHOICE

For each of the questions below select the *one best* answer.

1. What is the rash associated with German measles probably caused by?
   a. the virus infecting the cells of the skin
   b. the virus infecting the cells of the dermal layer beneath the skin
   c. an immunological reaction to the virus
   d. All of the above can cause this rash.

2. Who is the vaccine against rubella (German measles) recommended for?
   a. all individuals not previously exposed to the virus
   b. all children not previously exposed to the virus
   c. all women of childbearing age not previously exposed to the virus
   d. all children and all women of childbearing age not previously exposed to the virus

3. Which of the following is a complication of mumps?
   a. meningitis
   b. inflammation of the testes in the postpubescent male
   c. Both (a) and (b) are complications.
   d. Neither (a) nor (b) is a complication.

4. Which of the following contributed to the eradication of smallpox from the world?
   a. There are no hosts, reservoirs, or vectors, other than humans.
   b. There are no asymptomatic carriers.
   c. A vigorous worldwide vaccination program helped prevent spread of the virus.
   d. All of the above contributed to the eradication of smallpox.

5. Which of the following can stimulate a reactivation of the virus that causes cold sores?
   a. physical stress, such as fever, cold wind, trauma, and hormonal changes
   b. emotional stress
   c. Both (a) and (b) can stimulate reactivation.
   d. Neither (a) nor (b) can stimulate reactivation.

6. Which of the following is NOT true about the use of acyclovir to treat genital herpes?
   a. It decreases the duration of virus shedding.
   b. It kills the virus and thereby cures the disease if it is applied for a long enough time.
   c. It decreases the duration of pain.
   d. All of the above are true about the use of acyclovir to treat genital herpes.

7. Which of the following is NOT true of human retroviruses?
   a. They cause all known leukemias in humans.
   b. They cause adult T-cell leukemia in humans.
   c. They cause hairy-cell leukemia in humans.
   d. None of the above is true of retroviruses in humans.

8. Which of the following is NOT true about rabies?
   a. It can be transmitted to humans by the bite of an infected animal.
   b. It can be spread by aerosols in caves where bats roost.
   c. It can be transmitted by contamination of abrasions, wounds, scratches, or mucous membranes with saliva from an infected animal.
   d. All of the above are true about rabies.

9. Which of the following is the mode of transmission for hepatitis C?
   a. contaminated blood and blood products
   b. fecal-oral route
   c. Both (a) and (b) are correct.
   d. Neither (a) nor (b) is correct.

10. In which population is acute viral gastroenteritis the leading cause of death?
    a. children worldwide
    b. children in the United States
    c. children in developing countries where malnutrition is a major problem
    d. children in day-care centers

11. Which of the following is used for prevention and control of rabies?
    a. preexposure vaccination of dogs and cats
    b. postexposure vaccination of humans
    c. preexposure vaccination of humans at high risk
    d. All of the above are used in the prevention and control of rabies.

12. Drugs used to slow the progress of AIDS fall into which of the following categories?
    a. nucleoside analog reverse transcriptase inhibitors
    b. nonnucleoside reverse transcriptase inhibitors
    c. protease inhibitors
    d. all of the above

13. The agent that causes Creutzfeldt-Jakob disease is which of the following?
    a. virusoid protein
    b. misfolded brain protein
    c. viral envelope protein
    d. infectious mononucleoid

TRUE/FALSE

_____ 1. The same virus causes chickenpox and shingles.

_____ 2. Treatment of flu is symptomatic only, since there are no drugs that will affect the duration or severity of the disease.

_____ 3. Viral pneumonia is not specifically diagnosed, but is assumed when mycoplasmal pneumonia has been eliminated as a possibility.

_____ 4. Control of yellow fever depends on control of the insect vector.

_____ 5. Because HIV (the causative agent of AIDS) infects cells of the immune system, no antibodies are ever produced against this virus.

_____ 6. The drug azidothymidine (AZT) does not cure AIDS, but it does inhibit its transmission.

_____ 7. The most likely mode of transmission of the common cold is by hand-to-hand contact between a rhinovirus donor and a susceptible recipient.

_____ 8. Infectious mononucleosis is called the kissing disease because kissing is a primary mode of transmission.

_____ 9. Slow virus diseases are so named because the viruses that cause them have extremely slow replication cycles.

_____ 10. Herpetic keratitis is a herpes simplex virus (HSV)-related infection of the eye that can cause blindness.

_____ 11. Respiratory syncytial virus is the most dangerous cause of respiratory infection in young children.

_____ 12. The last case of natural infection of smallpox occurred in Somalia in 1977.

_____ 13. Flu epidemics vary greatly in severity and mortality.

_____ 14. Yellow fever is so named because of the jaundice it causes in severe cases.

_____ 15. In most cases, Lassa fever is mild.

_____ 16. Hantavirus is spread through rodent feces, urine, and saliva.

_____ 17. *Ixodes* ticks transmit encephalitis caused by *Flavivirus*.

_____ 18. Nipah virus is similar to the paramyxoviruses that cause mumps and measles.

_____ 19. Creutzfeldt-Jakob disease is not similar to mad cow disease.

CRITICAL THINKING

1. Discuss the pathology of AIDS giving the stages of progression. Why do people develop antibodies against HIV, the causative agent of AIDS, and yet are not protected against the disease by those antibodies? Does this mean that an effective vaccine against the virus is not possible? Why or why not?

2. There are at least two reasons why people continue to get the common cold throughout their lives. Explain what they are and why permanent immunity against this disease is not possible.

ANSWER KEY

Terms and Definitions

1. o, 2. i, 3. j, 4. m, 5. a, 6. e, 7.h, 8. f, 9. k, 10. l, 11. p, 12. q, 13. n, 14. r, 15. g, 16. b, 17. c, 18. d

Fill in the Blank

1. chickenpox; postherpetic neuralgia; shingles 2. hemorrhagic 3. measles; Koplik's spots; subacute sclerosing panencephalitis 4. AIDS 5. genital herpes; congenital (neonatal) 6. rubella; congenital rubella syndrome 7. hemorrhagic fevers; Marburg viral hemorrhagic fever; Ebola virus hemorrhagic fever; Korean hemorrhagic fever; hantavirus pulmonary syndrome; pulmonary syndrome hantavirus 8. hepatitis; hepatitis B; Dane; Delta agent; hepatitis D (HDV); hepatitis D 9. cytomegalovirus inclusion; intranuclear inclusion bodies 10. hemorrhagic fevers; Marburg; vascular; hemorrhaging 11. acute respiratory syndrome (SARS); contagious; pneumonia

True/False

1. T, 2. F, 3. T, 4. T, 5. F, 6. F, 7. T, 8. T, 9. F, 10. T, 11. T, 12. T, 13. T, 14. T, 15. F, 16. T, 17. T, 18. T, 19. F

Multiple Choice

1. c, 2. d, 3. c, 4. d, 5. c, 6. b, 7. a, 8. d, 9. c, 10. c, 11. d, 12. d, 13. b

38 Human Diseases Caused by Bacteria

CHAPTER OVERVIEW

This chapter discusses some of the more important bacterial diseases of humans.

CHAPTER OBJECTIVES

After reading this chapter you should be able to:

- describe bacterial diseases that are transmitted through the air
- discuss arthropod-borne bacterial diseases
- discuss bacterial diseases that require direct contact
- discuss food-borne and waterborne bacterial infections and bacterial intoxications
- discuss sepsis and septic shock
- discuss the bacterial odontopathogens involved in tooth decay and periodontal disease

CHAPTER OUTLINE

I. Airborne Diseases
   A. Chlamydial pneumonia—*Chlamydia pneumoniae*
      1. Mild upper respiratory infection (pharyngitis, bronchitis, sinusitis) with some lower respiratory tract involvement; symptoms include fever, productive cough, sore throat, hoarseness, and pain on swallowing
      2. Infections are common but sporadic; about 50% of adults have antibodies to *C. pneumoniae*; transmitted from human to human without a bird or animal reservoir
      3. Diagnosis is based on symptoms and a microimmunofluorescence test; treatment is with tetracycline and erythromycin
   B. Diphtheria—*Corynebacterium diphtheriae*
      1. Usually affects poor people living in crowded conditions
      2. Caused by an exotoxin (diphtheria toxin) produced by lysogenized bacteria; the toxin leads to inhibition of protein synthesis and cell death
      3. Symptoms include nasal discharge, fever, cough, and the formation of a pseudomembrane in the throat; can lead to paralysis and death; can infect skin at wounds or lesion leading to slow-healing ulcers (cutaneous diphtheria)
      4. Diagnosis is made by observation of pseudomembrane and bacterial culture; treatment is with antitoxin to remove exotoxins and with penicillin or erythromycin to eliminate the bacteria; prevention is by active immunization with the diphtheria-pertussis-tetanus vaccine (DPT)
   C. Legionnaires' disease and Pontiac fever—*Legionella pneumophila*
      1. Legionnaires' disease (legionellosis)
         a. Bacteria are normally found in soil and aquatic ecosystems; also found in air-conditioning systems and shower stalls
         b. Infection causes cytotoxic damage to lung alveoli; symptoms include fever, cough, headache, neuralgia, and bronchopneumonia
         c. Common-source spread; important nosocomial infection
         d. Diagnosis is based on isolation of the bacterium and serological tests; treatment is supportive but also includes administration of erythromycin or rifampin
         e. Prevention is accomplished by elimination of environmental sources

2. Pontiac fever—resembles an allergic disease more than an infection and is characterized by abrupt onset of fever, headache, dizziness and muscle pain; pneumonia does not occur; usually spontaneously resolves in two to five days

D. Meningitis—inflammation of brain or spinal cord meninges caused by a variety of organisms and conditions

    1. Bacterial (septic) meningitis

        a. Diagnosed by the presence of bacteria in the cerebrospinal fluid; transmitted by inhalation of respiratory secretions from carriers or active cases

        b. Symptoms include initial respiratory illness or sore throat interrupted by one of the following: vomiting, headache, lethargy, confusion, and stiffness in the neck and back

        c. Cause is determined by Gram stain, isolation of bacterium from cerebrospinal fluid, or rapid tests; treated with various antibiotics, depending on the specific bacterium involved

        d. Many bacteria can be causes, however, three dominate: *Streptococcus pneumoniae*, *Neisseria meningitidis*, and *Haemophilus influenzae*

            1) *N. meningitidis* causes meningococcal meningitis through airborne transmission; crosses the nasopharynx epithelium, entering the bloodstream; can be prevented with vaccination commonly given to college students in dormitories

            2) *H. influenzae* serotype b is a gram-negative bacterium transmitted by droplet nuclei and leading to pneumonia; bacteremia can ensue and then meningitis; primarily observed in children under age 5; has been dramatically reduced by active immunization with the HIB vaccine

    2. Aseptic (nonbacterial) meningitis syndrome is more difficult to treat and prognosis is poor

E. *Mycobacterium avium-M. intracellulare* and *M. tuberculosis* pulmonary diseases

    1. *M. avium-M. intracellulare* infections

        a. Organisms are normal inhabitants in soil, water, and home dust; these bacteria are closely related and referred to as *M. avium* complex (MAC); it is the most common mycobacterial infection in the U.S.

        b. Both the respiratory and the gastrointestinal tracts have been proposed as portals of entry; the gastrointestinal tract is thought to be the most common site of colonization and dissemination in AIDS patients, in whom the disease can have debilitating effects; pulmonary infection is similar to tuberculosis and is most often seen in elderly patients with preexisting pulmonary disease

        c. MAC can be isolated from sputum and other specimens and identified by acid-fast stain and other methods; treatment is usually multiple drug therapy

    2. *Mycobacterium tuberculosis* infections (tuberculosis; TB)

        a. Human-to-human transmission by droplet nuclei and food-borne transmission; mainly caused by *M. tuberculosis*, but worldwide, TB also is caused by *M. bovis* and *M. africanum*

        b. In lungs, bacterium forms nodules (tubercles) and the disease usually stops, but the bacterium remains alive; over time the tubercles can change into forms (caseous lesions) that lead to reactivation of the disease; *M. tuberculosis* does not have classic virulence factors, but uses toxic mycolic acids as a hydrophobic barrier and have several mechanisms for surviving within macrophages

        c. Infected individuals develop cell-mediated immunity that involves sensitized T cells; when exposed to tuberculosis antigens, these cells cause a delayed-type hypersensitivity; this reaction is the basis of skin tests that indicate prior exposure to *M. tuberculosis*

        d. Diagnosis is by isolation of organism, chest X-ray, skin test, or DNA probes; chemotherapeutic and prophylactic treatment is isoniazid and rifampin, and streptomycin and/or ethambutol

        e. Multidrug-resistant strains are appearing in the population

        f. Prevention and control are accomplished by treatment of infected individuals, vaccination, and better public health measures

F. Pertussis—whooping cough caused by *Bordetella pertussis*

    1. Highly contagious disease that primarily affects children

2. Transmission is by droplet inhalation; the exotoxin PTx causes increases in cellular cAMP levels while a tracheal cytotoxin and dermonecrotic toxin destroy epithelial tissue

3. Disease progresses in stages
   a. Catarrhal stage—inflamed mucous membranes; resembles a cold
   b. Paroxysmal stage—prolonged coughing sieges with inspiratory whoop
   c. Convalescent stage—may take months (some fatalities)

4. Diagnosis is by culture of the bacterium, fluorescent antibody staining, PCR and serological tests; treatment with erythromycin, tetracycline, or chloramphenicol; prevented by DPT vaccine

G. Mycoplasmal pneumonia—atypical pneumonia caused by *Mycoplasma pneumoniae*
   1. Spread by close contact and/or airborne droplets; common and mild in infants; more serious in older children and young adults
   2. Symptoms vary from none to serious pneumonia
   3. Diagnosis is considered if other bacteria cannot be isolated and viruses cannot be detected; rapid antigenic detection kits are now available; colony morphology also is helpful; treatment is usually tetracycline or erythromycin; no preventative measures

H. Streptococcal diseases
   1. Streptococci are a heterogeneous group of gram-positive bacteria, and one of the most important is *Streptococcus pyogenes*; streptococci have a variety of virulence factors, including extracellular enzymes that break down host molecules, streptokinases (destruction of blood clots), cytolysins (kill leukocytes), and capsules and M protein (retard phagocytosis)
   2. *S. pyogenes* is widely distributed in humans and many are asymptomatic carriers; transmission can occur through respiratory droplets, direct contact, or indirect contact
   3. Diagnosis is based on clinical and lab findings; rapid tests are available; treatment is with penicillin or erythromycin; vaccines are not available, except for streptococcal pneumonia
   4. Best control measure is prevention of transmission by isolation and treatment of infected persons
   5. Cellulitis and erysipelas
      a. Cellulitis—diffuse, spreading infection of subcutaneous tissue characterized by redness and swelling
      b. Impetigo—superficial cutaneous infection commonly seen in children
      c. Erysipelas—acute infection of the dermis characterized by reddish patches
   6. Invasive streptococcus A infections
      a. Dependent on specific strains and predisposing host factors; if bacterium penetrates a mucous membrane or takes up residence in a skin lesion, can cause necrotizing fasciitis (destruction of the sheath covering skeletal muscle) or myositis (inflammation and destruction of skeletal muscle and fat tissue)
      b. Rapid treatment with penicillin G reduces the risk of death
      c. Pyogenic exotoxins A and B are produced by 85% of the bacterial isolates; these evoke host defenses that destroy vascular tissues, and the surrounding tissues die from lack of oxygen; one of the toxins is a protease
      d. Can also trigger a toxic shock-like syndrome (TSLS) with a mortality rate over 30%
      e. Best preventative measures include covering food, washing hands, and cleansing and medicating wounds
   7. Poststreptococcal diseases—onset is one to four weeks after an acute streptococcal infection
      a. Glomerulonephritis (Bright's disease)—antibody-mediated inflammatory reaction (type III hypersensitivity); may spontaneously heal or may become chronic; for chronic illness a kidney transplant or lifelong renal dialysis may be necessary
      b. Rheumatic fever—autoimmune disease involving the heart valves, other parts of the heart, joints, subcutaneous tissues, and central nervous system; mechanism is unknown; occurs primarily in children ages 6 to 15 years old; therapy is directed at decreasing inflammation and fever, as well as controlling cardiac symptoms and damage
   8. Streptococcal pharyngitis (strep throat)—inflammatory response with lysis of leukocytes and erythrocytes; diagnosis by rapid tests; treatment is with penicillin, primarily to minimize the

possibility of subsequent rheumatic fever and glomerulonephritis; prevented by proper disposal and cleansing of contaminated objects

    9.   Streptococcal pneumonia

        a.   Endogenous (opportunistic) infection caused by *S. pneumoniae*, a member of normal microbiota; individuals usually have predisposing factors such as viral infection of the respiratory tract, physical injury to the respiratory tract, alcoholism, or diabetes; also may cause sinusitis, conjunctivitis, otitis media, bacteremia, and meningitis

        b.   Bacterium's capsular polysaccharides and a toxin are important virulence factors; diagnosis is by chest X-ray, biochemical tests, and culture

        c.   Treatment is with cefotaxime, ofloxacin, and ceftriaxone; a vaccine (Pneumovax) is available and preventative measures include vaccination and treatment of infectious persons

II.   Arthropod-Borne Diseases

   A.   Ehrlichiosis

      1.   First case was diagnosed in the United States in 1986; caused by a new species of *Rickettsiaceae, Ehrlichia chaffeensis,* which is transmitted from unknown animal vectors to humans by ticks; bacterium infects circulating monocytes and causes a nonspecific febrile illness (human monocytic ehrlichiosis; HME) that resembles Rocky Mountain spotted fever; diagnosis is by serological tests; treatment is with tetracycline

      2.   In 1994 a new form (human granulocytic ehrlichiosis; HGE) was discovered; caused by another species, as yet unidentified; has rapid onset of fever, headache, and muscle aches and is treated with antibiotics

   B.   Epidemic (louse-borne) typhus—*Rickettsia prowazekii*

      1.   Transmitted from person to person by the body louse (in the U.S., a reservoir is the southern flying squirrel); organism is found in insect feces, and feces are deposited when the insect takes a blood meal; as the person scratches, the bite becomes infected; the resulting vasculitis leads to headache, fever, muscle aches, and a characteristic rash; if untreated, recovery takes two weeks, but mortality rate is 50%; recovery gives a solid immunity that also cross-protects against endemic (murine) typhus

      2.   Diagnosis is by characteristic rash, symptoms, and the Weil-Felix reaction; treatment is usually tetracycline and chloramphenicol; control of body louse is important preventive measure; a vaccine also is available

   C.   Endemic (murine) typhus—*R. typhi*

      1.   Occurs in isolated areas around the world, including southeastern and Gulf Coast states, especially Texas; transmitted from rats by fleas

      2.   Similar to epidemic typhus, but milder with lower mortality rate (less than 5%); diagnosis and treatment are the same as for epidemic typhus; rat control and avoidance of rats are preventative measures

   D.   Lyme disease— (LD, Lyme borreliosis) caused by *Borrelia burgdorferi, B. garinii,* and *B. afzelii*

      1.   Tick-borne, with deer, mice, or the woodrat as the natural reservoir

      2.   Disease is complex and progressive; is divide into three stages

        a.   Initial localized stage—characteristic bull's eye rash and flulike symptoms

        b.   Disseminated stage—heart inflammation, arthritis, and neurological symptoms

        c.   Late stage—symptoms resembling Alzheimer's disease and multiple sclerosis with behavioral changes as well

      3.   Laboratory diagnosis is by isolation of the spirochete, PCR to detect DNA in the urine, or serological testing (ELISA or Western Blot); treatment with amoxicillin or tetracycline is effective if administered early; ceftriaxone is used if nervous system involvement is suspected

      4.   Prevention and control involves environmental modification to destroy tick habitat and use of anti-tick compounds

   E.   Plague—*Yersinia pestis*

      1.   Transmitted from rodent by bite of flea, direct contact with animals or animal products, or inhalation of airborne droplets; bacteria survive and proliferate inside phagocytic cells; use

type III secretion system to deliver YOPS (yersinal plasmid-encoded outer membrane proteins)

    2. Symptoms include subcutaneous hemorrhages, fever, and enlarged lymph nodes (buboes); mortality rate is 50 to 70% if untreated

    3. Diagnosis is by direct microscopic examination, culture of buboes, serological tests, PCR, and phage testing; treatment is with streptomycin or tetracycline

    4. Prevention and control involve ectoparasite and rodent control, isolation of human patients, prophylaxis, and vaccination of people at high risk

  F. Q fever—*Coxiella burnetii*

    1. Bacterium can survive outside host by forming endosporelike structures; transmitted by ticks between animals and by contaminated dust to humans; disease is an occupational hazard among slaughterhouse workers, farmers, and veterinarians

    2. Acute onset of severe headache, muscle pain, and fever; rarely fatal, but some develop endocarditis and hepatitis; diagnosis is serological and treatment is usually tetracycline and chloramphenicol; prevention and control measures consist of vaccinating researchers and others of high occupational risk, as well as pasteurization of cow and sheep milk in areas of endemic Q fever

  G. Rocky Mountain spotted fever—*R. rickettsii*

    1. Transmitted by the wood tick or the dog tick; also can be passed from generation to generation of ticks by transovarian passage

    2. Disease is characterized by sudden onset of headache, high fever, chills, and a characteristic rash; if untreated, can destroy blood vessels in the heart, lungs, or kidneys, and lead to death; treatment is usually chloramphenicol and chlortetracycline; diagnosis is through observation of rash and serological tests; best prevention is by avoidance of ticks

III. Direct Contact Diseases

  A. Gas gangrene or clostridial myonecrosis—*Clostridium perfringens, C. novyi,* and *C. septicum*

    1. Found in soil and intestinal tract microbiota; contamination of injured tissues by endospores in soil or fecal material is usual route of transmission

    2. If endospores germinate in anaerobic tissues, bacteria grow, generate gas, and produce toxin and enzymes that cause necrosis (gangrene)

    3. Diagnosis is through recovery of bacterium; treatment involves extensive surgical wound debridement, administration of antitoxins and antibiotics, and the use of hyperbaric oxygen

    4. Prevention and control measures include debridement of contaminated wounds plus antimicrobial prophylaxis and prompt treatment of all wound infections; amputation may be necessary to prevent spread

  B. Group B streptococcal disease—*Streptococcus agalactiae*

    1. Causes severe illness in newborns, including sepsis, meningitis, and pneumonia; may be exposed during birth

    2. Transmitted from person-to-person; many asymptomatic carriers; diagnosed by culturing or immunoassay; mothers often screened late in pregnancy; treated with penicillin or ampicillin

  C. Inclusion conjunctivitis—*Chlamydia trachomatis*

    1. Characterized by copious mucous discharge from eye, inflamed and swollen conjunctiva, and inclusion bodies in host cell cytoplasm; inclusion conjunctivitis of newborns is established from contact with an infected birth canal; in adults, disease spreads primarily by sexual contact

    2. Without treatment, recovery occurs spontaneously; therapy involves treatment with tetracycline, erythromycin, or a sulfonamide; diagnosis is by direct immunofluorescence, Giemsa stain, nucleic acid probes, and culture; prevention depends upon diagnosis and treatment of all infected individuals

  D. Leprosy—severely disfiguring skin disease caused by *Mycobacterium leprae*

    1. Usually requires prolonged exposure to nasal secretion of heavy bacteria shedders

    2. The incubation period may be three to five years, or even longer; starts as skin lesion and progresses slowly; most lesions heal spontaneously, those that don't develop into one of two types of leprosy:

        a.   Tuberculoid (neural) leprosy—mild, nonprogressive form associated with delayed-type hypersensitivity reaction

        b.   Lepromatous (progressive) leprosy—relentlessly progressive disfigurement

    3.   Diagnosis is by observation in biopsy specimens and by serodiagnostic tests

    4.   Treatment—long-term use of sulfa drugs (diacetyl/dapsone) and rifampin, sometimes in conjunction with clofazimine; use of vaccine in conjunction with the drugs shortens the duration of therapy

    5.   Control by identification and treatment of patients; children of contagious parents should be given prophylactic drug therapy until their parents are treated and have become noninfectious

E.   Peptic ulcer disease and gastritis—*Helicobacter pylori*

    1.   Bacterium colonizes gastric mucus-secreting cells, alters gastric pH to favor its own growth, and releases toxins that damage epithelial mucosal cells; strong correlations between gastric cancer rates and *H. pylori* infection rates have been demonstrated

    2.   Transmission is probably person-to-person, but common source has not been definitively ruled out

    3.   Diagnosis is by culture of gastric biopsy specimens, serological testing, stool antigen assays, and tests for urease production

    4.   Treatment includes acid-reducing drugs, bismuth subsalicylate (Pepto-Bismol) and antibiotics

F.   Staphylococcal diseases

    1.   Staphylococci are gram-positive, facultative anaerobes and are usually catalase positive

    2.   Staphylococci are very important human pathogens and also are part of normal human microbiota

    3.   Staphylococci can be divided into pathogenic species and relatively nonpathogenic species by the coagulase test

        a.   *S. aureus*—coagulase positive, pathogenic; causes severe chronic infections

        b.   *S. epidermidis*—coagulase negative, less invasive, opportunistic pathogens associated with nosocomial infections

    4.   Many of the pathogenic strains are slime producers; slime is a viscous extracellular glycoconjugate that allows the bacteria to adhere to smooth surfaces, such as medical prostheses and catheters, and form biofilms; slime also inhibits neutrophil chemotaxis, phagocytosis, and the antimicrobial agents vancomycin and teicoplanin

    5.   Can be spread by hands, expelled from respiratory tract, or transported in or on inanimate objects; staphylococci cause disease in any organ of the body; disease is most likely to occur in individuals whose defenses have been compromised

    6.   Staphylococci produce exotoxins and substances that promote invasiveness

    7.   They produce toxins that can cause disease ranging from food poisoning to bacteremia

        a.   Abscesses—related to coagulase production, which leads to formation of abscess; at core, tissue necrosis occurs

        b.   Impetigo—a superficial skin infection often observed in children

        c.   Toxic shock syndrome (TSS)—serious disease characterized by low blood pressure, fever, diarrhea, skin rash, and shedding of the skin; due to exotoxin that acts as a superantigen causing cytokine overproduction, circulatory collapse, and shock

        d.   Staphylococcal scalded skin syndrome—caused by strains of *S. aureus* that carry a gene for exfoliative toxin; common in infants and children

    8.   Diagnosis is by culture identification, catalase and coagulase tests, serology, DNA fingerprinting, and phage typing; no specific prevention; several antibiotics can be used for treatment, but isolates should be tested for sensitivity because of the existence of many drug-resistant strains; cleanliness, hygiene, and aseptic management of lesions are best control measures

G.   Sexually transmitted diseases

    1.   Introduction

        a.   A global health problem caused by viruses, bacteria, yeasts, and protozoa

        b.   Spread of sexually transmitted diseases (STDs) is currently out of control

c.   STDs are most frequent in the most sexually active group (15–30 years of age); the more sexual partners, the more likely that a person will acquire an STD
2.   Bacterial vaginosis
   a.   Disease is sexually transmitted with polymicrobic etiology; also may be an autoinfection (rectum is inhabited by these organisms)
   b.   Disease is mild but is a risk factor for obstetric infections, various adverse outcomes of pregnancy, and pelvic inflammatory disease
   c.   Diagnosis is based on fishy odor and microscopic observation of clue cells (sloughed-off vaginal epithelial cells covered with bacteria) in the discharge; treatment is with metronidazole
3.   Chancroid-genital ulcer disease—caused by the gram-negative bacillus, *Haemophilus ducreyi*
   a.   Bacterium enters the skin through a break in the epithelium; after 4 to 7 days a papular lesion develops with swelling and white blood cell infiltration; a pustule forms and ruptures leading to a painful ulcer on the penis or vagina; is a cofactor in the transmission of AIDS
   b.   Diagnosis is by isolating the bacterium; treatment is with erythromycin or ceftriaxone; prevention is by use of condoms or abstinence
4.   Genitourinary diseases—*Mycoplasma urealyticum* and *Ureaplasma hominis*
   a.   Transmission is related to sexual activity
   b.   Bacteria opportunistically cause inflammation of reproductive organs of males and females
   c.   Bacteria are difficult to recognize because they are not usually cultured in the clinical microbiology laboratory; diagnosis is usually by recognition of clinical syndromes; treatment is usually tetracycline or erythromycin
5.   Gonorrhea—*Neisseria gonorrhoeae* (gonococci)
   a.   Sexually transmitted disease of the genitourinary tract, eye, rectum, and throat
   b.   Bacteria invade mucosal cells, causing inflammation and formation of pus
   c.   In males there is urethral discharge and painful, burning urination; in females, disease can be asymptomatic, can cause some vaginal discharge, or may lead to pelvic inflammatory disease (PID); in both sexes, disseminated infection can occur; birth through infected vagina can result in neonatal eye infections (ophthalmia neonatorum or conjunctivitis of the newborn) that can lead to blindness
   d.   Diagnosis is by culture of the bacterium, oxidase reaction, Gram stain reaction, and colony and cell morphology; a DNA probe also is useful
   e.   Treatment—antibiotic treatment regimens have been found to be effective; silver nitrate is often used in the eyes of newborns to prevent infection
   f.   Prevention and control by public education, diagnosis, treatment of symptomatic and asymptomatic individuals, and use of condoms
6.   Lymphogranuloma venereum—sexually transmitted disease caused by *Chlamydia trachomatis*
   a.   Occurs in phases
      1)   Primary phase—ulcer on genitals that heals with no scar
      2)   Secondary phase—enlargement of lymph nodes (buboes); fever, chills, and anorexia are common
      3)   Late phase—fibrotic changes and abnormal lymphatic drainage leading to fistulas and/or urethral or rectal strictures; leads to untreatable fluid accumulation in the penis, scrotum, or vaginal area
   b.   Diagnosis by staining infected cells with iodine to observe inclusions, culture, nucleic acid probes, and serological tests; treated by aspiration of buboes and by antibiotics in early phases; by surgery in late phase; controlled by education, prophylaxis, and early diagnosis and treatment

7. Nongonococcal urethritis (NGU) —an inflammation of the urethra *not* caused by *Neisseria gonorrhoeae*
   a. Caused by a variety of agents including *C. trachomatis*; organisms are sexually transmitted; 50% are caused by chlamydia; NGU caused by chlamydia is the most common STD in the U.S.
   b. Infection may be asymptomatic in males or may cause urethral discharge, itching, and inflammation of genital tract; females may be asymptomatic or may develop pelvic inflammatory disease (PID), which can lead to sterility; disease is serious in pregnant females, where it may lead to miscarriage, stillbirth, inclusion conjunctivitis, and infant pneumonia
   c. Diagnosis is by observation of leukocyte exudates, Gram stain reaction, and culture; rapid diagnostic tests are now available; treatment is with various antibiotics
8. Syphilis—*Treponema pallidum*
   a. Sexually transmitted or congenitally acquired in utero
   b. Disease progresses in stages
      1) Primary stage—lesion (chancre) at infection site that can transmit organism during sexual intercourse
      2) Secondary stage—skin rash and other more general symptoms
      3) Latent stage—not communicable after two to four years except possibly congenitally
      4) Tertiary stage—degenerative lesions (gummas) in the skin, bone, and nervous system
   c. Diagnosed by clinical history, physical examination, microscopic examination of fluids from lesions, and serology
   d. Treatment—penicillin in early stages, tertiary stage is highly resistant to treatment; immunity is incomplete and subsequent infections can occur
   e. Prevention and control is by public education, treatment, follow-up on sources and contacts, sexual hygiene, and prophylaxis (use of condoms)
H. Tetanus—*Clostridium tetani*
   1. Found in soil, dust, hospital environments, and mammalian feces
   2. Transmission is associated with skin wounds; bacterium exhibits low invasiveness, but in deep tissues with low oxygen tension, its endospores germinate; when the vegetative cells lyse, they release tetanospasmin (an exotoxin)
   3. Toxin causes prolonged muscle spasms; a hemolysin (tetanolysin) also is produced and aids in tissue destruction
   4. Prevention is important and involves:
      a. Active immunization with toxoid (DPT)
      b. Proper care of wounds contaminated with soil
      c. Prophylactic use of antitoxin
      d. Administration of penicillin
I. Trachoma—*Chlamydia trachomatis*
   1. Greatest single cause of blindness in the world, although uncommon in the U.S.
   2. Transmitted by hand-to-hand contact, by contact with infected fomites, and by flies; first infection usually heals spontaneously with no lasting effects; with reinfection, vascularization of the cornea (pannus formation) and scarring of the conjunctiva occur
   3. Diagnosis and treatment is the same as for inclusion conjunctivitis; prevention and control is by health education, personal hygiene, and access to clean water for washing
IV. Food-Borne and Waterborne Diseases
   A. Food poisoning—gastroenteritis that can arise in two ways
      1. Food-borne infection—microorganism is transferred to host in food and then colonizes host
      2. Food intoxication—toxin is ingested in food; the toxins are called enterotoxins

B. Botulism—*Clostridium botulinum*
1. Frequently caused by canned goods containing endospores, which germinate and produce an exotoxin (neurotoxin) within the food; if food is eaten without adequate cooking, the toxin remains active
2. Can cause death by respiratory or cardiac failure
3. Diagnosis is by toxin detection in the patient's serum, stools, or vomitus; treatment is supportive and also involves antitoxin administration
4. Infant botulism is a disease of infants under 1 year of age; endospores germinate in infant's intestines and then produce toxin
5. Prevention and control involves safe food-processing practices in the food industry and in home canning; not feeding honey to babies under one year of age helps prevent infant botulism

C. *Campylobacter jejuni* gastroenteritis
1. Transmitted by contaminated food or water, contact with infected animals, or anal-oral sexual activity
2. Causes diarrhea, fever, intestinal inflammation and ulceration, and bloody stools
3. Diagnosis is by culture in reduced oxygen environment; disease is self-limited; treatment is supportive, with fluid and electrolyte replacement; erythromycin is used in severe cases

D. Cholera—*Vibrio cholerae*
1. Acquired by ingesting food or water contaminated with fecal material; shellfish and copepods are natural reservoirs
2. Bacteria adhere to the intestinal mucosa of the small intestine; are not invasive, but secrete cholera enterotoxin (choleragen), which stimulates hypersecretion of water and chloride ions, while inhibiting adsorption of sodium ions; leads to fluid loss; death may result from increased protein concentrations in blood, causing circulatory shock and collapse; recent evidence indicates that passage through human host enhances infectivity and may fuel cholera epidemics
3. Diagnosis is by culture of the bacterium from feces and by serotyping; treatment is rehydration therapy (fluid and electrolyte replacement) and administration of antibiotics; control is based on proper sanitation

E. Listeriosis—*Listeria monocytogenes*
1. *L. monocytogenes* is isolated from soil, vegetation, and many animal reservoirs; disease generally occurs in pregnant women or in immunosuppressed individuals; causes meningitis, sepsis, and stillbirth; *does not cause* gastrointestinal illness
2. Bacterium is an intracellular pathogen; can be part of normal gastrointestinal microbiota; pathogenicity is due to production of hemolysins and other enzymes
3. Diagnosis is by culture; treatment is intravenous administration of ampicillin or penicillin; the USDA and food manufacturers are developing food safety measures

F. Salmonellosis—*Salmonella* Typhimurium and other serovars
1. Food-borne, particularly in poultry, eggs, and egg products; also in contaminated water
2. Food infection; bacteria must multiply and invade the intestinal mucosa; as they reproduce they produce enterotoxin and cytotoxin, which destroy intestinal epithelial cells; this causes abdominal pain, cramps, diarrhea, and fever; fluid loss can be a problem, particularly for children and elderly people; treatment is fluid and electrolyte replacement; prevention depends on good food processing practices, proper refrigeration, and adequate cooking
3. Typhoid fever—*Salmonella typhi*
   a. Caused by ingestion of food or water contaminated with human or animal feces
   b. Symptoms are fever, headache, abdominal pain, and malaise, which last several weeks; paratyphoid fever is a milder form caused by different serovars
   c. Diagnosis is by demonstration of bacterium in blood, urine, or stools and by serology; treatment is with antibiotics
   d. Prevention and control involves purification of drinking water, pasteurization of milk, preventing carriers from handling food, and complete patient isolation; a vaccine is available for high-risk individuals

G. Shigellosis—*Shigella* spp.
    1. Shigellosis or bacterial dysentery is transmitted by fecal-oral route and is most prevalent in children 1 to 4 years old; bacterium has small infectious dose (10 to 100 bacteria); in U.S. shigellosis is a particular problem in daycare centers and custodial institutions where there is crowding
    2. Bacteria are facultative intracellular parasites phagocytosed by Peyer's patch cells, but do not usually spread beyond the colon epithelium; endotoxins and exotoxins (particularly shiga-toxin delivered by a type III secretion system) cause watery stools that often contain blood, mucus, and pus; in some cases colon becomes ulcerated
    3. Identification is based on biochemical characteristics and serology; disease is self-limiting in adults but may be fatal in children; treatment is fluid and electrolyte replacement; antibiotics may be used in severe cases; prevention is a matter of personal hygiene and maintenance of a clean water supply
H. Staphylococcal food poisoning—*Staphylococcus aureus*
    1. Caused by ingestion of improperly stored or prepared food in which the organism has grown
    2. Organism produces several enterotoxins that are heat stable
    3. Symptoms include severe abdominal pain, diarrhea, vomiting, and nausea; symptoms come quickly (one to six hours) and leave quickly (24 hours)
    4. Diagnosis is based on symptoms or identification of bacteria or enterotoxins in food; treatment is fluid and electrolyte replacement; prevention and control involves avoidance of contaminated food and control of personnel responsible for food preparation and distribution
I. Traveler's diarrhea and *Escherichia coli* infections
    1. Traveler's diarrhea is a rapidly acting, dehydrating condition caused by certain viruses, bacteria or protozoa normally absent from the traveler's environment; *E. coli* is one of the major causative agents
    2. Six categories or strains of diarrheagenic *E. coli* are now recognized
        a. Enterotoxigenic *E. coli* (ETEC) produces two enterotoxins that are responsible for symptoms, including hypersecretion of electrolytes and water into the intestinal lumen
        b. Enteroinvasive *E. coli* (EIEC) multiplies within the intestinal epithelial cells; may also produce a cytotoxin and an enterotoxin
        c. Enteropathogenic *E. coli* (EPEC) causes effacing lesions, destruction of brush border microvilli on intestinal epithelial cells
        d. Enterohemorrhagic *E. coli* (EHEC) causes attaching-effacing lesions leading to hemorrhagic colitis; it also releases toxins that kill vascular epithelial cells; *E. coli* 0517:H7 is a major form of EHEC and has caused many outbreaks of hemorrhagic colitis in the U.S.
        e. Enteroaggregative *E. coli* (EAggEC) forms clumps adhering to epithelial cells, toxins have not been identified but are suspected from the type of damage done
        f. Diffusely adhering *E. coli* (DAEC) adheres in a uniform pattern to epithelial cells and is particularly problematic in immunologically naive or malnourished children
    3. Diagnosis is based on past travel history and symptoms; lab diagnosis is by isolation of the specific type of *E. coli* from feces and identification using DNA probes, determination of virulence factors, and PCR; treatment is electrolyte replacement plus antibiotics; prevention and control involve avoiding contaminated food and water
V. Sepsis and Septic Shock
A. Cannot be categorized under a specific mode of transmission
B. Sepsis
    1. Systemic response to a microbial infection
    2. Manifested by fever or retrograde fever, heart rate > 90 beats per minute, respiratory rate > 20 breaths per minute, a pCO_2 < 32 mmHg, a leukocyte count > 12,000 cells per ml or < 4,000 cells per ml
C. Septic shock
    1. Sepsis associated with severe hypotension (low blood pressure)

2. Gram-positive bacteria, fungi, and endotoxin-containing gram-negative bacteria can initiate the pathogenic cascade of sepsis leading to septic shock
3. Lipopolysaccharide (LPS), an integral component of the outer membrane of gram-negative bacteria, has been implicated

D. Pathogenesis begins with localized proliferation of the microorganism
1. Bacteria may invade the bloodstream or may proliferate locally and release various products into the bloodstream
2. Products include structural components (endotoxins) and secreted exotoxins
3. These products stimulate the release of endogenous mediators of shock from plasma cells, monocytes, macrophages, endothelial cells, neutrophils, and their precursors
4. The endogenous mediators have profound effects on the heart, vasculature, and other body organs
5. Death ensues if one or more organ systems fail completely

VI. Zoonotic Diseases
A. Anthrax—*Bacillus anthracis*
1. Transmitted by direct contact with infected animals or their products
2. Three forms of disease
  a. Cutaneous anthrax results from contamination of cut or abrasion of the skin; most common form; results in formation of a characteristic skin lesion called an eschar; without antibiotic treatment, mortality can be as high as 20%; treated with either ciprofloxacin or doxycycline plus one or two additional antimicrobial agents; length of treatment depends on whether the disease is thought to be the result of bioterrorism (60 days) or naturally acquired (7 to 10 days)
  b. Pulmonary anthrax (woolsorter's disease) results from inhaling endospores; in the lungs, endospores are engulfed by alveolar macrophages and germinate within the endosomes; they then spread to regional lymph nodes and into the bloodstream; in bloodstream, the bacteria produce a tripartite exotoxin that interferes with host cell communication systems and kills macrophages; illness begins as a nonspecific illness but progresses rapidly to septic shock and death if treatment is not initiated quickly (same antibiotics as for cutaneous anthrax)
  c. Gastrointestinal anthrax occurs if endospores are ingested; illness begins with nausea, vomiting, fever, and abdominal pain; it progresses rapidly to severe, bloody diarrhea; mortality is greater than 50%
3. Diagnosis is by direct microscopic examination, culture of bacteria, and serology; vaccination, particularly of animals and persons with high occupational risks, is an important control measure

B. Brucellosis (undulant fever)— caused by a variety of *Brucella* species
1. Tiny gram-negative coccobacilli transmitted through contact with or consumption of infected animals or waters; can be transmitted by inhalation and through skin wounds; most commonly through infected (unapsteurized) milk products
2. Three forms of the illness:
  a. Acute form—flulike symptoms
  b. Undulant form—rising and falling fever, arthritis, and testicular inflammation; possible neurological effects
  c. Chronic form—chronic fatigue syndrome, depression, and arthritis
3. Treated with doxycycline and rifampin in combination

C. Psittacosis (ornithosis) —*Chlamydia psittaci*
1. Spread by handling infected birds or by inhalation of dried bird excreta; occupational hazard in the poultry industry (particularly to workers in turkey processing plants)
2. Infects respiratory tract, liver, spleen, and lungs, causing inflammation, hemorrhaging, and pneumonia
3. Diagnosis based on isolation of *C. psittaci* from blood or sputum, or by serology; treatment is with tetracycline; prevention is by chemoprophylaxis for pet birds and poultry (this practice can lead to antibiotic resistance and so is discouraged)

D. Tularemia—*Francisella tularensis*
  1. Is spread from animal reservoirs by a variety of mechanisms, including biting arthropods, direct contact with infected tissue, inhalation of aerosolized bacteria, and ingestion
  2. Characterized by ulcerative lesions, enlarged lymph nodes, and fever
  3. Diagnosis by PCR or culture and serological tests; treated with antibiotics; prevention and control involves public education, protective clothing, and vector control; a vaccine is available for high-risk laboratory workers
  4. *F. tularensis* is also a microorganism of concern as a biological threat agent
VII. Dental Infections—Caused by Various Odontopathogens
  A. Dental plaque
    1. Acquired enamel pellicle—a membranous layer produced by the selective absorption of saliva glycoproteins to the hard enamel surface of tooth; its net negative charge helps repel bacteria
    2. Dental plaque is initiated by the colonization of the acquired enamel pellicle by streptococci; this is followed by coaggregation due to cell-to-cell recognition between genetically distinct species; eventually an environment develops that allows *Streptococcus mutans* and *S. sobrinus* to colonize the tooth surface
    3. *S. mutans* and *S. sobrinus* produce glucans that cement plaque bacteria together and create anaerobic microenvironments; these are colonized by anaerobes
    4. After the plaque ecosystem develops, bacteria produce acids that can demineralize the enamel and initiate tooth decay
  B. Dental decay (caries)
    1. Production of fermentation acids after eating and the subsequent return to a neutral pH leads to a demineralization-remineralization cycle
    2. When diet is too rich in fermentable substrates, demineralization exceeds remineralization and leads to dental caries
    3. Drugs are not available to treat dental caries; prevention includes minimal ingestion of sucrose; daily brushing, flossing, and mouthwashes; and professional application of fluoride
  C. Periodontal disease—diseases of the periodontum
    1. Periodontium—supporting structure of tooth; includes the cementum, the periodontal membrane, the bones of the jaw, and the gingivae; disease begins by formation of subgingival plaque and leads to inflammatory reaction (periodontitis); periodontitis leads to formation of periodontal pockets that are colonized by bacteria, causing more inflammation; eventually bone destruction (periodontosis), inflammation of gingiva (gingivitis), and general tissue necrosis occur
    2. Can be controlled by plaque removal; by brushing, flossing, and mouthwashes; and at times by oral surgery

TERMS AND DEFINITIONS

Place the letter of each term in the space next to the definition or description that best matches it.

_____ 1. Inflammation of the membranes around the brain or spinal cord

_____ 2. Diffuse, spreading infection of subcutaneous skin tissue

_____ 3. Acute inflammation of the dermal layer of the skin in people with a history of streptococcal sore throat

_____ 4. Inflammatory disease of the membranous sites within the kidney where blood is filtered

_____ 5. Inflammatory process that destroys the sheath covering skeletal muscle

_____ 6. Inflammation and destruction of skeletal muscle and fat tissue

_____ 7. Inflammation of the throat

_____ 8. Inflammation of the tonsils

_____ 9. An infection caused the person's own microbiota

_____ 10. A hard nodule of bacteria in the lungs surrounded by lymphocytes, macrophages, and connective tissue

_____ 11. A lesion in the lungs that has a cheeselike consistency

_____ 12. Calcified caseous lesions that are easily observed in a chest X ray

_____ 13. An inflammation of the blood vessels

_____ 14. Enlarged lymph nodes associated with the plague

_____ 15. An ulcerated papule on the skin associated with anthrax

_____ 16. Sloughed-off vaginal epithelial cells that are covered with bacteria and are diagnostic for bacterial vaginosis

_____ 17. Tissue necrosis caused by the anaerobic growth of _Clostridium perfringens_ and related species in wounds

_____ 18. A viscous extracellular glycoconjugate that allows bacteria to adhere to smooth surfaces

_____ 19. A small, painless, reddened ulcer with a hard ridge that appears at the infection site in the early stages of syphilis

_____ 20. Degenerative lesions in the skin, bone, and nervous system that are associated with tertiary syphilis

_____ 21. Vascularization of the cornea

_____ 22. Toxins that disrupt the function of the intestinal mucosa

_____ 23. Cholera toxin

_____ 24. Lesions due to the destruction of intestinal brush border microvilli by enteropathogenic _E. coli_

_____ 25. Systemic response to a microbial infection

_____ 26. Dental pathogens

_____ 27. Sepsis accompanied by severe hypotension; often fatal

_____ 28. A covering of dextrans and microorganisms that forms on teeth

_____ 29. Binding of one bacterial species to another

_____ 30. Branched-chain polysaccharides composed of glucose

_____ 31. Cavitation of the teeth caused by lactic acid and acetic acid produced by microorganisms

_____ 32. Inflammation of the gums

a. buboes
b. caries
c. caseous lesion
d. cellulitis
e. chancre
f. choleragen
g. clue cells
h. coaggregation
i. dental plaque
j. effacing lesions
k. endogenous infection
l. enterotoxins
m. erysipelas
n. eschar
o. gas gangrene (clostridial myonecrosis)
p. Ghon complexes
q. gingivitis
r. glomerulonephritis (Bright's disease)
s. glucans
t. gummas
u. meningitis
v. myositis
w. necrotizing fasciitis
x. odontopathogens
y. pannus
z. pharyngitis
aa. sepsis
bb. septic shock
cc. slime
dd. tonsillitis
ee. tubercle
ff. vasculitis

BACTERIAL DISEASES

Select the organism that causes each disease listed below. If more than one organism is associated with a particular disease, select all of the appropriate alternatives.

_____ 1. anthrax
_____ 2. bacterial vaginosis
_____ 3. botulism
_____ 4. brucellosis
_____ 5. chancroid (genital ulcer disease)
_____ 6. cholera
_____ 7. diphtheria
_____ 8. endemic typhus
_____ 9. epidemic typhus
_____ 10. ehrlichiosis
_____ 11. gas gangrene (clostridial myonecrosis)
_____ 12. gastroenteritis
_____ 13. gonorrhea

_____ 14. hemolytic uremic syndrome
_____ 15. impetigo
_____ 16. inclusion conjunctivitis
_____ 17. legionellosis
_____ 18. leprosy
_____ 19. listeriosis
_____ 20. Lyme disease
_____ 21. lymphogranuloma venereum
_____ 22. nongonococcal urethritis
_____ 23. opthalmia neonatoram
_____ 24. pertussis
_____ 25. peptic ulcer disease
_____ 26. plague

_____ 27. pneumonia
_____ 28. Pontiac fever
_____ 29. psittacosis (ornithosis)
_____ 30. Q fever
_____ 31. rheumatic fever
_____ 32. Rocky Mountain spotted fever
_____ 33. shigellosis
_____ 34. syphilis (venereal syphilis and congenital syphilis)
_____ 35. tetanus
_____ 36. trachoma
_____ 37. traveler's diarrhea
_____ 38. tuberculosis
_____ 39. tularemia
_____ 40. typhoid fever

a. *Bacillus anthracis*
b. *Bordetella pertussis*
c. *Borrelia burgdorferi*
d. *Brucella* spp.
e. *Campylobacter jejeuni*
f. *Chlamydia pneumoniae*
g. *Chlamydia psittaci*
h. *Chlamydia trachomatis*
i. *Clostridium botulinum*
j. *Clostridium novyi*
k. *Clostridium perfringens*
l. *Clostridium tetani*
m. *Corynebacterium diphtheriae*
n. *Coxiella burnetii*
o. *Ehrlichia chaffeensis*
p. *Escherichia coli*
q. *Franciscella tularensis*
r. *Gardnerella vaginalis*
s. *Haemophilus ducreyi*
t. *Helicobacter pylori*
u. *Legionella pneumophila*
v. *Listeria monocytogenes*
w. *Mycobacterium leprae*
x. *Mycobacterium tuberculosis*
y. *Mycoplasma hominis*
z. *Mycoplasma pneumoniae*
aa. *Neisseria gonorrhoeae*
bb. *Rickettsia prowazekii*
cc. *Rickettsia rickettsii*
dd. *Rickettsia typhi*
ee. *Salmonella* spp.
ff. *Salmonella typhi*
gg. *Shigella* spp.
hh. *Staphylococcus aureus*
ii. *Streptococcus pyogenes*
jj. *Treponema pallidum*
kk. *Ureaplasma urealyticum*
ll. *Vibrio cholerae*
mm. *Yersinia pestis*

FILL IN THE BLANK

1. The DPT vaccine protects against _____, _____, and _____.

2. Bacterial meningitis is also called _____ meningitis, while meningitis caused by other types of agents, such as viruses, is called _____ meningitis.

3. The lesions caused by *Mycobacterium tuberculosis* can take various forms. If they are hard nodules surrounded by lymphocytes, macrophages, and connective tissue, they are called _____. Later they may take on a cheeselike consistency and are referred to as _____ lesions; if they calcify, they are called _____ _____, and show up prominently in a chest X ray. Sometimes the lesions liquefy and form air-filled cavities called _____ cavities. From these, the bacteria can spread to new foci of infections. This spreading is referred to as either _____ tuberculosis or as _____ tuberculosis.

4. Staphylococci are responsible for a wide variety of diseases because of their numerous virulence factors. One important virulence factor observed in some strains is the ability to produce _____, a viscous extracellular glycoconjugate that is important in the formation of _____ on catheters and prosthetic medical devises. Another important virulence factor is the secretion of exotoxins, which are responsible for diseases such as staphylococcal _____ _____, _____ _____ _____ (TSS), and staphylococcal _____ _____ _____. The latter involves the action of _____ toxin, also called _____.

5. The plague is sometimes referred to as _____ plague because of the presence of enlarged lymph nodes called _____. If the plague bacteria invade the lungs, they may cause a more serious disease referred to as _____ plague.

6. Leprosy is sometimes referred to as _____ disease. It can take two distinct forms: _____ (neural) leprosy, a mild, nonprogressive hypersensitivity to surface antigens of the organism; and _____ (progressive) leprosy, a relentlessly progressive form that kills skin cells and leads to the loss of facial features, fingers, toes and other structures.

7. Two toxins released by *C. tetani* cause the disease called _____ (more commonly lockjaw). One toxin, called _____, causes uncontrolled stimulation of skeletal muscles (spasms). The other toxin, called _____, is a hemolysin that aids in tissue destruction.

8. Streptococci cause a wide variety of diseases including one that resembles toxic shock syndrome. The streptococcal disease, called _____ _____ _____ is caused by a strain that produces one or more pyrogenic exotoxins.

9. Streptococcal pneumonia is considered an _____ _____ because it is caused by one's own microbiota.

10. In the primary phase of the disease _____ _____, an ulcer appears in the genital area; it heals quickly and generally leaves no scar. In the secondary phase, the regional lymph nodes become enlarged and tender, and are referred to as _____. If not treated, a late phase with fibrotic changes and abnormal lymphatic drainage occurs, which can lead to untreatable fluid accumulation in the penis, scrotum, or vagina.

11. Any inflammation of the urethra not caused by *Neisseria gonorrhoeae* is referred to as _____ _____. Males have few, if any, manifestations of the disease. Females may be asymptomatic, or they may have a severe infection called _____ _____ _____.

12. *Chlamydia psittaci* causes a disease that was first recognized in parrots and parakeets (psittacine birds); it was referred to as _____. However, it is now recognized in pigeons, chickens, ducks, turkeys, and other birds, and has been given the more general name _____.

13. In _____, the first infection site usually heals spontaneously. Reinfection leads to vascularization of the cornea (called _____ formation), which leads to scarring of the conjunctiva and may result in blindness.

14. A group of diseases called _____ disease affect the _____ (the supporting structure of a tooth). These diseases are initiated by the formation of _____ plaque, plaque that forms at the dentogingival margin and extends down into the gingival tissue. The result of this is an initial inflammatory reaction known as _____. It leads to the formation of pockets that can be colonized, leading to more inflammation. If the new inflammation is restricted to the gums, it is referred to as _____; if it involves bone destruction, it is called _____.

399

15. Teeth have a membranous layer called the _____ _____ _____, which acts as a natural defense against colonization of their surfaces. However, this defense breaks down when _____ _____ formation begins. After initial colonization by streptococci, other bacteria attach by a process called _____, which results from cell-to-cell recognition between genetically distinct bacteria.

16. The disease _____, caused by the bacterium *Bacillus anthracis*, takes several different forms in humans. If the bacterium gains access through a cut or abrasion, _____ _____ occurs. If endospores are inhaled, the resulting illness is _____ _____, also know as woolsorter's disease. If endospores reach the intestine, the resulting illness is called _____ _____.

17. Although primarily thought of as a respiratory pathogen, *C. diphtheriae* can also infect the skin causing _____ _____.

18. Six categories of diarrheagenic *E. coli* are now recognized: *E. coli* _____, *E. coli* _____ *E. coli* _____, *E. coli* _____, *E. coli* _____, and _____ *E. coli.*

19. Some diseases are caused by a variety of different microorganisms some examples are _____, an inflammation of the meninges; _____, inflammation of the intestines; and urethritis, inflammation of the urethra. The latter can be caused by _____, mycoplasmas, chlamydiae, and other bacteria.

20. Gastroenteritis caused by *Salmonella* spp. is called _____ and that caused by *C. jejuni* is called_____.

MULTIPLE CHOICE

For each of the questions below select the *one best* answer.

1. Which of the following is NOT a streptococcal disease?
    a. cellulitis
    b. erysipelas
    c. rheumatic fever
    d. All of the above are streptococcal diseases.
2. Which term refers to diseases of the gastrointestinal tract that are caused by ingestion of preformed bacterial exotoxins?
    a. food intoxications (poisonings)
    b. food infections
    c. Both (a) and (b) are correct.
    d. Neither (a) nor (b) is correct.
3. Which term refers to diseases of the gastrointestinal tract that require colonization by a viable microorganism to produce the symptoms?
    a. food intoxications (poisonings)
    b. food infections
    c. Both (a) and (b) are correct.
    d. Neither (a) nor (b) is correct.

4. Which of the following is NOT considered a predisposing factor for streptococcal pneumonia?
    a. diabetes
    b. alcoholism
    c. viral respiratory infections
    d. All of the above are predisposing factors for streptococcal pneumonia.
5. Which of the following is used in the treatment of gas gangrene?
    a. surgical removal of foreign material and dead tissue from the wound
    b. antitoxins against the necrosis-causing toxins released by the organism
    c. hyperbaric chambers containing pressurized oxygen to poison the anaerobic organism that causes the condition
    d. All of the above are used in the treatment of gas gangrene.
6. Which of the following is a mode of transmission for *Campylobacter jejuni*?
    a. contaminated food or water
    b. contact with infected animals
    c. anal-oral sexual activity
    d. All of the above are correct.

7. Which of the following factors determine whether an invasive streptococcus A infection will occur?
   a. which strain infects the host
   b. predisposing factors in the host, such as diabetes and wounds.
   c. Both (a) and (b) are correct.
   d. Neither (a) nor (b) is correct.

8. Which term refers to sloughed-off vaginal epithelial cells that are covered with bacteria and are associated with bacterial vaginosis?
   a. nurse cells
   b. clue cells
   c. vaginotal cells
   d. diagnostic cells

9. Which of the following procedures is used to prevent and control syphilis?
   a. follow-up on sources and contacts
   b. sexual hygiene
   c. use of condoms
   d. All of the above help control syphilis.

10. Which of the following is NOT used to control botulism?
    a. safe food-processing practices in the food industry
    b. safe food-processing practices in home canning
    c. not feeding honey to babies less than one year of age
    d. All of the above are used to control botulism.

11. In which setting is shigellosis (bacterial dysentery) a particular problem?
    a. daycare centers
    b. crowded custodial institutions
    c. Both (a) and (b) are correct.
    d. Neither (a) nor (b) is correct.

12. Which diagnostic approach is used to diagnose traveler's diarrhea and identify the specific type of *E. coli* involved?
    a. DNA probes
    b. determination of virulence factors
    c. PCR
    d. All of the above are used.

13. Which of the following is sensitive to sulfonamides?
    a. *C. trachomatis*
    b. *C. psittaci*
    c. Both (a) and (b) are correct.
    d. Neither (a) nor (b) is correct.

14. Which of the following does NOT normally cause genitourinary disease?
    a. *Mycoplasma hominis*
    b. *Ureaplasma urealyticum*
    c. *Neisseria gonorrhoeae*
    d. *Helicobacter pylori*

15. Which of the following can survive outside the host by forming endosporelike structures?
    a. *Rickettsia prowazekii*
    b. *Rickettsia typhi*
    c. *Coxiella burnetii*
    d. *Rickettsia rickettsii*

16. Which of the following commonly uses an avian reservoir?
    a. *C. psittaci*
    b. *C. pneumoniae*
    c. *C. trachomatis*
    d. All of the above use an avian reservoir.

17. Psittacosis is generally considered to be an occupational hazard for which of the following groups of people?
    a. hog farmers
    b. turkey processing plant workers
    c. students working on a microbiology study guide
    d. All of the above are correct.

18. Prevention and control of Q fever consists of
    a. vaccination of high-risk individuals.
    b. pasteurization of cow and sheep milk in areas of endemic Q fever.
    c. Both (a) and (b) are correct.
    d. Neither (a) nor (b) is correct.

19. Late in pregnancy, mothers are often screened for which of the following?
    a. *Brucella*
    b. *Campylobacter*
    c. Group B streptococcus
    d. *Clostridia perfringens*

20. Which of the following is also called undulant fever?
    a. brucellosis
    b. plague
    c. necrotizing fasciitis
    d. Q fever

TRUE/FALSE

_____ 1. *Corynebacterium diphtheriae* can only cause diphtheria when it is lysogenized by a phage responsible for the production of diphtheria toxin.

_____ 2. Once an initial case has occurred, Legionnaires' disease spreads as a propagated epidemic.

_____ 3. Bacterial meningitis is more difficult to treat than meningitis caused by other agents.

_____ 4. Although permanent immunity is not established, outbreaks of streptococcal sore throat are less frequent among adults because of their accumulation of antibodies to many different serotypes.

_____ 5. There are more male than female asymptomatic carriers of gonorrhea.

_____ 6. Gastroenteritis caused by *Campylobacter jejuni* has a high mortality rate (30 to 50%) if untreated.

_____ 7. *Legionella pneumophila* causes both legionellosis and Pontiac fever.

_____ 8. Whooping cough is divided into three stages: the catarrhal stage, the paroxysmal stage, and the convalescent stage.

_____ 9. Mycobacterial (MAC) pneumonia has been seen in 15 to 40% of AIDS patients and is becoming an increasing problem.

_____ 10. Because leprosy is not very contagious, it usually is not necessary to treat uninfected children of infected parents.

_____ 11. The evidence is very strong, if not overwhelming, that *Helicobacter pylori* is the primary cause of peptic ulcer disease.

_____ 12. Bacterial vaginosis is frequently the result of autoinfection from the rectum where the causative organisms usually reside.

_____ 13. *Staphylococcus epidermidis* is more invasive than *Staphylococcus aureus*.

_____ 14. Shigellosis is often fatal in both adults and children.

_____ 15. Sepsis and septic shock cannot be categorized under a specific mode of transmission.

_____ 16. Septic shock is actually the result of endogenous mediators released from the patient's own cells in response to bacterial endotoxins or exotoxins.

_____ 17. *E. coli* 0157:H7 is a major cause of hemorrhagic colitis in the U.S.

_____ 18. Tularemia is primarily spread from animal reservoirs.

_____ 19. Streptococcal sore throat is one of the most common bacterial infections.

_____ 20. Six categories of diarrheagenic *E. coli* are now recognized.

_____ 21. Different serotypes of *C. trachomatis* cause different disease states.

_____ 22. *Chlamydia trachomatis* is the greatest single cause of blindness throughout the world.

_____ 23. Mycoplasmas that cause genitourinary disease can be readily cultured and identified so that specific treatment can be initiated.

_____ 24. Mycoplasmal pneumonia is considered only when other bacterial and viral etiologies have been eliminated as possibilities.

_____ 25. Recovery from epidemic typhus confers permanent immunity and also protects against endemic typhus.

_____ 26. Rocky Mountain spotted fever is so named because it only occurs in the Rocky Mountain area.

_____ 27. Nongonococcal urethritis (NGU) is the most common STD in the U.S.

_____ 28. *C. pneumoniae* infections are common but sporadic; about 50% of adults have antibody to the organism.

_____ 29. Transmission of *C. pneumoniae* requires an avian reservoir.

_____ 30. Most active cases of tuberculosis are due to the reactivation of old dormant infections.

_____ 31. *R. rickettsia* can be transmitted form one generation of the tick vector to the next by transovarian passage.

_____ 32. The Weil-Felix reaction is used to diagnose infections caused by members of the genus *Proteus*.

CRITICAL THINKING

1. In 1976, an outbreak of a respiratory disease occurred at a hotel in Philadelphia where a convention of the American Legion was being held. The previously unrecognized disease was named Legionnaires' disease. Assume you were the epidemiologist involved in this. How would you identify the source and/or reservoir for the causative agent, and how would you determine if it was a common-source or propagated epidemic?

2. What is the difference between a food infection and a food intoxication (poisoning)? Give at least two examples of each and discuss how the treatments vary for the different types of food-borne diseases.

3. Discuss the steps involved in the development of dental caries. At each step, name the microorganisms involved and describe how they contribute to tooth decay.

4. When characterizing the identifying the cause of an illness, clinical microbiologists often narrow the field of potential pathogens and the diagnostic procedures to use by considering the specimen obtained and the *most likely* pathogens with that illness and the normal microbiota usually associated with that specimen. For each of the following specimens and illnesses, make a list of the bacterial pathogens most likely to cause that disease. For each pathogen, indicate the mode of transmission and any characteristics that might be useful in identifying the pathogen. For each specimen, also indicate the normal microbiota that might also be present.

- Urine specimen: urethritis

- Throat swab: sore throat

- Stool sample; diarrhea

ANSWER KEY

Terms and Definitions

1. u, 2. d, 3. m, 4. r, 5. w, 6. v, 7. z, 8. dd. 9. k, 10. ee, 11. c, 12. p, 13. ff, 14. a, 15. n, 16. g, 17. o, 18. cc, 19. e, 20. t, 21. y, 22. l, 23. f, 24, j, 25. aa, 26. x, 27. bb, 28. i, 29. h, 30. s, 31. b, 32. q

Bacterial Diseases

1. a, 2. r, y, 3. i, 4. d, 5. s, 6. ll, 7. m, 8. dd, 9. bb, 10. o, 11. j, k, 12, e, ee 13. aa, 14. p, 15. hh, ii 16. h, 17. u, 18, w, 19. v, 20. c, 21. h, 22. h, y, kk, 23. aa, 24. b, 25. t, 26, mm, 27. f, z, hh, ii 28. u 29. g, 30. n, 31, ii, 32, cc, 33. gg, 34. jj, 35. l, 36. h, 37. p, 38. x, 39. q, 40. ff

Fill in the Blank

1. diphtheria; pertussis; tetanus 2. septic; aseptic 3. tubercles; caseous; Ghon complexes; tuberculosis; miliary; reactivation 4. slime; biofilms; food poisoning, toxic shock syndrome, scaled skin syndrome; exfoliative; exfolitin 5. bubonic; buboes; pneumonic 6. Hansen's; tuberculoid; lepromatous 7. tetanus; tetanospasmin; tetanolysin 8. toxic shock-like syndrome (TSLS) 9. endogenous infection 10. lymphogranuloma venereum; buboes 11. nongonococcal urethritis (NGU); pelvic inflammatory disease (PID) 12. psittacosis; ornithosis 13. pannus 14. periodontal; periodontium; subgingival; periodontitis; gingivitis; periodontosis 15. acquired enamel pellicle; dental plaque; coaggregation 16. anthrax; cutaneous anthrax; pulmonary anthrax; gastrointestinal anthrax 17. cutaneous diphtheria 18. enterotoxigenic (ETEC); enteroinvasive (EIEC); enteropathogenic (EPEC); enterohemorrhagic (EHEC); enteroaggregative (EaggEC); diffusely adhering (DAEC) 19. meningitis; gastroenteritis; gonococci 20. salmonellosis; campylobacteriosis

Multiple Choice

1. d, 2. a, 3. b, 4. d, 5. d, 6. d, 7. c, 8. b, 9. d, 10. d, 11. c, 12. d, 13. a, 14. d, 15. c. 16. a, 17. b, 18. c, 19. c, 20. a

True/False

1. T, 2. F, 3. F, 4. T, 5. F, 6. F, 7. T, 8. T, 9. T, 10. F, 11. T, 12. T, 13. F, 14. F, 15. T, 16. T, 17. T, 18. T, 19. T, 20. T, 21. T, 22. T, 23. F, 24. T, 25. T, 26. F, 27. T, 28. T, 29. F, 30 T, 31. T, 32. F

39 Human Diseases Caused by Fungi and Protists

CHAPTER OVERVIEW

This chapter discusses some of the more important fungal and protist diseases of humans. The clinical manifestations, diagnosis, epidemiology, pathogenesis, and treatment of selected diseases are presented.

CHAPTER OBJECTIVES

After reading this chapter you should be able to:

- discuss the five types of diseases caused by fungi (mycoses) and give examples of each
- discuss some of the more important diseases caused by protists

CHAPTER OUTLINE

I. Pathogenic Fungi and Protists
    A. Medical mycology—discipline that deals with fungi that cause human disease; fungal diseases are called mycoses
    B. Systemic mycoses are caused by dimorphic fungi (except for *Cryptococcus neoformans*, which has only a yeast form); usually acquired by inhalation of spores from soil; infection begins as lung lesions, becomes chronic, and disseminates through the bloodstream to other organs
    C. Protists are transmitted by arthropod vectors or by food and water vehicles; fewer than 20 protist genera cause human disease; however, malaria, trypanosomes, and amoeba account for hundreds of millions of cases
II. Airborne Diseases
    A. Blastomycosis—*Blastomyces dermatitidis*
        A. Occurs in three clinical forms: cutaneous, pulmonary, and disseminated
        B. Diagnosis is aided by serological tests; antifungal agents are effective; surgery may be necessary to drain large abscesses; no prevention or control measures
    B. Coccidiomycosis—*Coccidioides immitis*
        A. Acquired by inhalation of spores
        B. Usually an asymptomatic or mild respiratory infection that spontaneously resolves in a few weeks; occasionally progresses to chronic pulmonary disease
        C. Diagnosis is by culturing; serological tests also are available; treatment with several antifungal agents; prevention involves reduction of exposure to dust in endemic areas
    C. Cryptococcosis—*Cryptococcus neoformans*
        A. Aged, dried pigeon droppings are a source of infection; fungus enters by the respiratory tract
        B. Minor transitory pulmonary infection that can disseminate and cause meningitis
        C. Diagnosis is by microscopic examination of specimens and immunological procedures; treatment includes amphotericin B or intraconazole; no prevention or control measures
    D. Histoplasmosis—*Histoplasma capsulatum*
        A. A facultative fungus that grows intracellularly
        B. Found worldwide in soils; spores are easily spread by air currents and inhaled; the spores are most prevalent where bird droppings have accumulated
        C. Inhaled microsporidia are transformed into budding yeasts within alveolar spaces and are eventually observed within the cells of the monocyte-macrophage system; symptoms are usually those of mild respiratory involvement; it rarely disseminates

D. Diagnosis by immunological tests and culture; most effective treatment is amphotericin B, ketoconazole, or intraconazole; prevention and control by using protective clothing and masks and by soil decontamination where feasible

III. Arthropod-Borne Diseases
    A. Malaria—*Plasmodium falcipurum, P. malariae, P. vivax, P. ovale*
        A. Transmitted to humans by bite of an infected female *Anopheles* mosquito; sporozoites in blood move to and reproduce in the liver (asexual schizogony), producing merozoites; these enter erythrocytes and form large trophozoites that undergo schizogony and the schizonts produce merozoites; the cycle repeats every 48 to 72 hours
        B. Periodic sudden release of merozoites, toxins, cell debris from the infected erythrocytes; and TNF-α and interleukin-1 from macrophages trigger the characteristic attack of chills and fever; anemia can result, and the spleen and liver often hypertrophy
        C. Diagnosis is by microscopic examination of blood smears; serological tests are also available; treatment is by chloroquine or related drugs; a potential vaccine (Mosquirix) provides partial protection in children
    B. Leishmaniasis
        A. Caused by flagellated protists (hemoflagellates) transmitted by sandflies from canines and rodents
        B. Can be mucocutaneous, cutaneous, or visceral; symptoms vary with the particular etiological organism involved
        C. Treated with pentavalent antimonial compounds; recovery usually confers permanent immunity; vector and reservoir control and epidemiological surveillance are the best options for control
    C. Trypanosomiasis
        A. *T. brucei,* a hemoflagellate, causes African trypanosomiasis; transmitted by tsetse flies; causes interstitial inflammation and necrosis of the lymph nodes, brain, and heart; causes sleeping sickness (uncontrollable lethargy)
        B. *T. cruzi* causes American trypanosomiasis (Chagas' disease); transmitted when bite of triatomid (kissing) bug is contaminated with insect feces; symptoms are similar to those caused by *T. brucei*
        C. Trypanosomiasis is diagnosed by microscopic examination of blood and by serological tests; drugs are available only for treatment of African trypanosomiasis; vaccines are not useful because the parasite can change its coat to avoid the immune response

IV. Direct Contact Diseases
    A. Superficial mycoses
        A. Most occur in the tropics
        B. The fungi that cause the disease are limited to the outer surface of the hair and the skin
            1. Piedras are infections of hair shaft that result in formation of a hard nodule
            2. Tineas are infections of the outer layer of skin, nails, and hair
        C. Treatment involves removal of skin scales and infected hairs; prevention is by good personal hygiene
    B. Cutaneous mycoses—dermatomycoses, ringworms, tineas
        A. Occur worldwide; most common fungal diseases
        B. Three genera, *Epidermophyton, Microsporum,* and *Trichophyton,* are involved
        C. Diagnosed by microscopic examination of skin biopsies and by culture on Sabouraud's glucose agar
        D. Treatment—topical ointments, oral griseofulvin, or oral itraconazole (sporanox)
        E. Different diseases are distinguished according to the causative agent and the area of the body affected (tinea barbae—beard hair, tinea capitis—scalp hair, tinea corporis—any part of skin, tinea cruris—groin, tinea pedis—athlete's foot, tinea mannum—hands, tinea unguium —nail bed)
    C. Subcutaneous mycoses
        A. The fungi that cause these diseases are saprophytes in soil; they gain entry by puncture wounds

      B.   Disease develops slowly over a period of years, during which time nodule develops and then ulcerates; organisms spread along lymphatic channels, producing more nodules at other locations

      C.   Treatment is with 5-fluorocytosine, iodides, amphotericin B, and surgical excision; diagnosis is by culture of the infected tissue

      D.   Examples include chromoblastomycosis, maduromycosis, and sporotrichosis

D.   Toxoplasmosis—*Toxoplasma gondii*

      A.   Fecal-oral transmission from infected animals; also transmitted by ingestion of undercooked meat and by congenital transfer, blood transfusion, or tissue transplant

      B.   Most cases are asymptomatic or resemble mononucleosis; can be fatal in immunocompromised individuals; leads to severe congenital effects for pregnant women

      C.   Acute disease is characterized by lymphadenopathy, enlargement of reticular cells, pulmonary necrosis, myocarditis, hepatitis, and retinitis; a major cause of death in AIDS patients

      D.   Diagnosis is by serological tests; chemotherapeutic agents are available for treatment; prevention and control require minimizing exposure by not eating raw meat and eggs, washing hands after working in soil, cleaning cat litter boxes daily, keeping cats indoors, and feeding cats commercial food

E.   Trichomoniasis—*Trichomonas vaginalis;* a sexually transmitted disease; host accumulates leukocytes at the site of infection; in females, this leads to a yellow purulent discharge and itching; in males, most infections are asymptomatic; treatment is with metronidazole

V.   Food-Borne and Waterborne Diseases

A.   Amebiasis (amebic dysentery) —*Entamoeba histolytica*

      A.   Ingested cysts excyst in the intestine and proteolytically destroy the epithelial lining of the large intestine

      B.   Disease severity ranges from asymptomatic to fulminating dysentery, exhaustive diarrhea, and abscesses of the liver, lungs, and brain

      C.   Diagnosis is based on finding trophozoites in fresh, warm stools and cysts in ordinary stools; serological testing also should be done; treatment with several drugs is possible; prevention and control involves avoiding contaminated water; hyperchlorination or iodination can destroy waterborne cysts

B.   Amebic meningoencephalitis—caused by *Naegleria* and *Acanthamoebae;* facultative parasites that cause primary amebic meningoencephalitis and keratitis (particularly among wearers of soft contact lenses); found in freshwater and soil; diagnosis is by microscopic examination of clinical specimens; most are resistant to common antimicrobial agents

C.   Cryptosporidiosis—*Cryptosporidium parvum*

      A.   Found in the intestines of many birds and mammals, which shed oocysts into the environment in fecal material; when oocysts are ingested, they excyst in the small intestine; the released sporozoites parasitize intestinal epithelial cells

      B.   Major symptom of infection is diarrhea; diagnosis is by microscopic examination of feces; treatment is supportive; patients will usually recover, but the disease can be fatal in late stage AIDS patients

D.   Cyclospora—*Cyclospora cayetanensis*

      A.   Mainly in tropical regions; linked to contaminated produce

      B.   Frequent, explosive diarrhea with fever, fatigue, and weight loss; infects intestines and shed cyst oocysts in feces; these must differentiate into sporozoites to be infectious

      C.   Diagnosis is by observation of oocysts in feces; treatment is with trimethoprim and sulfamethoxazole and fluid replacement; prevention is through avoidance of contaminated food and water

E.   Giardiasis—*Giardia intestinalis*

      A.   Most common cause of waterborne epidemic diarrheal disease; commonly found in daycare facilities with diapered children

      B.   Transmission is usually by cyst-contaminated water supplies, and disease is common in wilderness areas where animal carriers shed cysts into otherwise "clean" water

C. Disease varies in severity; asymptomatic carriers are common; may be chronic or acute

D. Diagnosis is by identification of trophozoites; immunological tests also are available; treatment is usually metronidazole (Flagyl); prevention involves avoiding contaminated water and the use of slow-sand filters in the processing of drinking water

VI. Opportunistic Diseases

A. Opportunistic organisms are normally harmless but can cause disease in a compromised host

B. Aspergillosis—*Aspergillus fumigatus* or *A. flavus*

1. Portal of entry is respiratory tract; inhalation can lead to several types of pulmonary aspergillosis; the fungus can spread to other tissues and organs; in immunocompromised patients, invasive aspergillosis (mycelia in lungs) may occur

2. Diagnosis depends on examination of specimens or isolation and characterization of fungus; treated with intraconazole

C. Candidiasis—*Candida albicans*

1. *C. albicans* is part of normal microbiota and can be transmitted sexually

2. Exhibits a diverse spectrum of disease:

a. Oral candidiasis (thrush) —mouth; common in newborns

b. Paronychia—subcutaneous tissues of the digits

c. Onychomycosis—subcutaneous tissues of the nails

d. Intertriginous candidiasis—warm, moist areas such as axillae, groin, and skin folds (e.g., diaper candidiasis, candidal vaginitis, and balanitis)

3. Diagnosis is difficult; no satisfactory treatment; cutaneous lesions can be treated with topical agents; oral antibiotics are used for systemic candidiasis

D. Microsporidia—obligate intracellular *Microspora*

1. Infect a wide range of animal hosts that act as reservoirs; highly resistant spores (with characteristic polar tube structure) survive long periods in the environment; mainly seen in AIDS patients

2. Spores inject intraspore contents (sporoplasm) into host cells using polar body; the sporoplasm multiplies asexually and generates more spores; symptoms include hepatitis, pneumonia, skin lesions, weight loss, diarrhea, and wasting syndrome

3. Diagnosis is based on stained microsporidia, if possible with electron microscopy, and via PCR; treatment is not well defined

E. *Pneumocystis* pneumonia

1. Caused by a fungus (*Pneumocystis jeroveci*) that was once thought to be a protozoan

2. Disease occurs almost exclusively in immunocompromised hosts including more than 80% of AIDS patients; the fungus remains localized in the lungs, even in fatal cases

3. Definitive diagnosis involves demonstrating the presence of the organism in infected lung material or PCR analysis; treatment is by oxygen therapy and combination drug therapy; prevention and control is through prophylaxis with drugs in susceptible persons

TERMS AND DEFINITIONS

Place the letter of each term in the space next to the definition or description that best matches it.

_____ 1. The discipline that deals with fungi that cause human disease

_____ 2. General term for diseases caused by fungi

_____ 3. Fungal infections of the hair shaft

_____ 4. Fungi that cause cutaneous infections

_____ 5. A tumorlike deformity that results from a subcutaneous fungal infection

_____ 6. A microorganism that is generally harmless in its own environment but becomes pathogenic in a compromised host

_____ 7. Candidal infection of the digits

_____ 8. Candidal infection of the nails

_____ 9. An inflammation of the cornea

_____ 10. Transmitted by the bites of infected arthropods; protozoans that infect the blood and tissues of humans

a. dermatophytes
b. eumycota mycetoma
c. hemoflagellates
d. keratitis
e. medical mycology
f. mycoses
g. onychomycosis

h. opportunistic microorganism
i. paronychia
j. piedras

FILL IN THE BLANK

1. Cutaneous mycoses, which are also called _____, are distinguished in part by the area of the body infected. Infections of beard hair are called tinea _____; those of scalp hair are called tinea _____; infections any part of the skin are called tinea _____; those of the groin are called tinea _____ (commonly called _____ _____); and those of the feet, hands, and nail beds are called tinea _____ (commonly called _____ _____), tinea _____, and tinea _____, respectively.

2. An _____ microorganism is one that is generally harmless in its normal environment but that can become pathogenic in a _____ host.

3. The most important opportunistic mycoses are _____ and _____. A third important opportunistic mycosis is _____ _____, which is caused by an organism that was thought to be a protozoan but is now classified as a fungus.

4. Candidiasis takes many forms. If the disease involves the mouth, it is referred to as _____ candidiasis or _____. Infections in areas of the body that are warm and moist (e.g., axillae and groin) are called _____ candidiasis. One example of this type of candidiasis is _____ candidiasis, which is observed in infants whose diapers are changed infrequently. Candidiasis can also involve the female genital tract, in which case it is referred to as _____ _____. If *C. albicans* is sexually transmitted to a male, the resulting infection of the glans penis is referred to as _____.

5. Flagellated protozoa that infect the blood and tissues of humans fall into two major groups of organism: _____ and _____. A major disease caused by *Plasmodium* is _____.

MULTIPLE CHOICE

For each of the questions below select the *one best* answer.

1. Which of the following terms is used to refer to cutaneous mycoses?
   a. dermatomycoses
   b. ringworms
   c. tineas
   d. All of the above refer to cutaneous mycoses.

2. Which of the following is NOT a subcutaneous mycosis?
   a. chromoblastomycosis
   b. coccidiomycosis
   c. maduromycosis
   d. sporotrichosis

3. Which term refers to sporotrichosis that has spread throughout the body?
   a. extracutaneous sporotrichosis
   b. disseminated sporotrichosis
   c. systemic sporotrichosis

   d. fulminating sporotrichosis

4. Which of the following systemic mycoses is caused by a fungus that is NOT dimorphic?
   a. *Blastomyces dermatitidis*
   b. *Coccidioides immitis*
   c. *Cryptococcus neoformans*
   d. *Histoplasma capsulatum*

5. Which of the following statements about *Pneumocystis* pneumonia (PCP) is NOT true?
   a. It occurs almost exclusively in immunocompromised hosts.
   b. It remains localized in the lungs, even in fatal cases.
   c. It occurs in 80% of AIDS patients.
   d. All of the above are true about PCP.

6. Which of the following diseases is caused by trypanosomes?

a. sleeping sickness
b. Chagas' disease
c. Both (a) and (b) are caused by trypanosomes.
d. Neither (a) nor (b) is caused by trypanosomes.

7. Why aren't vaccines effective against trypanosomiasis?
   a. because the organisms are never exposed to the immune system
   b. because the organism is only weakly immunogenic
   c. because the organism can change its protein coat and thereby evade the immune response
   d. All of the above are correct.

8. What is one of the leading causes of death in AIDS patients?
   a. giardiasis
   b. toxoplasmosis
   c. trichomoniasis
   d. sporotrichosis

9. Which of the following is NOT caused by a protozoan?
   a. cryptosporidiosis
   b. cryptococcosis
   c. Both (a) and (b) are caused by a protozoan.
   d. Neither (a) nor (b) is caused by a protozoan.

10. Which of the following is a superficial mycosis?
    a. black piedra
    b. white piedra
    c. tinea versicolor
    d. All of the above are superficial mycoses.

FUNGAL AND PROTOZOAL DISEASES

Disease	Causative Agent(s) and Mode of Transmission	Type of Organism (Fungus or Protozoan)
amebiasis (amebic dysentery)		
American trypanosomiasis (Chaga's disease)		
aspergillosis		
blastomycosis		
coccidiomycosis		
cryptococcosis		
cryptosporidiosis		
giardiasis		
histoplasmosis		
leishmaniasis		
malaria		
Pneumocystis pneumonia		
primary amebic meningoencephalitis		
sporotrichosis		
toxoplasmosis		
trichomoniasis		
African trypanosomiasis		

TRUE/FALSE

_____ 1. Jock itch and athlete's foot are caused by the same set of fungi.

_____ 2. Subcutaneous mycoses are caused by soil-inhabiting fungi that can easily penetrate the skin.

_____ 3. *Candida albicans* is normally found in the vagina but does not usually cause disease because the acidic pH created by *Lactobacilli* prevents its overgrowth.

_____ 4. The impact of protozoan diseases worldwide is great even though there are relatively few genera associated with human diseases.

_____ 5. Asymptomatic cyst shedders of *Entamoeba histolytica* do not need treatment.

_____ 6. *Entamoeba histolytica* cysts can be destroyed by hyperchlorination and iodination.

_____ 7. *Giardia lamblia* is the most common cause of epidemic waterborne diarrheal disease in the U. S.

_____ 8. Trichomoniasis is frequently asymptomatic in males, but seldom is asymptomatic in females.

411

CRITICAL THINKING

1. One of the apparent paradoxes associated with giardiasis is that outbreaks occur more frequently in the Rocky Mountain and New England states where the raw water is considered to be of fairly high quality, whereas outbreaks are less frequent in the South, Southwest, and Midwest, where quality of raw water is actually considered much poorer. Explain. (Consider the types of treatment used in the two types of areas.)

2. Describe the life cycle of *Plasmodium vivax* and discuss how it is related to the symptoms associated with malaria.

ANSWER KEY

Terms and Definitions

1. e, 2. f, 3. j, 4. a, 5. b, 6. h, 7. i, 8. g, 9. d, 10. c

Fill in the Blank

1. dermatomycoses; barbae; capitis; corporis; cruris; jock itch; pedis; athlete's foot; mannum, unguium 2. opportunistic; compromised; 3. aspergillosis; candidiasis; *Pneumocystis* pneumonia 4. oral; thrush; intertriginous; napkin (diaper); candidal vaginitis; balanitis 5. leishmanias; trypanosomes; malaria

Multiple Choice

1. d, 2. b, 3. a, 4. c, 5. d, 6. c, 7. c, 8. b, 9. b, 10. d

True/False

1. T, 2. F, 3. T, 4. T, 5. F, 6. T, 7. T, 8. T

40 Microbiology of Food

CHAPTER OVERVIEW

This chapter discusses the microorganisms associated with foods. Some of these microorganisms cause food spoilage, some are pathogens that are transmitted via foods, and some are used in the production of foods.

CHAPTER OBJECTIVES

After reading this chapter you should be able to:

- discuss the interaction of intrinsic (food-related) and extrinsic (environmental) factors related to food spoilage
- describe the various physical, chemical, and biological processes used to preserve foods
- discuss the various diseases that can be transmitted to humans by foods
- differentiate between food infections and food intoxications
- discuss the detection of disease-causing organisms in foods
- describe the fermentation of dairy products, grains, meats, fruits, and vegetables
- discuss the toxins produced by fungi growing in moist corn and grain products
- discuss the direct use of microbial cells as food by humans and animals
- list foods that are made with the aid of microorganisms and indicate the types of microorganisms used in their production
- describe probiotics

CHAPTER OUTLINE

I. Microorganism Growth in Foods
   A. Intrinsic factors
      1. Food composition
         a. Carbohydrates—do not result in major odors
         b. Proteins and/or fats result in a variety of foul odors (e.g., putrefactions)
      2. pH—low pH allows yeasts and molds to become dominant; higher pH allows bacteria to become dominant; higher pH favors putrefaction (the anaerobic breakdown of proteins that releases foul-smelling amine compounds)
      3. Physical structure affects the course and extent of spoilage
         a. Grinding and mixing (e.g., sausage and hamburger) increases surface area, alters cellular structure, and distributes microorganisms throughout the food
         b. Vegetables and fruits have outer skins that protect against spoilage; spoilage microorganisms have enzymes that weaken and penetrate such protective coverings
      4. Presence and availability of water
         a. Drying (removal of water) controls or eliminates food spoilage
         b. Addition of salt or sugar decreases water availability and reduces microbial spoilage

413

        c.   Even under these conditions spoilage can occur by certain kinds of microorganisms
            1)   Osmophilic—prefer high osmotic pressure
            2)   Xerophilic—prefer low water availability
        d.   Oxidation-reduction potential can be affected (lowered) by cooking, making foods more susceptible to anaerobic spoilage

    5.   Many foods contain natural antimicrobial substances (e.g., fruits and vegetables, milk and eggs, herbs and spices, and unfermented green and black teas)

B.   Extrinsic factors

    1.   Temperature and relative humidity—at higher relative humidity, microbial growth is initiated more rapidly, even at lower temperatures

    2.   Atmosphere—oxygen usually promotes growth and spoilage even in shrink-wrapped foods since oxygen can diffuse through the plastic; high CO_2 tends to decrease pH and reduce spoilage; modified atmosphere packaging (MAP) involves the use of modern shrink wrap materials and vacuum technology to package foods in a desired atmosphere (e.g., high CO_2 or high O_2)

II.   Microbial Growth and Food Spoilage

A.   Meats and dairy products are ideal environments for spoilage by microorganisms because of their high nutritional value and the presence of easily utilizable carbohydrates, fats, and proteins; proteolysis (aerobic) and putrefaction (anaerobic) decompose proteins; in spoilage of unpasteurized milk, a four-step succession of microorganisms occurs

B.   Fruits and vegetables have much lower protein and fat content then meats and dairy products and undergo different kind of spoilage; the presence of readily degradable carbohydrates in vegetables favors spoilage by bacteria; high oxidation–reduction potential favors aerobic and facultative bacteria; molds usually initiate spoilage in whole fruits

C.   Frozen citrus products are minimally processed and can be spoiled by lactobacilli and yeasts

D.   Grains, corn, and nuts can spoil when held under moist conditions; this can lead to production of toxic substances

    1.   Ergotism is caused by hallucinogenic alkaloids produced by fungi in corn and grains

    2.   Aflatoxins—planar molecules that intercalate into DNA and act as frameshift mutagens and carcinogens; if consumed by dairy cows, aflatoxins can appear in milk; also have been observed in beer, cocoa, raisins, and soybean meal; aflatoxin sensitivity can be influenced by prior disease exposure (e.g., hepatitis B infection increases sensitivity)

    3.   Fumonisins—contaminants of corn; cause disease in animals and esophageal cancer in humans; disrupt synthesis and metabolism of sphingolipids

E.   Shellfish and finfish can be contaminated by algal toxins, which cause a variety of illnesses in humans

III. Controlling Food Spoilage

A.   Removal of microorganisms—filtration of water, wine, beer, juices, soft drinks, and other liquids can keep bacterial populations low or eliminate them entirely

B.   Low temperature—refrigeration and/or freezing retards microbial growth but does not prevent spoilage

C.   High temperature

    1.   Canning

        a.   Canned food is heated in special containers called retorts to 115°C for 25–100 minutes to kill spoilage microorganisms

        b.   Canned foods can undergo spoilage despite safety precautions; spoilage can be due to spoilage prior to canning, underprocessing during canning, or leakage of contaminated water through can seams during cooling

    2.   Pasteurization—kills pathogens and substantially reduces the number of spoilage organisms

        a.   Low-temperature holding (LTH)—62.8°C for 30 minutes

        b.   High-temperature short-time (HTST)—71°C for 15 seconds

        c.   Ultra-high temperature (UHT)—141°C for 2 seconds

        d.   Shorter times result in improved flavor and extended shelf-life

3. Heat treatments are based on a statistical process involving the probability that the number of remaining viable microorganisms will be below a certain level after a specified time at a specified temperature

D. Water availability—dehydration procedures (e.g., freeze-drying) remove water and increase solute concentration

E. Chemical-based preservation
   1. Regulated by the U.S. Food and Drug Administration (FDA); preservatives are listed as "generally recognized as safe" (GRAS); include simple organic acids, sulfite, ethylene oxide as a gaseous sterilant, sodium nitrite, and ethyl formate
   2. Effectiveness depends on pH; nitrites protect against *Clostridium botulinum*, but are of some concern because of their potential to form carcinogenic nitrosamines when meats preserved with them are cooked

F. Radiation—nonionizing (ultraviolet or UV) radiation is used for surfaces of food-handling utensils, but does not penetrate foods; ionizing (gamma radiation) penetrates well but must be used with moist foods to produce peroxides, which oxidize sensitive cellular constituents (radappertization); ionizing radiation is used for seafoods, fruits, vegetables, and meats; electron beams also can be used irradiate foods

G. Microbial product-based inhibition
   1. Bacteriocins—bactericidal proteins produced by bacteria; active against only closely related bacteria (e.g., nisin)
   2. Bacteriocins function by several mechanisms, including dissipation of proton motive force, formation of hydrophobic pores in membranes, or inhibition of protein and RNA synthesis

IV. Food-borne Diseases
   A. Food-borne illnesses impact the entire world; are either infections or intoxications; are associated with poor hygiene practices
   B. Food-borne infections
      1. Due to ingestion of microorganisms, followed by growth, tissue invasion and/or release of toxins
      2. Salmonellosis—caused by a variety of *Salmonella* serovars; commonly transmitted by meats, poultry, and eggs; can arise from contamination of food by workers in food-processing plants and restaurants
      3. *Campylobacter jejuni*—transmitted by uncooked or poorly cooked poultry products, raw milk and red meats; thorough cooking prevents transmission
      4. Listeriosis—transmitted by dairy and meat products
      5. Enteropathogenic, enteroinvasive, and enterotoxigenic *Escherichia coli*
         a. Spread by fecal-oral route; found in meat products, in unpasteurized fruit drinks, and on fruits and vegetables
         b. Prevention requires prevention of food contamination throughout all stages of production, handling, and cooking
      6. Viral pathogens—usually transmitted by water or by direct contamination by food processors and handlers; recently Noroviruses have been involved in major outbreaks on several large cruise ships
      7. Variant Creutzfeld-Jakob disease—transmitted by ingestion of beef from infected cattle; transmission between animals is due to the use of mammalian tissue in ruminant animal feeds; prevention and control are difficult
      8. Foods transported and consumed in uncooked state are increasingly important sources of food-borne infection, especially as there is increasingly rapid movement of people and products around the world
         a. Sprouts can be a problem if germinated in contaminated water
         b. Shellfish and finfish can be contaminated by pathogens (e.g., *Vibrio* and viruses) found in raw sewage
         c. Raspberries often are transported by air to far-away markets; if contaminated, outbreak occurs far from source of pathogen

C. Food intoxications
1. Ingestion of microbial toxins in foods
2. Staphylococcal food poisoning is caused by exotoxins released by *Staphylococcus aureus,* which is frequently transmitted from its normal habitat (nasal cavity) to food by person's hands; improper refrigeration leads to growth of bacterium and toxin production
3. *Clostridium botulinum, C. perfringens,* and *B. subtilis* also cause food intoxication
   a. Botulism, caused by *C. botulinum,* often results from spores not destroyed in cooking or canning process; these germinate and quickly produce deadly toxin
   b. *C. perfringens* is a common inhabitant of food, soil, water, spices, and intestinal tract; upon ingestion, endospores germinate and produce enterotoxins within the intestine; this causes food poisoning; often occurs when meats are cooked slowly
   c. *Bacillus cereus* food poisoning is associated with starchy foods
V. Detection of Food-borne Pathogens
   A. Methods need to be rapid; therefore, traditional culture methods that might take days to weeks to complete are too slow; identification also is complicated by low numbers of pathogens compared to normal microflora; chemical and physical properties of food can make isolation of food-borne pathogens difficult
   B. Molecular methods are valuable for three reasons
      1. They can detect the presence of a single, specific pathogen
      2. They can detect viruses that cannot be conveniently cultured
      3. They can identify slow-growing or nonculturable pathogens
   C. Some examples
      1. DNA probes can be linked to enzymatic, isotopic, chromogenic, or luminescent/ fluorescent markers; are very rapid
      2. PCR can detect small numbers of pathogens (e.g., as few as 10 toxin-producing *E. coli* cells in a population of 100,000 cells isolated from soft cheese samples; as few as two colony-forming units of *Salmonella*); PCR systems are being developed for *Campylobacter jejuni* and *Arcobacter butzleri*
      3. Food-borne pathogen fingerprinting is an integral part of an initiative by the Centers for Disease Control (CDC) to control food-borne pathogens; the CDC has established a procedure (PulseNet) in which pulsed-field gel electrophoresis is used under carefully controlled and standardized conditions to detect the distinctive DNA patterns of nine major food pathogens; these pathogens are being followed by a surveillance network (FoodNet)
VI. Microbiology of Fermented Foods
   A. Fermented milks—at least 400 different fermented milks are produced throughout the world; fermentations are carried out by mesophilic, thermophilic, and therapeutic lactic acid bacteria, as well as by yeasts and molds
      1. Lactic acid bacteria—used for majority of fermented milks; gram-positive, acid-tolerant bacteria with a strictly fermentive metabolism
      2. Mesophilic—acid produced from microbial activity at temperatures lower than 45°C causes protein denaturation (e.g., cultured buttermilk and sour cream)
      3. Thermophilic—fermentations carried out at about 45°C (e.g., yogurt)
      4. Probiotics—fermented milks may have beneficial therapeutic effects
         a. Acidophilus milk contains *L. acidophilus;* improves general health by altering intestinal microflora; may help control colon cancer
         b. *Bifidbacterium* spp. fermented milk products improve lactose tolerance, possess anticancer activity, help reduce serum cholesterol levels, assist calcium absorption, and promote the synthesis of B-complex vitamins; may also reduce or prevent the excretion of rotaviruses, a cause of diarrhea among children
      5. Yeast-lactic fermentations—include kefir, which is made by the action of yeasts, lactic acid bacteria, and acetic acid bacteria
      6. Mold-lactic fermentation—used to make viili, a Finnish beverage; carried out by the mold *Geotrichium candidum* and lactic acid bacteria

B. Cheeses—produced by coagulation of curd by the stomach enzyme renin after an initial fermentation with starter lactic acid bacteria, squeezing out of watery whey, and ripening by continued microbial fermentation by nonstarter lactic acid bacteria and fungi; cheese can be internally inoculated or surface ripened; water content (ripening time) determines the hardness of the cheese

C. Meat and fish
1. Meat products include sausages, country-cured hams, bologna, and salami; these fermentations frequently involve *Pediococcus cerevisiae* and *Lactobacillus plantarum*
2. Fish products include izushi (fresh fish, rice, and vegetables incubated with *Lactobacillus* spp.) and katsuobushi (tuna incubated with *Aspergillus glaucus*)

D. Production of alcoholic beverages
1. Wines and champagnes
   a. Grapes are crushed and liquids that contain fermentable substrates (musts) are separated; musts can be fermented immediately, but the results can be unpredictable; usually must is sterilized by pasteurization or with sulfur dioxide fumigant; to make a red wine, the skins of a red grape are left in contact with the must before the fermentation process; if must was sterilized, the desired strain of *Saccharomyces cerevisiae* or *S. ellipsoideus* is added, and the mixture fermented (10 to 18% alcohol)
   b. Another important fermentative process that occurs is the malo-lactic fermentation carried out by *Leuoconostoc* spp.; this fermentation reduces the amount of organic acids (e.g., malic acid) in the wine, improving its flavor, stability, and "mouth feel"
   c. For dry wine (no free sugar), the amount of sugar is limited so that all sugar is fermented before fermentation stops; for sweet wine (free sugar present), the fermentation is inhibited by alcohol accumulation before all sugar is used up; in the aging process flavoring compounds accumulate
   d. Racking—removal of sediments accumulated during the fermentation process
   e. Brandy (burned wine) is made by distilling wine to increase alcohol concentration; wine vinegar is made by controlled microbial oxidation (by *Acetobacter* or *Gluconobacter*) to produce acetic acid from ethanol
   f. For champagnes, fermentation is continued in bottles to produce a naturally sparkling wine
2. Beers and ales
   a. Malt is produced by germination of the barley grains and the activation of their enzymes; mash is produced from malt by enzymatic starch hydrolysis to accumulate utilizable carbohydrates; mash is heated with hops (dried flowers of the female vine *Humulus lupulis*) to provide flavor and clarify the wort; hops inactivate hydrolytic enzymes so that wort can be pitched (inoculated with yeast)
   b. Beer is produced with a bottom yeast, such as *Saccharomyces pastorianus*, and ale is produced with a top yeast, such as *S. cerevisiae*; freshly fermented (green) beers are lagered (aged), bottled, and carbonated; beer can be pasteurized or filtered to remove microorganisms and minimize flavor changes
3. Distilled spirits—beerlike fermented liquid is distilled to concentrate alcohol; type of liquor depends on composition of starting mash; flavorings also can be added; a sour mash involving *Lactobacillus delbrueckii*-mediated fermentation is often used

E. Production of breads
1. Aerobic yeast fermentation is used to increase carbon dioxide production and decrease alcohol production; other metabolic products add flavors
2. Other microorganisms make special breads, such as sourdough
3. Bread products can be spoiled by *Bacillus* species that produce ropiness

F. Other fermented foods
1. Sufu, fermented tofu (a chemically coagulated soybean milk product) and tempeh, made from soybean mash, are made by the action of molds

2. Sauerkraut—fermented cabbage; involves a microbial succession mediated by *Leuconostoc mesenteroides*, *Lactobacillus plantarum*, and *Lactobacillus brevis*
3. Pickles are cucumbers fermented in brine by a variety of bacteria; fermentation process involves a complex microbial succession
4. Silages—animal feeds produced by anaerobic, lactic-type mixed fermentation of grass, corn, and other fresh animal feeds

VII. Microorganisms as Foods and Food Amendments
   A. Microbes that are eaten include a variety of bacteria, yeasts, and other fungi (e.g., mushrooms, *Spirulina*)
   B. Prebiotics—oligosaccharide polymers that are not processed until reaching the large intestine; often combined with probiotics to create a synbiotic system
   C. Probiotics are being used with poultry to increase body weight and feed conversion; also reduce coliforms and *Campylobacter*; may be useful in preventing *Salmonella* from colonizing gut due to competitive exclusion

TERMS AND DEFINITIONS

Place the letter of each term in the space next to the definition or description that best matches it.

_____ 1. The food spoilage process that results in the release of foul-smelling amine compounds

_____ 2. Microorganisms that grow in media with high salt or sugar concentrations

_____ 3. Microorganisms that grow in media with low water activity

_____ 4. A commercial process that preserves food by heating it in containers at about 115°C for up to 100 minutes

_____ 5. A process that reduces the total microbial population and usually eliminates all pathogenic microorganisms

_____ 6. Chemicals used to preserve foods that are generally recognized as safe

_____ 7. The process of irradiating moist foods to kill contaminating microorganisms

_____ 8. Small, labeled nucleic acids used to detect specific sequences of DNA or RNA

_____ 9. Fruit juices that contain readily fermentable substrates

_____ 10. The process by which complex polysaccharides found in grains are hydrolyzed to form fermentable substrates

_____ 11. A clear liquid containing fermentable sugars derived from grains

_____ 12. The science of wine production

_____ 13. The process by which sediments in wine are removed

_____ 14. The product formed when *Acetobacter* and *Gluconobacter* are allowed to oxidize the ethanol in wine to acetic acid

_____ 15. A mash that has been inoculated with a homolactic acid bacterium in order to decrease the pH of the mash

_____ 16. An animal feed created by lactic acid fermentation of grass, chopped corn and other plant materials

a. canning
b. enology
c. GRAS
d. mashing
e. musts
f. osmophilic microorganisms
g. pasteurization
h. probes
i. putrefaction
j. racking
k. radappertization
l. silage
m. sour mash
n. wine vinegar
o. wort
p. xerophilic microorganisms

FILL IN THE BLANK

1. Food spoilage is dependent on food-related or _____ factors, such as pH, moisture, water activity or availability, oxidation-reduction potential, physical structure of the food, available nutrients, and the possible presence of antimicrobial compounds; and on environmental, or _____, factors, such as temperature, relative humidity, gases present and the types and levels of microorganisms added to the food.

2. Recently, the CDC has established a program called _____, in which pulsed-field gel electrophoresis is used under carefully controlled conditions to determine the distinctive DNA patterns of bacterial pathogens. Using data generated in this way, an active surveillance network called _____ is monitoring disease outbreaks caused by nine major food-borne pathogens.

3. If a food-borne disease requires ingestion of the pathogen, growth of the organism, and release of toxins in the intestine, it is referred to as a food-borne _____; if no growth of the microorganism is required after ingestion because the toxic substances are already present in the food as a result of previous growth, then it is referred to as a food-borne _____.

4. Infection of grains by the ascomycete *Claviceps purpura* can cause _____, in which hallucinogenic alkaloids produced by the fungus lead to altered behavior, abortion, and death after their consumption. Another group of fungi that can develop in grains and nuts produces powerful carcinogens known as _____. Another group of fungal toxins called _____ can contaminate corn, and finfish and shellfish can be contaminated with _____ toxins.

5. The liquid extracted from crushed grapes has readily fermentable substrates and is referred to as _____. However, if cereals and other starchy materials are used, the complex carbohydrates must be depolymerized by a process known as _____ to produce a liquid that has fermentable substrates. In this case, the liquid is referred to as a _____.

6. The production of beer and ale begins with the germination of barley grains and the activation of their enzymes to produce a _____. This is then mixed with water and the desired grains and the activated enzymes hydrolyze starch, producing utilizable carbohydrates. This mixture is called a _____. Heating this mixture with hops yields the _____, which is inoculated with the desired yeast. This is called _____. If a _____ yeast is used, the product is beer, and if a top yeast is used, the product is ale. Freshly fermented beers are then aged or _____.

7. A patented blend of bacteria is currently being used as a _____ for poultry. The blend is sprayed on young chicks, and as they preen, they ingest the mixture. Once ingested the bacterial blend helps establish a functional microbial community in the cecum, which limits the colonization of the gut by *Salmonella*. The basis of this effect is called _____ _____.

8. Many fermented milks are made worldwide, using numerous types of microorganisms. Thermophilic microorganisms such at *Streptococcus thermophilus* and *Lactobacillus bulgaricus* are used to make _____. Lactic acid producing yeasts are used to make _____, a frothy, foamy beverage that originated in the Caucasus Mountains.

9. Large quantities of sugar or salt added to foods cause microorganisms to dehydrate. However, food spoilage can still result from the action of _____ microorganisms that grow best at high osmotic concentration, and from _____ microorganisms that grow best in a low water activity environment.

10. One new approach to packaging foods is _____ _____ packaging, which creates a specific environment within the package that extends shelf life of the food.

MULTIPLE CHOICE

For each of the questions below select the *one best* answer.

1. Which of the following helps preserve foods by controlling availability of water?
   a. drying
   b. addition of salt
   c. addition of sugar
   d. All of the above help preserve foods by controlling the availability of water.

2. Which of the following reduces spoilage of the foods involved?
   a. grinding and mixing of foods such as sausage and hamburger
   b. peeling off the skins of fruits and vegetables
   c. Both (a) and (b) are correct.
   d. Neither (a) nor (b) is correct.

3. Which of the following products seems to aid a variety of intestinal processes and promote general health?
   a. acidophilus milk
   b. bifid-amended fermented milk products
   c. Both (a) and (b) are correct.
   d. Neither (a) nor (b) is correct.

4. Gamma irradiation has been used to sterilize certain types of foods. There are, however, certain limitations on its effectiveness. One of these is that it will only work on what types of food?
   a. moist foods
   b. dry foods
   c. foods that are not in metal cans
   d. Actually, it will work well on all foods.

5. Microbial fermentations are used in the production of cheeses in order to do which of the following steps?
   a. coagulate the milk solids to form a curd
   b. ripen the cheese to give a characteristic texture and flavor
   c. Microbial fermentations are used to accomplish both (a) and (b).
   d. Microbial fermentations are not used to accomplish either (a) or (b).

6. Which term refers to the addition of microorganisms to the diet in order to provide health benefits beyond basic nutritive value.
   a. probiotics
   b. microbial dietary adjuvants
   c. prebiotics
   d. synbiotics

7. Which term refers to oligosaccharide polymers in the diet that are not processed until reaching the large intestine?
   a. probiotics
   b. microbial dietary adjuvants
   c. prebiotics
   d. synbiotics

8. Which of the following determines whether or not a wine is considered to be dry?
   a. the percentage of alcohol after fermentation
   b. the amount of free sugar after fermentation
   c. the amount of carbon dioxide after fermentation
   d. the type of grape used for starting material

9. Why are molecular methods replacing traditional culture methods to detect disease-causing organisms in foods?
   a. Culture methods are too slow.
   b. Culture methods are not sensitive enough to detect low levels of pathogens against a high background of normal microflora.
   c. Both (a) and (b) are correct.
   d. Neither (a) nor (b) is correct.

10. Bacteriocins are important for controlling which microorganism or disease?
    a. *Streptococcus lactis*
    b. *Listeria monocytogenes*
    c. variant Creutzfeld-Jakob disease
    d. all of the above

11. Which of the following is true about molecular methods of detecting microorganisms in foods?
    a. They can detect the presence of a single specific pathogen.
    b. They can be used to detect viruses that are not easily cultured.
    c. They can be used to detect slow-growing pathogens.
    d. All of the above are true about the use of molecular detection methods.

12. Which of the following represents ways in which canned foods may undergo spoilage?
    a. spoilage before canning
    b. underprocessing during canning
    c. leakage of contaminated water through can seams during cooling
    d. All of the above are ways in which canned foods may undergo spoilage.
13. At which step of food handling and processing can food spoilage occur?
    a. during harvesting
    b. during transport
    c. during final preparation
    d. during storage
    e. Spoilage can occur at all of the above steps.
14. Which organism causes a food-borne infection?
    a. *Staphylococcus*
    b. *Bacillus*
    c. *Salmonella*
    d. all of the above
15. In which food products has *E. coli* O157:H7 been found?
    a. in meat products such as hamburger
    b. in unpasteurized fruit drinks
    c. on fruits and vegetables
    d. all of the above
16. What factor is contributing to increases in food-borne illness?
    a. the popularity of raw foods
    b. international shipment of raw foods
    c. Both (a) and (b) are contributing.
    d. Neither (a) nor (b) is contributing.

TRUE/FALSE

_____ 1. Food grinding and mixing tend to reduce spoilage by causing mechanical damage to the contaminating microorganisms.

_____ 2. Even at freezer temperatures, some microorganisms may grow and cause food spoilage, but the process is greatly slowed.

_____ 3. Microorganisms can be removed from some foods by filtration, resulting in sterilized food preparations.

_____ 4. Cooking can lead to increased anaerobic spoilage by lowering the food's oxidation-reduction potential.

_____ 5. Nitrites are used to protect against microorganisms but their use is of some concern because of the formation of carcinogenic nitrosamines upon cooking preserved foods.

_____ 6. The same microbial fermentation processes produce sour cream and cultured buttermilk, but the starting material is different.

_____ 7. Some fermented dairy products have been suggested to have beneficial effects, including a reduction in the incidence of colon cancer and the minimization of lactose intolerance.

_____ 8. White wines are produced from white (or green) grapes. If a red grape is used, the result will always be a red wine.

_____ 9. In bread making, yeast growth is usually carried out under aerobic conditions, which results in more carbon dioxide production and less alcohol accumulation.

_____ 10. The cause of most food-borne illnesses is known.

_____ 11. Prior illness with hepatitis B decreases susceptibility to aflatoxin-induced liver cancer.

_____ 12. Brandy is produced by distilling wine in order to increase the concentration of alcohol.

_____ 13. Vinegar is produced by microbial oxidation of the ethanol in wine to acetic acid.

_____ 14. PCR can detect as few as 10 toxin-producing *E. coli* in a population of 100,000 microorganisms isolated from soft-cheese samples.

_____ 15. Norwalk-type viruses, *Campylobacter*, and *Salmonella* are thought to be the most important causes of food-borne illness.

CRITICAL THINKING

1. Go through your refrigerator and kitchen shelves. List all the food products made by the activities of microorganisms.

2. Discuss the making of wines, champagnes, and other fermented grape products. In your discussion give particular consideration to the actions of the microorganisms involved. What steps are normally taken to control the quality of the final product?

ANSWER KEY

Terms and Definitions

1. i, 2. f, 3. p, 4. a, 5. g, 6. c, 7. k, 8. h, 9. e, 10. d, 11. o, 12. b, 13. j, 14. n, 15. m, 16. l

Fill in the Blank

1. intrinsic; extrinsic 2. PulseNet; FoodNet 3. infection; intoxication 4. ergotism; aflatoxins; fumonisins; algal
5. must; mashing; wort 6. malt; mash; wort; pitching; bottom; lagered 7. probiotic; competitive exclusion
8. yogurt; kefir 9. osmophilic; xerophilic 10. modified atmosphere

Multiple Choice

1. d, 2. d, 3. c, 4. a, 5. c, 6. a. 7. c, 8. b, 9. c, 10. b, 11. d, 12. d, 13. e, 14. c, 15. d, 16. c

True/False

1. F, 2. T, 3. T, 4. T, 5. T, 6. T, 7. T, 8. F, 9. T, 10. F, 11. T, 12. T; 13. T, 14. T, 15. T

41 Applied and Industrial Microbiology

CHAPTER OVERVIEW

This chapter discusses the uses of microorganisms in processes that are grouped under the heading of applied and industrial microbiology. Water is an excellent vehicle for the transmission of diseases, and the chapter explores the measures taken to ensure the availability of safe drinking water. The chapter continues with a discussion of the contamination of groundwaters by domestic and industrial wastes and the use of home wastewater treatment systems. Next, the use of genetically engineered microorganisms to increase the efficiency of the industrial processes and to produce new or modified products is discussed, as is the integration of biological and chemical processes to achieve a desired objective. The chapter concludes with discussions of biodegradation, some recent biotechnological applications, and the impact of microbial biotechnology on ecology and human society.

CHAPTER OBJECTIVES

After reading this chapter you should be able to:

- list the major pathogens transmitted by water
- describe the steps used to purify drinking water
- describe the ideal characteristics of indicator organisms and how they are used to measure the microbiological quality of water
- describe the methods commonly used to measure the level of organic material in wastewater
- discuss wastewater treatment systems
- compare and contrast wastewater treatment systems to the natural purification processes observed in waters
- discuss the ways humans impact groundwater and surface water
- describe home wastewater treatment systems
- discuss the sources of microorganisms for use in industrial microbiology and biotechnology
- discuss the genetic manipulation of microorganisms to construct strains that better meet the needs of an industrial or biotechnological process
- discuss the preservation of microorganisms
- describe the design or manipulation of environments in which desired processes will be carried out
- discuss the management of growth characteristics to produce the desired product
- list the major products or uses of industrial microbiology and biotechnology
- discuss the use of microorganisms in manufacturing nanotechnology, biosensors, and biopesticides
- discuss the manipulation of microorganisms in the environment to control biodegradation

CHAPTER OUTLINE

I. Water Purification and Sanitary Analysis
    A. Waterborne pathogens and water purification
        1. Many human pathogens are transmitted by water (e.g., *Vibrio* spp., *Giardia, Cryptosporidium*)
        2. Water purification is critical for public health and safety; common steps in water purification are:
            a. Sedimentation in a sedimentation basin removes sand and large particles

        b.   Coagulation with alum, lime, and/or organic polymers is followed by clarification in a settling basin; removes many of the microorganisms (including many of the viruses), organic matter, toxic contaminants, and suspended particles

        c.   Rapid sand filtration—physically traps particles and microbes

        d.   Disinfection with chlorine or ozone; chlorination can lead to formation of disinfection by-products (DBPs) that may be carcinogens

    3.   *Giardia* cysts, *Cryptosporidium* oocysts, and viruses are not consistently and reliably removed by the above procedures; slow sand filtration, which involves the slow passage of water over a bed of sand, more consistently removes *Giardia*

  B. Sanitary analysis of water

    1.   Since intestinal pathogens gradually lose their ability to form colonies after release into aquatic environments, microbiologists have generally monitored the presence and amount of indicator organisms as an index of fecal contamination of water; ideal indicator organisms have the following characteristics:

        a.   Should be suitable for the analysis of all types of water

        b.   Should be present whenever enteric pathogens are present

        c.   Should survive longer than the hardiest enteric pathogen

        d.   Should not reproduce in the contaminated water

        e.   Should be detected by a highly specific assay

        f.   Should be detected by a test that is easy to perform and sensitive

        g.   Should be harmless to humans (ensuring safety for laboratory personnel)

        h.   Concentration of indicator should directly reflect the degree of fecal pollution

    2.   Coliforms—the most commonly used indicator organisms

        a.   All are facultative anaerobic, gram-negative, nonsporing, rod-shaped bacteria that ferment lactose with gas formation within 48 hours at 35°C (e.g., *Escherichia coli, Enterobacter aerogenes,* and *Klebsiella pneumoniae*)

        b.   Detected by the following tests

            1)   Most probable number (MPN) —statistical estimation; does not distinguish coliforms from fecal coliforms (those derived from intestines of homeothermic animals; can grow at 44.5°C)

            2)   Membrane filtration technique—water is filtered, filter is placed on an absorptive pad containing liquid medium; this is incubated and colonies are counted; detects total coliforms, fecal coliforms, and fecal streptococci

            3)   Presence-absence (P-A) test—detects both coliforms and fecal coliforms

            4)   Defined substrate tests (e.g., Colilert) involve the production of a colored product (for total coliforms) or a fluorescent product (for *E. coli*) from a specific growth substrate

            5)   Molecular techniques are now being routinely used to detect *E. coli*

    3.   In the United States, general guidelines for microbiological quality of drinking water have been set, including standards for coliforms, viruses, and *Giardia*

    4.   Other indicator microorganisms used to test water safety include fecal enterococci; they are being used as indicators for brackish and marine waters because they survive longer than coliforms under these conditions

II. Wastewater Treatment

  A. Wastewater can contain high levels of organic matter and human pathogens; these can be removed (or their amount decreased) by wastewater treatment; such treatment is one of the most important factors in maintaining public health

  B. Measuring water quality—the tests described below monitor organic carbon in wastewater but do not address the levels of nitrate, phosphate, and sulfate; these also are of concern

    1.   Total organic carbon (TOC)—quantifies carbon concentration by oxidizing organic matter at high temperatures and measuring the amount of carbon dioxide produced

    2.   Chemical oxygen demand (COD)—quantifies the amount of organic matter present (except lignin) by reacting organic material with a strong acid

3. Biochemical oxygen demand (BOD) —amount of oxygen needed to utilize organic material as growth substrates; indirectly measures the amount of organic material in a sample; can be affected by presence of ammonia, so nitrification (nitrogen oxygen demand—NOD) is inhibited by addition of nitrapyrin to the sample

C. Water treatment processes
1. A controlled self-purification; usually involves the use of large basins where mixing and gas exchange are carefully controlled
2. Conventional wastewater treatment
   a. Primary (physical) treatment—removal of particulates (20 to 30% of the BOD); resulting solid material is called sludge
   b. Secondary (biological) treatment—removal of dissolved carbonaceous materials (90 to 95% of the BOD) and many bacterial pathogens; produces a sludge, which must be further processed or disposed of; if not carefully monitored, these processes can produce bulking sludge, which is not easily removed
      1) Aerated activated sludge systems—involve horizontal flow of materials and the addition of sludge, which acts as a source of microorganisms; the sludge microorganisms oxidize the organic matter; the resulting biomass is later removed
      2) Trickling filters—vertical flow over gravel on which microorganisms have developed in surface films; the microorganisms degrade the organic matter
      3) Extended aeration systems—reduce the amount of sludge produced by the process of biological self-consumption (endogenous respiration)
      4) All of the above secondary treatment processes as well as the primary treatment produce sludge; anaerobic sludge digestion reduces the amount of sludge that must be disposed of in landfills or by other means; also produces methane, which can be used as a fuel for generation of electrical power; involves three steps: fermentation of sludge components to form organic acids, production of methanogenic substrates (acetate, carbon dioxide, and hydrogen), and methanogenesis
   c. Tertiary treatment (physical, chemical and/or biological)—removes inorganic nitrogen, phosphorus, recalcitrant organics, viruses, etc.
3. Constructed wetlands use floating emergent and/or submerged plants to provide nutrients for microbial growth in their root zone; they help remove organic matter, inorganic matter, and metals from waters

D. Home Treatment Systems
1. Groundwater—water in gravel beds and fractured rocks below the surface of soil; it is an important source of water but the microbiological processes occurring in groundwater are not well understood; it is known that disease-causing organisms and organic matter are removed by adsorption and trapping as they move through the subsurface; microbial predators use trapped pathogens as food
2. Home treatment in a conventional septic tank system mimics the natural adsorption-biological predation process; septic systems are now being designed with nitrogen and phosphorus removal steps; a conventional system is described below
   a. Anaerobic liquefaction and digestion occurs in a septic tank
   b. Aerobic digestion, adsorption, and filtration of organic material are accomplished by drainage through suitable soil in a leach (drain) field; if drainage is too rapid, there is little adsorption and filtration, with subsequent contamination of well waters and groundwaters
3. Groundwater also can be contaminated by land disposal of sewage sludges, illegal dumping of septic tank pumpage, improper toxic waste disposal, agricultural runoff, and deep-well injection of industrial wastes
4. *In situ* treatment procedures for groundwater are under investigation; microorganisms are critical in many of these remediation efforts

III. Microorganisms for Industrial Microbiology

A. The characteristics of microbes that are desirable to the industrial microbiologist are: genetic stability, easy maintenance and growth, and amenability to procedures for extraction and purification of desired product

B. Finding microorganisms in nature—major sources of microorganisms for use in industrial processes are soil, water, and spoiled bread and fruits; only a minor portion of microbial species in most environments have been identified and cultured

C. Genetic manipulation of microorganisms

    1. Mutagenesis—once a promising culture is found, it can be improved by mutagenesis with chemical agents and UV light; site-directed mutagenesis is used to insert short lengths of DNA into specific sites in genome of a microorganism; leads to small changes in amino acid sequence that can result in unexpected changes in protein characteristics; site-directed mutagenesis is important to the field of protein engineering

    2. Protoplast fusion

        a. Widely used with yeasts and molds, especially if the microorganism is asexual or of a single mating type; involves removal of cell walls (to create protoplasts), mixing two different solutions of protoplasts, and growth in selective media

        b. Can be done using species that are not closely related

    3. Transfer of genetic information between different organisms

        a. Combinatorial biology—transfer of genes from one organism to another

        b. Can improve production efficiency and minimize purification of the product

        c. Numerous vectors are available for transfer of genes

    4. Modification of gene expression

        a. Can involve modifying gene regulation to overproduce a product

        b. Pathway architecture and metabolic pathway engineering—intentional alteration of pathways by inactivating or deregulating specific genes

        c. Metabolic control engineering—intentional alteration of controls for synthesis of a product

    5. Protein evolution—employs forced evolution and adaptive mutations; specific environmental stresses are used to force microorganism to mutate and adapt, this creates microorganism with new biological capabilities; also can be performed in vitro using cycles of transcription, reverse-transcription, and selection to generate novel and valuable gene sequences; high-throughput screening systems enable rapid selection of desirable mutants

D. Preservation of microorganisms—strain stability is of concern; methods that provide this stability are lyophilization (freeze-drying) and storage in liquid nitrogen

IV. Microorganism Growth in Controlled Environments

A. Industrial microbiologists use the term "fermentation" primarily to refer to the mass culture of microorganisms; the term has many other meanings to other microbiologists

B. Medium development

    1. Low-cost crude materials are frequently used as sources of carbon, nitrogen, and phosphorus; these include crude plant hydrolysates, whey from cheese processing, molasses, and by-products of beer and whiskey processing

    2. The balance of minerals (especially iron) and growth factors may be critical; it may be desirable to supply some critical nutrient in limiting amounts to cause a programmed shift from growth to production of desired metabolites

C. Growth of microorganisms in an industrial setting

    1. Physical environment must be defined (i.e., agitation, cooling, pH, oxygenation); oxygenation can be a particular problem with filamentous organisms as their growth creates a non-Newtonian broth (viscous) that is difficult to stir and aerate

    2. Attention must be focused on the above physical factors to ensure that they are not limiting when small-scale laboratory operations are scaled up to industrial-sized operations

    3. Culture tubes, shake flasks, and stirred fermenters of various sizes are used to culture microorganisms

        a.   In stirred fermenters, all steps in growth and harvesting must be carried out aseptically and computers often are used to monitor microbial biomass, levels of critical metabolic products, pH, input and exhaust gas composition, and other parameters

        b.   Continuous feed of a critical nutrient may be necessary to prevent excess utilization, which could lead to production and accumulation of undesirable metabolic waste products

        c.   Newer methods include air-lift fermenters, solid-state fermentation, and fixed and fluidized bed reactors, where the media flows around the attached or suspended microorganisms, respectively

        d.   Dialysis culture systems allow toxic wastes to diffuse away from microorganisms and nutrients to diffuse toward microorganisms

    4.   Microbial products often are classified as primary or secondary metabolites

        a.   Primary metabolites are related to the synthesis of microbial cells in the growth phase; they include amino acids, nucleotides, fermentation end products, and exoenzymes

        b.   Secondary metabolites usually accumulate in the period of nutrient limitation or waste product accumulation that follows active growth; they include antibiotics and mycotoxins

V.   Major Products of Industrial Microbiology

   A.   Antibiotics

      1.   Penicillin—careful adjustment of medium composition is used to slow growth and to stimulate penicillin production; side chain precursors can be added to stimulate production of particular penicillin derivatives; harvested product can then be modified chemically to produce a variety of semisynthetic penicillins

      2.   Streptomycin is a secondary metabolite that is produced after microorganism growth has slowed due to nitrogen limitation

   B.   Amino acids

      1.   Amino acids such as lysine and glutamic acid are used as nutritional supplements and as flavor enhancers

      2.   Amino acid production is usually increased through the use of regulatory mutants or through the use of mutants that alter pathway architecture

   C.   Organic acids

      1.   These include citric, acetic, lactic, fumaric, and gluconic acids

      2.   Citric acid, which is used in large quantities by the food and beverage industry, is produced largely by *Aspergillus niger* fermentation in which trace metals are limited to regulate glycolysis and the TCA cycle, thereby producing excess citric acid

      3.   Gluconic acid also is produced in large quantities by *A. niger,* but only under conditions of nitrogen limitation; gluconic acid is used in detergents

   D.   Specialty compounds for use in medicine and health—include sex hormones, ionophores, and compounds that influence bacteria, fungi, amoebae, insects, and plants

   E.   Biopolymers—microbially produced polymers

      1.   Polysaccharides are used as stabilizers, agents for dispersing particulates, and as film-forming agents; they also can be used to maintain texture in ice cream, as blood expanders and absorbents, to make plastics, and as food thickeners; also used to enhance oil recovery from drilling mud

      2.   Cyclodextrins can modify the solubility of pharmaceuticals, reduce their bitterness, and mask their chemical odors; also can be used to selectively remove cholesterol from eggs and butter, to protect spices from oxidation, or as stationary phases in gas chromatography

   F.   Biosurfactants

      1.   Biosurfactants are biodegradable agents used for emulsification, increasing detergency, wetting and phase dispersion, as well as for solubilization

      2.   The most widely used biosurfactants are glycolipids, which are excellent dispersing agents; many have antimicrobial properties due to their amphipathic nature (disrupt membranes)

   G.   Bioconversion processes—microbial transformations or biotransformations

      1.   Microorganisms are used as biocatalysts; bioconversions are frequently used to produce the appropriate stereoisomer; are very specific, and can be carried out under mild conditions

2. When bioconversion reactions require ATP or reductants, an energy source must be supplied

VI. Biodegradation and Bioremediation by Natural Communities
  A. Microorganisms can be used to carry out desirable processes in natural environments; in these environments, complete control of the process is not possible; processes carried out in natural environments include:
    1. Biodegradation, bioremediation, and environmental maintenance processes
    2. Addition of microorganisms to soils or plants for improvement of crop production
  B. Biodegradation and bioremediation processes
    1. Biodegradation has at least three definitions
      a. A minor change in an organic molecule, leaving the main structure still intact
      b. Fragmentation of a complex organic molecule in such a way that the fragments could be reassembled
      c. Complete mineralization
    2. Bioremediation is the use of microbes to transform contaminants into nontoxic degradation products
    3. Degradation of a complex compound such as a halogenated compound occurs in stages
      a. Dehalogenation often occurs faster under anaerobic conditions; humic substances may facilitate this stage
      b. Subsequent steps usually proceed more rapidly in the presence of oxygen
    4. Structure and stereochemistry impact rate of biodegradation (e.g., *meta* effect and preferential degradation of one isomer)
    5. Microbial communities change in response to addition of inorganic and organic substrates; these can impact rate and extent of biodegradation (e.g., repeated contact with a herbicide leads to the adaptation of the microbial community and a faster rate of degradation—acclimation)
    6. Waste materials can be degraded after incorporation into soil or as they flow across soil surface
    7. Biodegradation does not always reduce environmental problems (e.g., partial degradation can produce equally hazardous or more hazardous substances)
    8. Biodegradation can cause damage and financial losses (e.g., corrosion of metal pipes in oil fields)
  C. Stimulating biodegradation
    1. Engineered bioremediation—addition of oxygen or nutrients to stimulate degradation activities of microorganisms
    2. Stimulating hydrocarbon degradation in waters and soils—usually involves addition of nutrients and substances that increase contact between microorganisms and substrate to be degraded; also can involve aeration or creating anoxic conditions
    3. Stimulating degradation with plants—phytoremediation is the use of plants to stimulate the extraction, degradation, adsorption, stabilization, or volatilization of contaminants; transgenic plants can be used
    4. Stimulation of metal bioleaching from minerals—involves the use of acid-producing bacteria to solubilize metals in ores; may require addition of nitrogen and phosphorus if they are limiting
  D. Bioaugmentation—addition of microorganisms to complex microbial communities
    1. Impact of protective microhabitats
      a. Often fails to produce long-lasting increases in rates of biodegradation; this may be due to three factors:
        1) Attractiveness of laboratory grown microbes as a food source for predators
        2) Inability of microorganisms to contact the compounds to be degraded
        3) Failure of the microorganisms to survive
      b. "Toughening" microorganisms by starvation before they are added has increased microbial survival somewhat, but has not solved the problem

2. More success is possible when considering protective microhabitats when adding microorganisms by including materials that provide protection and/or supply nutrients
   a. Living microhabitats—include surfaces of a seed, a root, or a leaf
   b. Inert microhabitats—include microporous glass or "clay hutches"

VII. Microbes as Products
   A. Nanotechnology
      1. Diatom shells have precise structures at the micrometer scale; these can be grown and the silicon oxides replaced by magnesium oxides
      2. Magnetosomes formed by bacteria are minute, perfectly formed magnetic beads with a membrane envelope that can be used for drug delivery or diagnostic techniques
   B. Biosensors
      1. Biosensors make use of microorganisms or microbial enzymes that are linked to electrodes in order to detect specific substances by converting biological reactions to electric currents
      2. Biosensors have been or are being developed to measure specific components in beer, to monitor pollutants, to detect flavor compounds in foods, and to study environmental processes such as changes in biofilm concentration gradients; they also are being used to detect glucose and other metabolites in medical situations and to combat bioterrorism
      3. New immunochemical-based biosensors are being developed; these are used to detect pathogens, herbicides, toxins, proteins, and DNA
   C. Biopesticides
      1. Bacteria (e.g., *Bacillus thuringiensis*) —are being used to control insects; accomplished by inserting toxin-encoding gene into the plant or by production of a wettable powder that can be applied to agricultural crops; the toxin gene also has been introduced into crop plants
      2. Viruses—nuclear polyhedrosis viruses (NPV), granulosis viruses (GV), and cytoplasmic polyhedrosis viruses (CPV) have potential as bioinsecticides
      3. Fungi—fungal biopesticides are increasingly being used in agriculture

VIII. Impacts of Microbial Biotechnology
   A. Ethical and ecological considerations are important in the use of biotechnology
   B. Industrial ecology—discipline concerned with tracking the flow of elements and compounds through biosphere and anthrosphere

TERMS AND DEFINITIONS

Place the letter of each term in the space next to the definition or description that best matches it.

_____ 1. Microbial biomass produced during wastewater treatment that does not settle properly; it results from the development of massive amounts of certain filamentous organisms

_____ 2. Organisms that are used to indicate fecal contamination of water

_____ 3. Methods for treating groundwaters where they are located

_____ 4. Chemicals such as trihalomethanes (THMs) produced during water treatment as a result of the reaction of chlorine with organic matter in the water

_____ 5. The active biomass that is formed when organic matter is oxidized and degraded by microorganisms in a wastewater treatment system

_____ 6. Use of site-directed mutagenesis and other methods to produce enzymes and bioactive peptides with desired characteristics

_____ 7. The analysis, design, and modification of biochemical pathways for the purpose of increasing the efficiency of the process

_____ 8. The process whereby water is removed from a preparation at freezing temperatures while under a vacuum

_____ 9. A process whereby a critical nutrient is added periodically so that the organism will not have excess substrate available to it at any time

_____ 10. Molecules produced by a microorganism that are directly related to synthesis of cell material during the growth phase

_____ 11. Molecules produced by a cell that are not directly related to the synthesis of cell material and that are usually produced after the growth phase has ended
_____ 12. Microbial polymers used to modify the flow characteristics of liquids and to serve as gelling agents
_____ 13. Minor modifications in molecules that are carried out by nongrowing microbes
_____ 14. Microorganisms used to carry out biotransformations
_____ 15. A process that removes chlorine from compounds such as PCBs under anaerobic conditions
_____ 16. The decrease in the rate of biodegradation observed when a constituent of a molecule is in the meta position rather than the ortho position
_____ 17. Describes molecules with the characteristic of being asymmetric
_____ 18. A process in which waste material is incorporated in soil or allowed to flow across the soil surface, where degradation occurs
_____ 19. Stimulation of degradative activities of microorganisms by modifying the water or soil in which biodegradation is occurring
_____ 20. The stimulation of the degradation of recalcitrant molecules by the addition of easily degraded organic molecules
_____ 21. The use of plants to stimulate the degradation, transformation, or removal of compounds
_____ 22. The use of natural microbial communities in the environmental management of pollutants
_____ 23. Living microorganisms that are linked with electrodes to convert biological reactions to electrical currents
_____ 24. Bacteria, viruses, and fungi, or their products, that can be used to control insect pests
_____ 25. The addition of microbes with known activities to soils, waters, or other complex systems in an attempt to speed up existing microbiological processes

a.   activated sludge
b.   bioaugmentation
c.   biocatalysts
d.   bioinsecticides (biopesticides)
e.   biopolymers
f.   biosensors
g.   biotransformations
h.   bulking sludge
i.   chiral
j.   coliforms
k.   cometabolism
l.   continuous feed
m.   disinfection by-products (DPBs)
n.   engineered bioremediation
o.   in situ treatment
p.   land farming
q.   lyophilization
r.   meta effect
s.   natural attenuation
t.   pathway architecture
u.   phytoremediation
v.   primary metabolites
w.   protein engineering
x.   reductive dehalogenation
y.   secondary metabolites

FILL IN THE BLANK

1. Three methods can be used to monitor the amount of organic matter present in wastewater as it is processed in a wastewater treatment facility. One method determines the amount of a strong acid neutralized by the organic matter in a sample; it is referred to as the _____ _____ _____. Another method reacts the organic matter with oxygen at a high temperature and measures the amount of carbon dioxide produced; this is referred to as _____ _____ _____. The third method is the most commonly used method. It measures the amount of oxygen consumed when the organic material is used as a growth substrate; it is referred to as the_____ _____ _____). BOD is impacted by nitrification (_____ _____ _____), which can be inhibited by adding nitrapyrin to the sample.

2. The addition of chemicals such as alum and lime during water purification precipitates material in the water. These settle to the bottom of _____ basins. This process is called _____.

3. The solid material collected by primary treatment of wastewater is called _____. This term is also applied to the microbial biomass produced by aerobic secondary treatment of wastewater. Although the amount of excess microbial biomass generated can be decreased by a process called _____ _____, it ultimately is further treated by _____ _____, which yields methane.

4. Home systems for treatment of wastewater include an anaerobic liquefaction and digestion step that occurs in a _____ _____, followed by aerobic degradation in the leach field.

5. Wastewater treatment is a complex process that resembles the natural purification processes that occur in streams and other bodies of water. The first step, called _____ treatment, physically removes particulate material by screening, precipitation, and settling. This is followed by _____ treatment in which dissolved organic matter is removed by biological activity. There are several forms of this treatment. One commonly used system is the _____ _____ system, which involves a horizontal flow of materials and the introduction of sludge to the wastewater; this sludge is the source of microorganisms that will degrade the organic matter, creating more sludge that later is removed in settling basins. A second commonly used method is the _____ _____ method. In this method, wastewater is passed over rocks or other solid materials upon which microbial films have developed; the microbial community in these films degrades the organic waste. The last type of treatment is more expensive and is not used by all wastewater treatment facilities. It is _____ treatment, and it can use physical, chemical, or biological methods to remove nitrogen or phosphorus.

6. Water that is safe to drink (_____ water) is the product of water purification systems. These systems typically use _____ to partially clarify water, _____ to further clarify the water by adding alum or other chemicals that cause impurities to precipitate out, _____ to trap fine particles and remove most bacteria, and _____ with chlorine or ozone to destroy pathogens.

7. Rather than test directly for pathogens in drinking water, microbiologists monitor the presence and numbers of _____ organisms that serve as an index for fecal contamination of water. The _____ are the most commonly used _____ organisms. They are defined as facultatively anaerobic, gram-negative, nonsporing, rod-shaped bacteria that ferment lactose with gas formation within 48 hours at 35° C. Unfortunately, this description includes a wide variety of bacteria, including those that are not inhabitants of the intestinal tract. Therefore in testing drinking water quality, the presence of _____ _____ is also determined. These differ from the above definition in that they can grow at the more restrictive temperature of 44.5° C.

8. The growth of filamentous organisms results in a viscous, plastic medium called a _____ _____, which offers great resistance to stirring and aeration.

9. To assure that physical factors are not limiting microbial activity, physical conditions used in industrial processes must be considered at the level of the individual microbe. This is most critical in _____, where a process developed in a small shake flask, if successful, must be carried out in a large fermenter.

10. Compounds related to the synthesis of microbial cells and often involved in the growth phase are called _____ _____; those that have no direct relationship to the synthesis of cell materials and that are usually produced after the growth phase are referred to as _____ _____.

11. Commercial production of amino acids is typically carried out using _____ mutants, which over produce the desired amino acid.

12. A rapidly developing area of biotechnology concerns the linking of microorganisms to electronic components to create _____ that convert biochemical reactions into electrical current.

13. The newest approach for creating new metabolic capabilities in a given microorganism is _____ _____ _____. This approach uses specific environmental stresses to force microorganisms to mutate and adapt, processes described as _____ _____ and _____ _____.

14. Microbiology is a critical part of the area of _____ _____, which is concerned with tracking the flow of elements and compounds thru the natural world (_____) and the social world (_____).

15. Increasingly microorganisms used in industrial processes are being improved by molecular methods. For instance, _____ _____ involves the transfer of genes from one organism to another. Another approach is _____ _____, in which small synthetic DNA molecules are inserted into a microorganism at a specific site. Such techniques can be used to engineer metabolic pathways [i.e., _____ _____ engineering and _____ _____ engineering]

MULTIPLE CHOICE

For each of the questions below select the *one best* answer.

1. Secondary wastewater treatment removes organic material by which of the following?
   a. biological processes
   b. physical processes
   c. chemical processes
   d. All of the above are correct.

2. During one type of secondary wastewater treatment, the wastewater flows horizontally through an agitated aeration tank. What is this called?
   a. lagooning
   b. activated sludge treatment
   c. trickling filter processing
   d. endogenous respiration

3. Which of the following is NOT an advantage of anaerobic digestion?
   a. Most of the biomass produced aerobically is utilized for methane production.
   b. The remaining sludge can be dried easily before disposal.
   c. Heavy metals are concentrated in the sludge.
   d. All of the above are advantages to anaerobic sludge digestion.

4. Which of the following does NOT contribute to the removal of organic material in leach fields of home sewage treatment systems?
   a. aerobic digestion as the waste percolates through the soil
   b. adsorption of organic material to soil particle surfaces
   c. entrapment of microbes in the pores of the leach field
   d. All of the above contribute to the removal of organic material.

5. Which of the following is true about the Colilert defined substrate test?
   a. It can detect total coliforms but not *E. coli.*
   b. It can detect *E. coli* but not total coliforms.
   c. It can detect total coliforms and *E. coli* simultaneously and independently.
   d. None of the above is correct.

6. Constructed wetlands use plants and their associated microorganisms to remove which of the following?
   a. organic material
   b. inorganic material
   c. metals
   d. All of the above are correct.

7. Which microorganism can usually be effectively removed by slow sand filtration?
   a. *Giardia lamblia*
   b. *Cryptosporidium parvum*
   c. Both (a) and (b) are correct.
   d. Neither (a) nor (b) is correct.

8. During which step of wastewater treatment are bacterial pathogens primarily removed?
   a. primary sewage treatment
   b. secondary sewage treatment
   c. Both (a) and (b) contribute equally to pathogen removal.
   d. Neither (a) nor (b) significantly remove bacterial pathogens.

9. Which of the following is used during water purification to remove large particles such as sand from raw water?
   a. sedimentation basins
   b. settling basins
   c. rapid sand filters
   d. slow sand filters

10. Which of the following organisms is not readily removed or inactivated to acceptable levels by conventional water purification and chlorination?
    a. *Cyclospora*
    b. *Cryptosporidium*
    c. viruses
    d. None of the above is effectively removed or inactivated to acceptable levels.

11. Why is chlorination of water during water purification increasingly being replaced by ozonation?
    a. because ozonation is particularly effective at destroying *Cryptosporidium* oocysts
    b. because chlorination can lead to the formation of disinfection by-products (e.g., trihalomethanes), which may be carcinogenic
    c. Both (a) and (b) are correct.
    d. Neither (a) nor (b) is correct.

12. Which step of wastewater treatment physically removes particulate material from wastewater?
    a. primary treatment
    b. secondary treatment
    c. tertiary treatment
    d. All of the above physically remove particulate matter.

13. Which step of wastewater treatment uses biological processes to remove dissolved organic matter from wastewater?
    a. primary treatment
    b. secondary treatment
    c. tertiary treatment
    d. All of the above use biological processes to remove dissolved organic matter.

14. Which of the following is NOT an accepted meaning for the term fermentation?
    a. any process involving the mass culture of microorganisms, either anaerobic or aerobic
    b. the production of alcoholic beverages
    c. food spoilage
    d. All of the above are accepted meanings for the term fermentation.

15. One culture system used in industrial settings allows toxic waste products to diffuse through a membrane away from microorganisms, and allows substrates to diffuse toward the microorganisms. What is this culture system called?
    a. fixed bed reactor
    b. dialysis culture system
    c. fluidized bed reactor
    d. continuous feed reactor

16. Agitation is used to maintain proper oxygen availability. However, this can be a difficult when trying to culture which of the following organisms?
    a. bacteria
    b. filamentous fungi
    c. yeasts
    d. It is a major problem with all of the above organisms.
17. Which of the following methods is used to improve strains used to carry out microbial fermentations?
    a. mutation and selection
    b. recombinant DNA modification of gene expression
    c. protoplast fusion
    d. All of the above are used to improve strains used for microbial fermentations.

18. Which of the following would NOT be considered a secondary metabolite?
    a. ethanol or some other fermentation end product
    b. mycotoxins
    c. antibiotics
    d. All of the above are considered secondary metabolites.
19. Which term refers to the use of microbes to make minor modifications to chemical compounds?
    a. bioconversions
    b. microbial transformations
    c. biotransformations
    d. All of the above are correct.
20. Which of the following is NOT being used as a bioinsecticide?
    a. protozoa
    b. bacteria
    c. viruses
    d. fungi
21. Which of the following best describes the production of semisynthetic penicillins?
    a. addition of a side-chain precursor molecule to a *Penicillium* culture to produce a desired modification
    b. chemical modification of a penicillin product after it has been released from the *Penicillium* culture
    c. Both (a) and (b) are correct.
    d. Neither (a) nor (b) is correct.
22. Why are bioconversions increasingly replacing chemical synthesis of certain compounds?
    a. Bioconversions yield only the desired stereoisomer.
    b. Bioconversions can be done under mild reaction conditions.
    c. Both (a) and (b) are correct.
    d. Neither (a) nor (b) is correct.
23. Why have genetically engineered microorganisms used in natural environments not been as effective as was originally anticipated?
    a. They are used as a food source by protozoa.
    b. They frequently do not come in contact with the materials to be degraded.
    c. They do not survive well once released and cannot compete with indigenous microbes.
    d. All of the above are correct.
24. Natural biodegradative processes are usually viewed as beneficial. However, they can have adverse effects. Which of the following is an adverse outcome of biodegradation?
    a. corrosion of metal pipes in oil fields
    b. creation of equally hazardous or more hazardous biodegradation end products
    c. unwanted degradation of jet fuel
    d. All of the above are adverse effects.

METHODS FOR EVALUATING MICROBIAL CHARACTERISTICS OF WATER

For each of the descriptions below, indicate which of the following tests fits the description: most probable number test (MPN), membrane filtration technique (MFT), presence-absence test (P-AT), defined substrate test (DST), and polymerase chain reaction tests (PCR).

1. Uses a filter to trap microbial contaminants in the water sample	
2. Can be used to detect total coliforms	
3. Can be used to detect fecal coliforms or *E. coli*	
4. Can be used to detect fecal streptococci	
5. Can be used to detect fecal viruses	
6. Tubes containing test medium are inoculated with different volumes of water	
7. Test medium is inoculated with 100 ml of water	
8. Test medium contains ONPG and MUG as the only nutrients	
9. Can be used to differentiate nonpathogenic and enterotoxigenic strains of *E. coli*	
10. Results are adversely impacted by high turbidity, large populations of noncoliform bacteria, metals, and phenol	
11. Relatively rapid (i.e., < 4 days to complete)	

TRUE/FALSE

_____ 1. Wastewater treatment uses unique processes that are unlike any that occur in aquatic environments.

_____ 2. Anaerobic sludge digestion resulting in methane production is dependent on the presence of carbon dioxide, hydrogen, and organic acids to initiate the reaction.

_____ 3. Because of our great dependence on groundwater as a drinking water supply, we have developed a tremendous understanding of the microorganisms and microbiological processes occurring in this environment.

_____ 4. If a leach field floods, it becomes anaerobic and effective treatment ceases.

_____ 5. Inadequately treated municipal waste is considered to be a nonpoint pollution source.

_____ 6. Potable water is unfit for consumption or recreation because of the high levels of microbial contaminants present in it.

_____ 7. *Cryptosporidium* has recently become of greater concern than *Giardia* because it is harder to remove from water.

_____ 8. Industrial processes involving microorganisms require that the culture be maintained in an active growth phase in order to maximize product formation.

_____ 9. As with laboratory scale operations, industrial fermentations rely primarily on purified media components in order to maintain better control of the process and of the final product.

_____ 10. Continuous feed processes involve periodic additions of a critical nutrient so that the organism will not have excess substrate available at any time.

_____ 11. Protoplast fusion can be used only between members of the same species.

_____ 12. Environmental conditions often have to be adjusted to switch from those conducive to microbial growth to those conducive to product formation. For example, *Streptomyces griseus* is grown in a high nitrogen medium, but the availability of nitrogen must be limited before the antibiotic streptomycin will accumulate.

_____ 13. Biocatalysts are often immobilized by attaching them to ion exchange resins or by entrapping them in a polymerized matrix to enable their recovery and repeated use.

_____ 14 Engineered proteins can be used to degrade previously recalcitrant molecules that were not considered to be amenable to biological processing.

_____ 15. Plants have been made pest-resistant by introducing a bacterial gene for an insect toxin into the plant's DNA.

_____ 16. Non-Newtonian broths are created when filamentous microorganisms grow and prevent adequate mixing of the growth media.

_____ 17. Although the terms can be used interchangeably, many differentiate industrial microbiology from biotechnology because the former uses natural isolates modified by mutation and selection, whereas the latter used molecular techniques to modify and improve microorganisms.

_____ 18. Most microorganisms of potential use in industrial microbiology have already been identified.

_____ 19. Because growth of microorganisms in controlled environments is inexpensive, it is preferable to carry out all microbial processes under such conditions.

_____ 20. Microorganisms can be used in controlled environments to make a variety of products. They can also be used in natural complex environments to carry out environmental management processes.

_____ 21. The survival of microorganisms added to complex microbial communities in soil or water can be improved somewhat if they are added with materials that provide microhabitats.

CRITICAL THINKING

1. Describe the primary, secondary, and tertiary (if any) processes used by the sewage treatment facility in your community. At each stage, describe the method used and the basis for its activity. If your residence utilizes a septic tank, then describe the functioning of that system.

2. Discuss the ideal characteristics of an indicator organism for fecal contamination of water. Using these characteristics, discuss the use of total coliforms, fecal coliforms, and fecal streptococci as indicators. Under what circumstances is one a better choice than the others? Why are none of these organisms particularly useful as indicators of enteric viruses or protozoans?

3. Discuss how regulatory mutants are used (and why they are necessary) in the production of amino acids. Cite a specific example and describe the nature of the regulatory changes. Why is it desirable to increase membrane permeability in these organisms as well?

4. Genetically engineered microorganisms have been developed for use as pesticides. What steps should be taken to test the safety of these organisms before releasing them for use in a natural environment? Justify your choices. If you do not think they should ever be used, state your reasons.

5. Search the cabinets and shelves in your home and identify products containing compounds made by microorganisms. List the product, the microbial contribution, and describe the function of the compound in the product.

ANSWER KEY

Terms and Definitions

1. h, 2. j, 3. o, 4. m, 5. a, 6. w, 7. t, 8. q, 9. l, 10. v, 11. y, 12. e, 13. g, 14. c, 15. x, 16. r, 17. i, 18. p, 19. n, 20. k, 21. u, 22. s, 23. f, 24. d, 25. b

Fill in the Blank

1. chemical oxygen demand (COD); total organic carbon (TOC); biochemical oxygen demand (BOD); nitrogen oxygen demand 2. settling; coagulation 3. sludge; extended aeration; anaerobic digestion 4. septic tank 5. primary; secondary; activated sludge; trickling filters; tertiary 6. potable; sedimentation; coagulation; filtration; disinfection 7. indicator; coliforms; indicator; fecal coliforms 8. non-Newtonian broth 9. scaleup 10. primary metabolites; secondary metabolites 11. regulatory 12. biosensors 13. natural genetic engineering; forced evolution; adaptive mutations 14. industrial ecology; biosphere; anthrosphere 15. combinatorial biology; site-directed mutagenesis; metabolic pathway (MPE); metabolic control (MCE)

True/False

1. F, 2. T, 3. F, 4. T, 5. F, 6. F, 7. T, 8. F, 9. F, 10. T, 11. F, 12. T, 13. T, 14. T, 15. T, 16. T, 17. T, 18. F, 19. F, 20. T, 21. T

Multiple Choice

1. a, 2. b, 3. c, 4. d, 5. c, 6. d, 7. a, 8. b, 9. a, 10. d, 11. b, 12. a, 13. b, 14. d, 15. b, 16. b, 17. d, 18. a, 19. d, 20. a, 21. b, 22. c, 23. d, 24. d

Methods for Evaluating Microbial Characteristics of Water

1. MFT 2. MPN; MFT; P-AT; DST 3. MFT; P-AT; DST; PCR 4. MFT 5. None 6. MPN 7. P-AT; DST 8. DST 9. PCR 10. MFT 11. MFT; P-AT; DST; PCR

Notes

Notes

Notes

Notes